Advanced Models for Manufacturing Systems Management

CRC Mathematical Modelling Series

Series Editor
Nicola Bellomo
Politecnico di Torino, Italy

Advisory Editorial Board

Giuseppe Geymonat
ENS de Cachan, France

Konstantin Markov
University of Sofia, Bulgaria

Carlo Naldi
Politecnico di Torino, Italy

Andrzej Palczewski
Warsaw University, Poland

Vladimir Protopopescu
Oak Ridge National Laboratory, USA

Johan Grasman
Wageningen University, The Netherlands

Agostino Villa
Politecnico di Torino, Italy

Wolfgang Kliemann
Iowa State University, USA

Titles included in the series:

Movchan: **Mathematical Modelling of Solids with Nonregular Boundaries**

Mittnik: **System Theoretic Methods in Empirical Economic Modelling**

Villa/Brandimarte: **Advanced Models for Manufacturing Systems Management**

Kusiak/Bielli: **Designing Innovations in Industrial Logistics Modelling**

Kliemann/Namachchivaya: **Nonlinear Dynamics and Stochastic Mechanics**

Bellomo: **Modelling Mathematical Methods and Scientific Computation**

Advanced Models for Manufacturing Systems Management

Paolo Brandimarte
Agostino Villa
*Department of Manufacturing Systems
and Production Economics
Technical University of Torino
Torino, Italy*

CRC Press
Boca Raton   New York   London   Tokyo

Library of Congress Cataloging-in-Publication Data

Brandimarte, Paolo.
  Advanced models for manufacturing systems management /Paolo Brandimarte and Agostino Villa
    p.   cm.
  Includes bibliographical references and index.
  ISBN 0-8493-8332-3
  1. Production management—Mathematical models.   I. Villa, Agostino.   II. Title.
  TS155.B629 1995.
  658.5'01'5118—dc20

                                                                95-32355
                                                                CIP

This book contains information obtained from authentic and highly regarded sources. Reprinted material is quoted with permission, and sources are indicated. A wide variety of references are listed. Reasonable efforts have been made to publish reliable data and information, but the author and the publisher cannot assume responsibility for the validity of all materials or for the consequences of their use.

Neither this book nor any part may be reproduced or transmitted in any form or by any means, electronic or mechanical, including photocopying, microfilming, and recording, or by any information storage or retrieval system, without prior permission in writing from the publisher.

CRC Press, Inc.'s consent does not extend to copying for general distribution, for promotion, for creating new works, or for resale. Specific permission must be obtained in writing from CRC Press for such copying.

Direct all inquiries to CRC Press, Inc., 2000 Corporate Blvd., N.W., Boca Raton, Florida 33431.

© 1995 by CRC Press, Inc.

No claim to original U.S. Government works
International Standard Book Number 0-8493-8332-3
Library of Congress Card Number 95-33164
Printed in the United States of America 1 2 3 4 5 6 7 8 9 0
Printed on acid-free paper

Contents

1 Manufacturing systems modeling   **1**
   1.1 The nature of mathematical models of manufacturing systems   1
   1.2 Dynamic models of manufacturing systems   6
   1.3 An overview of management problems in manufacturing systems   10
      1.3.1 A hierarchy of management problems   10
      1.3.2 Production scenarios and manufacturing systems layout   11
      1.3.3 Performance measures   14
   1.4 Plan of the book   17
   For further reading   18

2 Optimization models and model solving   **20**
   2.1 Classes of optimization models   21
      2.1.1 Finite/infinite-dimensional problems   23
      2.1.2 Linear/nonlinear problems   25
      2.1.3 Continuous/discrete problems   26
      2.1.4 Deterministic/stochastic problems   28
      2.1.5 Single/multiobjective problems   29
   2.2 An overview of optimization methods   30
   2.3 Complexity of optimization problems   33
   2.4 Good and bad model formulations   38
   2.5 Developing heuristics from mathematical models   43
   S2.1 The theory of NP–completeness   45
   S2.2 An outlook on multiobjective optimization   49
   For further reading   51

3 Discrete-time models   **52**
   3.1 Aggregate Production Planning   53
      3.1.1 The basic APP model   54

	3.1.2	Extensions of the basic APP model	55
	3.1.3	A decomposition approach for the basic APP model	56
3.2	The Capacitated Lot-Sizing Problem	60
	3.2.1	Formulation of the CLSP	60
	3.2.2	Solving the CLSP with standard MILP codes	61
	3.2.3	A dual approach to the CLSP	63
	3.2.4	The single-item uncapacitated lot-sizing problem	64
	3.2.5	A modal formulation for the CLSP problem	68
	3.2.6	Hierarchical approaches for CLSP	70
	3.2.7	Multilevel lot-sizing	72
3.3	The Discrete Lot-Sizing and Scheduling Problem	74
	3.3.1	The DLSP formulation	75
	3.3.2	A dual decomposition approach for the multi-item DLSP problem	77
3.4	Continuous-flow models for production scheduling	77
	3.4.1	Discrete-time dynamic models and trajectory tracking	78
	3.4.2	A discrete-time continuous-flow model for batch production	82
3.5	Discussion: the flexibility of discrete-time models	89
	3.5.1	Periodic scheduling in a press shop	90
S3.1	Strong formulations for lot-sizing problems	95
S3.2	The single item DLSP problem: a dynamic programming approach	98
For further reading	100

4 DEDS models for scheduling problems   102

4.1	Classical machine scheduling theory	103
4.2	Classification of scheduling problems	106
	4.2.1 Machine environment	106
	4.2.2 Job characteristics	109
	4.2.3 Objective functions	109
	4.2.4 Semiactive, active and nondelay schedules	113
	4.2.5 Nonstandard scheduling problems	115
4.3	Polynomial complexity scheduling problems	118
	4.3.1 The SPT sequencing rule	119
	4.3.2 The EDD sequencing rule	121
	4.3.3 The Johnson algorithm	122
	4.3.4 Complexity of preemptive scheduling	123
4.4	Dynamic programming approaches	125
	4.4.1 Applying Dynamic Programming to single-machine problems	125
	4.4.2 Solving $1/s_{ij}/*$ problems by Dynamic Programming	129
	4.4.3 The curse of dimensionality	129
4.5	A modeling framework based on node potential assignment	130

	4.5.1	Project networks	130

- 4.5.1 Project networks 130
- 4.5.2 The disjunctive graph representation of job shop problems 132
- 4.6 MILP models for scheduling problems 135
  - 4.6.1 Permutation modeling 135
  - 4.6.2 Disjunctive graph based models 137
  - 4.6.3 TSP-like modeling 139
  - 4.6.4 Modeling with time-indexed variables 140
  - 4.6.5 Modeling with assignment and selection variables . . 142
- 4.7 Branch and Bound methods 144
  - 4.7.1 Integer programming approaches to machine scheduling 145
  - 4.7.2 A Branch and Bound algorithm for $1/r_i/L_{\max}$ . . . 146
  - 4.7.3 A Branch and Bound algorithm for the permutation $F//C_{\max}$ problem 148
  - 4.7.4 Lagrangian relaxation approaches to machine scheduling 149
- 4.8 Heuristic scheduling methods 156
  - 4.8.1 List scheduling and greedy methods 157
  - 4.8.2 Dual heuristics for scheduling problems 161
  - 4.8.3 Scheduling by simulated annealing 163
  - 4.8.4 Scheduling by tabu search 167
  - 4.8.5 LPT and Multifit algorithms for load balancing . . . 171
  - 4.8.6 Decomposition approaches to FJS and FMS scheduling 175
  - 4.8.7 The Shifting Bottleneck procedure for $J//C_{\max}$ . . . 177
- 4.9 Discussion 179
- S4.1 Dynamic Programming for scheduling a batch processor . . 180
- S4.2 Periodic scheduling problems 182
- S4.3 A multiobjective approach to machine loading 185
- For further reading 189

5 Evaluative models   191

- 5.1 Introduction to queueing models 192
  - 5.1.1 Little's law 195
  - 5.1.2 Elementary Markovian queueing models 195
  - 5.1.3 Approximations for the $GI/G/1$ queue 198
- 5.2 Queueing networks 199
  - 5.2.1 Two-machine queueing networks 200
  - 5.2.2 Product form open queueing networks 202
  - 5.2.3 Closed queueing networks 207
- 5.3 Computational methods for closed networks 209
  - 5.3.1 The convolution algorithm 210
  - 5.3.2 The Mean Value Analysis algorithm 214

    5.4   Approximate analysis of nonproduct form queueing
           networks 216
           5.4.1   Approximate Mean Value Analysis 216
           5.4.2   A node decomposition approach for general queueing
                  networks 217
    5.5   Petri net models 219
    5.6   Discussion 225
    S5.1 Product forms and local balance equations 226
    For further reading 228

6 Putting things together   230

    6.1   Integrating optimization methods and evaluative models .. 230
           6.1.1   The Response Surface Methodology 231
           6.1.2   Perturbation Analysis 233
           6.1.3   Optimization with queueing models 235
    6.2   Optimal control of failure-prone manufacturing systems .. 237
    6.3   Model management and modeling languages 240
    For further reading 244

APPENDICES   245

A Fundamentals of Mathematical Programming   246

    A.1   Convex sets and convex functions 247
           A.1.1   Qualitative properties of mathematical programming
                  problems 250
    A.2   Unconstrained optimization problems 251
    A.3   Numerical methods for unconstrained optimization 253
    A.4   Constrained optimization problems 254
           A.4.1   The penalty function approach 255
           A.4.2   Multiplier methods: optimization with equality
                  constraints 256
           A.4.3   Multiplier methods: optimization with inequality
                  constraints 260
           A.4.4   Kuhn-Tucker conditions for the general constrained
                  problem 261
           A.4.5   Interpreting dual variables 263
    A.5   Duality theory 264
    A.6   Decomposition of large-scale optimization problems 270
    A.7   Nonsmooth optimization 272
    For further reading 277

B Linear Programming and Network Optimization   278

    B.1   Background 279
           B.1.1   Convex polyhedra and polytopes 279

	B.1.2	Graph Theory	281
B.2		The standard and canonical forms of LP problems	283
B.3		Geometric and algebraic features of Linear Programming	287
B.4		The Simplex method	289
	B.4.1	Getting a starting basis	294
	B.4.2	Computational issues and interior point methods	294
B.5		Duality in Linear Programming	296
	B.5.1	Applications of LP duality	297
B.6		Network Optimization	299
	B.6.1	The minimum cost flow problem	300
	B.6.2	The transportation problem	301
	B.6.3	The maximum flow problem	302
	B.6.4	The Shortest and Longest Path problems	303
	B.6.5	The Traveling Salesman Problem	303
B.7		Total unimodularity	304
For further reading			306

C Enumerative and Heuristic Methods for Discrete Optimization   307

C.1		Classical discrete optimization problems	308
C.2		Modeling with binary variables	314
	C.2.1	Expressing logical constraints	314
	C.2.2	Modeling piecewise linear objective functions	315
	C.2.3	Modal formulations	317
	C.2.4	Linearizing nonlinear 0/1 problems	319
C.3		Branch and Bound methods	319
C.4		Relaxation methods for bounding	323
	C.4.1	Constraints elimination	323
	C.4.2	Continuous relaxation	323
	C.4.3	Lagrangian relaxation	324
	C.4.4	Surrogate relaxation	326
	C.4.5	Comparing alternative relaxations	326
C.5		Branching strategies	328
	C.5.1	Branching for MILP models	328
	C.5.2	Branching in special purpose Branch and Bound algorithms	330
	C.5.3	Selecting the branching variable	330
	C.5.4	Obtaining feasible solutions	332
C.6		Search strategies	332
C.7		Dominance conditions	333
C.8		Heuristic methods for discrete optimization	334
	C.8.1	Heuristics from mathematical models	335
	C.8.2	Greedy algorithms	335
	C.8.3	Truncated exponential schemes	336

C.8.4 Local Search	337
For further reading	344

D Dynamic Programming   346
- D.1 The shortest path problem 346
- D.2 Sequential decision processes 349
  - D.2.1 The optimality principle and solving the functional equation 351
- D.3 Dynamic Programming for infinite-dimensional problems .. 354
- For further reading 356

E Stochastic modeling   357
- E.1 Motivation 357
  - E.1.1 A refresher on Probability Theory and random variables 360
  - E.1.2 Memoryless random variables 362
- E.2 Stochastic processes and Markov chains 364
  - E.2.1 The Poisson process 364
  - E.2.2 Discrete-time Markov chains 366
  - E.2.3 Continuous-time Markov chains 368
- E.3 Steady-state analysis of Markov chains 369
  - E.3.1 Steady-state analysis of DTMCs 369
  - E.3.2 Steady-state analysis of CTMCs 372
  - E.3.3 An example: steady-state analysis of a birth-death process 374
- For further reading 375

Problems   376

Bibliography   384

Index   399

Preface

> *Mathematical modeling of managerial problems does not produce universal laws as in the natural sciences, but tentative representations of relationships which are felt to be relevant for the analysis of a given problem, in a particular period of time, and in a particular context.*[1]

> *Practical applications of the above ideas have been disappointing, thus supporting the claim above that there is no advantage in translating scheduling problems into integer programming problems.*[2]

> *Inasmuch as at the end of the sixties, practical integer programming problems were commonly considered intractable, now, in the mid-eighties, it is difficult to find a practical problem, not an artificially constructed one, which cannot be solved at least near-optimally by existing codes.*[3]

This book contains a tutorial introduction to mathematical modeling for managing manufacturing systems, designed for a course at the graduate level. The motivations for writing it, and the reasons behind the selection and organization of the covered topics are roughly contained in the above three quotations. Mathematical modeling in this book has a different meaning than modeling in physics: the models presented here are *computational* in nature and aim at helping a decision making process; they are not aimed at stating universal law, nor to help in understanding manufacturing related problems. When formulating a model, a basic tradeoff must be carefully considered between the need for representing the problem faithfully and the need for formulating a computationally tractable model.

---

[1] F.A. Lootsma. 1994. Alternative Optimization Strategies for Large-Scale Production-Allocation Problems. *European Journal of Operational Research* 75, p. 13.

[2] S. French. 1982. *Sequencing and Scheduling: an Introduction to the Mathematics of the Job Shop.* Ellis Horwood, p. 135.

[3] S. Walukiewicz. 1991. *Integer Programming.* Kluwer, p. 167.

This is particularly true for Mixed-Integer Linear Programming (MILP) models, which are the most common type of model in the book. The second and third quotation are seemingly contrasting views, which illustrate the fundamental role of computational aspects. Indeed, there are computationally intractable MILP models of rather modest size, which can be contrasted to well-solvable, large-scale models: we believe it is fundamental to gain a deep understanding of why certain models are easy to solve and others are not.

Apart from computational issues, the very role of mathematics in management is the subject of debate: on the one hand, there are concerned views about the distortions induced by the academic character of much research on this topic[4]; on the other hand, impressive progress is reported in the application to real world problems [5]. As one may expect, the truth lies somewhere in between: this book is aimed at showing what can be accomplished despite the limitations of mathematical models.

Most of the book deals with generative models solved by exact or heuristic optimization algorithms; evaluative models for performance evaluation are also covered, with emphasis on analytical models based on queueing theory. This is not a book on *problems*. The topics are organized from a model-oriented perspective: for instance, scheduling problems are treated in two different chapters, depending on the type of model adopted. This choice has been dictated by the character of this book, but it should not be taken as a suggestion that a deep understanding of practical problems is ancillary to the knowledge of mathematical modeling. Some fundamental readings are suggested in the first chapter.

The book is organized in six chapters and five appendices: the latter contain prerequisite material that has been isolated for the reader's convenience. Some optional material is contained in supplements at the end of the chapters: these sections may be safely omitted. The book prerequisites are at the level of undergraduate courses in Mathematics, in particular, Linear Algebra and Probability Theory. An Optimization background is helpful, but not necessary. Preliminary versions, in the form of classroom notes have been used for a course on Production Planning and Control, held at Politecnico di Torino for students in Industrial Engineering.

Feedback. Any comment or suggestion for improvement will be greatly appreciated. The best way to contact us is by e-mail: please send your message to brandimarte@polito.it.

---

[4]See e.g. A. Reisman, F. Kirschnick. 1994. The devolution of OR/MS. *Operations Research* **42**, 577-588, and references therein.

[5]G. Nemhauser. 1994. The Age of Optimization: Solving Large-Scale Real-World Problems. *Operations Research* **42**, 1-12.

1
Manufacturing systems modeling

Mathematical modeling of manufacturing systems has not been a tradition as long as modeling in physics; nevertheless, it is older than one would probably think. From a methodological point of view, the background of mathematical models is a body of knowledge which can be included within Operations Research. However, while the name "Operations Research" was born after 1940, the well known Economic Order Quantity formula, balancing fixed order and inventory costs, dates back to 1915. The worried reader may relax: we will not go that far. Nevertheless, it may be worthwhile to use the rather "historic" Economic Order Quantity problem as an example to set the stage for a few considerations on the nature of mathematical models of manufacturing systems. This is done in section 1.1. In section 1.2 we turn our attention to the *dynamics* of manufacturing systems, introducing different types of models based on continuous-time, discrete-time, and discrete-event representations. An overview of management problems related to manufacturing systems is given in section 1.3: we illustrate the different hierarchical levels, production scenarios, layouts and performance measures involved. Finally, in section 1.4 we present the organization of the book.

1.1 The nature of mathematical models of manufacturing systems

To illustrate some important issues about mathematical modeling, a good starting point is a simple and classical problem concerning an inventory control system. The aim of an inventory control system is to determine "when" and "how much" to order of a certain item in order to meet the demand at minimum cost. The inventory acts as a buffer decoupling a low-frequency replenishment pattern (infrequent replenishment orders for

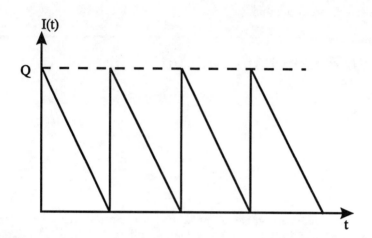

FIGURE 1.1
The time evolution of the inventory level for the EOQ problem

large quantities) from a high-frequency depletion pattern (many frequent demands for small quantities). If we assume a stationary demand, expressed by a rate of d units per unit time, the inventory level will vary in time as illustrated in Figure 1.1. Note that the inventory is replenished exactly when it reaches level zero: this is possible if there is no uncertainty in the demand and in the lead time, i.e., the time elapsing between the release of an order and the replenishment.

Meeting demand at minimum cost requires finding the right compromise between two conflicting needs. On the one hand we would like to release as few orders as possible, since for each order we have to pay a fixed cost s; this cost may be related to paperwork or shipping costs in the case of a purchase order, or to setup costs in the case of a production order. On the other hand, we would like to keep inventory to a minimum, since there is a holding cost h per item per unit time: this calls for frequent small orders. In sum, we have to find the "optimal" order size Q^* yielding the minimum overall cost: such a quantity is known as the Economic Order Quantity (EOQ). The ordering period is clearly Q^*/d, and it is easy to see that we must release the order when the inventory level is $d \cdot LT$, i.e. the demand during the lead time LT (provided that the lead time is smaller than the ordering period): this critical inventory level is called the **reorder level** (ROL). A ROL and an optimal order size define a **fixed quantity** inventory control policy whereby we launch a replenishment order whenever the ROL is reached; a second approach may be considered, called **fixed period** policy, whereby the inventory level is checked at fixed time intervals and a replenishment order is launched in order to reach a target inventory level. Note that the two policies are the same if the demand rate is constant, but they differ if demand variations are considered.

To determine the optimal lot size, we have to determine the cost per unit time. By looking at Figure 1.1, we can find the cost per ordering cycle:

$$\frac{1}{2}hQ\frac{Q}{d} + s.$$

The first term is simply h times the area of each triangle; the second term is the fixed cost incurred during each ordering cycle. Dividing by the cycle length Q/d, we get the cost per unit time:

$$C(Q) = \frac{1}{2}hQ + \frac{sd}{Q}. \quad (1.1)$$

The shape of $C(Q)$ is shown in Figure 1.2: it is the sum of a term linearly increasing with Q and a term decreasing with Q. To solve the problem we must minimize the cost:

$$\min_{Q \geq 0} C(Q).$$

In this case, we have only to compute the first derivative of $C(Q)$ and set it to 0, yielding[1]

$$Q^* = \sqrt{\frac{2ds}{h}}.$$

The EOQ model is very simple to formulate and to solve, but it suffers from many flaws:

- it assumes a constant demand d, whereas we may have to deal with a time-varying demand $d(t)$;
- it neglects any possible source of uncertainty, both in the demand rate and the lead time;
- it does not take into account interactions among different items, which are treated independently: in practice, production capacity constraints limit the possibility of ordering arbitrary quantities with arbitrary timing (this is relevant for *production* orders); furthermore, if an item is used to assemble another item, there is an obvious relationship between their demands: we must distinguish *independent* and *dependent* demands;
- it assumes a continuous demand for items that could be discrete in nature;

---

[1] The following chapters will discuss technical details concerning when and why setting the first derivative to zero solves an optimization problem.

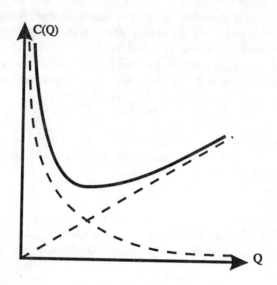

FIGURE 1.2
A plot of the total cost per unit time as a function of the lot size Q

- it does not consider the cost of purchasing the item, which could be affected by quantity discounts;
- it considers a steady-state situation, neglecting the transient phase due to the initial inventory level: this may lead to infeasibilities in the case of finite capacity problems.

Such criticisms about the EOQ approach led, on the one hand, to the development of more complex mathematical models for lot-sizing (see Chapter 3), and, on the other hand, to the development of computerized production management tools like Material Requirements Planning (MRP) systems. From our point of view, it is interesting to discuss why the EOQ formulation is so well-known despite its apparent disadvantages.

A first important point is that there is no such thing as an "exact" model. It has been argued that

modeling is the art of selective simplification of reality.

The EOQ model is not actually wrong per se; with some improvement it may be used to obtain the parameters of an ROL-based inventory management policy considering demand and lead time uncertainty. What may be actually wrong is the adoption of an ROL policy. There are many production scenarios in which this would lead to poor performance; on the other hand, an ROL policy is quite easy to implement, and it does not necessarily require a computer. The reason behind this simplicity is that each item is treated independently from the others; more sophisticated approaches require the use of a computer.

We will see that, indeed, computational issues play a major role: *the models considered in this book have a computational nature*. We do not deal with *explicative* models; the EOQ model can be considered a **generative** model, in the sense that its solution yields a set of decision variables (in our case the optimal lot size). The EOQ model is an example of an optimization model; it is important to realize that the solution may be optimal from the point of view of the mathematical model, but this does not imply that it is the truly optimal solution for the real problem. Computational issues limit our ability to cope with difficult optimization models.

In some cases the manufacturing system is so complex that it is practically impossible to find an analytical relationship between decision variables and performance measures. In such cases, we need to find a way to *evaluate* the performance measure for a given decision policy. For instance, we may want to evaluate the service level for a complex inventory control system in a stochastic environment; an obvious approach would be to carry out a simulation experiment on a computer. In such a case, we have an **evaluative** model. In the book we deal both with generative and evaluative models, but, among evaluative models, we will give emphasis to **analytical**, as opposed to **experimental**, models. Analytical models (such as the queueing models discussed in Chapter 5) rely on a mathematical representation of the manufacturing system, whereas experimental models are oriented to building simulation programs and do not necessarily require a mathematical formalism (still, they must be considered as formal models since they must be executable on a computer).

A further point to be made is that the selection of a certain model depends on the *scope* of the decisions we must make. For instance, we have not considered the cost of purchasing an item in the EOQ model. Let c be the unit cost, and assume that no quantity discount applies; then the total cost per unit time would be

$$C(Q) = \frac{1}{2}hQ + \frac{sd}{Q} + c.$$

Clearly, the optimal solution does not change if we add a constant to the cost function. However, if we must also select a supplier, the model may change: assuming that we have two suppliers such that $s_1 > s_2$, but $c_1 < c_2$, we would need to include the purchasing cost in the model.

In summary, building mathematical models involves finding a suitable tradeoff between the need for faithfully representing the manufacturing system and the production scenario and the need for keeping data and computation requirements as low as possible. Model building is not simply

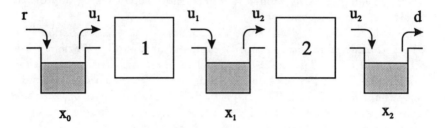

FIGURE 1.3
A simple two-stage manufacturing system

the problem of taking a snapshot of an objective reality: the right blend of knowledge and commonsense is needed in order to build a model, solve it, validate it, and successfully exploit the results.

1.2 Dynamic models of manufacturing systems

The EOQ formulation assumes a static demand pattern. To cope with a dynamic environment, we can borrow the idea of **state equations** from physics. State equations describe the evolution of a set of quantities, called states, selected in such a way that the knowledge of their values at certain instants is sufficient to predict the future evolution of the system, whereas the past history can be forgotten. A common form of state equation is a differential equation.

Example 1.1
Consider the two-stage manufacturing system depicted in Figure 1.3. We have a raw material storage, a Work in Process (WIP) storage between stage S_1 and S_2, and an end item storage: the storage levels are denoted, respectively, by $x_0(t)$, $x_1(t)$ and $x_2(t)$. Let $d(t)$ be the time-varying demand rate; we control the system by determining the purchase rate of raw material $r(t)$ and the production rates $u_1(t)$, and $u_2(t)$.

To model the dynamics of the system, we can write differential equations such as

$$\dot{x}_0(t) = r(t) - u_1(t)$$

$$\dot{x}_1(t) = u_1(t) - u_2(t)$$
$$\dot{x}_2(t) = u_2(t) - d(t),$$

where $\dot{x}$ denotes, as usual, the derivative of x with respect to time. We see that the storage levels are the state variables of the system, i.e. the only information we need to keep memory of the past history of the system. The production and purchase rates are used to control the system.

We must also state some non-negativity constraints such as

$$x_0(t), x_1(t), x_2(t), u_1(t), u_2(t), r(t) \geq 0.$$

Another fundamental constraint is the capacity constraint:

$$u_1(t) \leq \bar{u}_1 \qquad u_2(t) \leq \bar{u}_2,$$

i.e. each machine cannot produce at a higher rate than $\bar{u}_i$. We might also have upper bounds on the state variables if the storages have a limited capacity. Finally, we are given the initial inventory levels $x_0(0)$, $x_1(0)$, and $x_2(0)$.

In controlling the system, we might want to minimize not only the end item inventory level, but the overall level over a given planning horizon T:

$$\min \int_0^T [h_0 x_0(t) + h_1 x_1(t) + h_2 x_2(t)] \, dt. \quad \square$$

We have just shown an example of a **continuous-time** model. From a managerial point of view, such a model may be unsuitable. In general, it requires instantaneous changes in the production rates, which are difficult to implement. We will see in section 6.2 that this needs not be the case, but, for now, let us pursue the idea of keeping production rates constant over certain time periods. This idea leads to **discrete-time** models; we discretize the time horizon in equal time periods of length Δ, and we consider only time instants of the form $k\Delta$ ($k = 0, 1, 2, 3, \ldots$). Production rates are kept constant during each time period. In this case we have difference rather than differential equations.

Example 1.2
The discrete-time version of the previous model would read like:

$$\min \sum_{t=1}^T (h_0 x_{0t} + h_1 x_{1t} + h_2 x_{2t})$$
$$\text{s.t.} \quad x_{0t} = x_{0,t-1} + r_t - u_{1t}$$
$$x_{1t} = x_{1,t-1} + u_{1t} - u_{2t}$$

$$x_{2t} = x_{2,t-1} + u_{2t} - d_t,$$

to which we should add the same constraints as in the continuous-time case. Note that here t is a discrete-time index. □

On the one hand, discrete-time formulations may simplify the task of devising solution procedures; on the other hand, they are quite powerful in that they enable us to model several interesting features. To illustrate this point, in the next example we formulate a dynamic version of the EOQ problem, including setup issues.

Example 1.3
Let d_t and u_t denote the demand and the production during period t, and x_t the inventory level at the end of period t. A mathematical model for the minimization of inventory plus setup costs is

$$\min \sum_{t=1}^{T} (hx_t + s\delta(u_t))$$
$$\text{s.t.} \quad x_t = x_{t-1} + u_t - d_t \qquad t = 1, \ldots, T \qquad (1.2)$$
$$x_t, u_t \geq 0, \qquad t = 1, \ldots, T$$

where h and s are the inventory and setup costs and

$$\delta(u) = \begin{cases} 1 & \text{if } u > 0 \\ 0 & \text{if } u = 0. \end{cases}$$

Eq. (1.2) simply states that the inventory level at the end of a period is the level at the end of the previous period, plus the production, minus the demand. □

Both continuous-time and discrete-time models are common in mathematical modeling of a wide range of systems, including physical, economical, and biological systems. In manufacturing systems it is often necessary to introduce another type of dynamic system, a **discrete-event dynamic system** (DEDS for short), common when we must model operations at a detailed level. In the previous examples, we have assumed that the inventory levels were represented by real numbers; this is equivalent to the assumption that items are continuously divisible. In the following example we see a case where this is not appropriate.

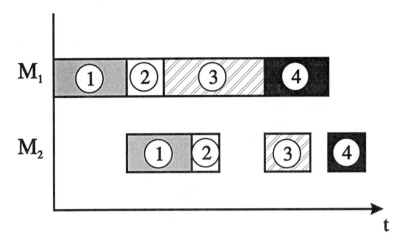

FIGURE 1.4
A Gantt chart for a two-stage manufacturing system

Example 1.4
Consider again the two stage system of example 1.1, but assume that we have to produce *discrete* parts. We have N single parts or batches of parts to produce; if the batches are consecutively processed and it is not possible to move a fraction of a batch to the next machine, the two cases are formally identical, and we may speak of *jobs*. Let the processing times of each job J_i ($i = 1, \ldots, N$) on each machine be denoted by p_{i1} and p_{i2}. Each machine can work on one job at a time, and the processing cannot be interrupted. To start processing a job on the second machine, we must have completed processing on the first machine. Assuming a given sequence of jobs, the type of solution we get is shown in Figure 1.4, where a **Gantt chart** is depicted. In this case, we are concerned with *events* occurring at discrete instants, namely the start and completion times of each operation on each machine. Actually, we need to consider only the completion time of each job i on each machine j, denoted by C_{ij}, since the start times are obviously given by $S_{ij} = C_{ij} - p_{ij}$.

Let us assume that the job sequence is common for both machines and that the storage between the machines is unlimited (no blocking occurs). The performance measure could be the makespan, i.e., the time needed to process the whole set of jobs. Then our problem boils down to a sequencing problem; we must find an optimal permutation σ of the jobs. Let $\sigma(k)$ denote the kth job in the sequence; under the above hypotheses, the first machine will have no idle time:

$$C_{\sigma(k+1),1} = C_{\sigma(k),1} + p_{\sigma(k+1),1}.$$

For the second machine, a job is started when the machine has finished processing the previous job and when the job has finished processing on the first machine:

$$C_{\sigma(k+1),2} = \max\{C_{\sigma(k+1),1}, C_{\sigma(k),2}\} + p_{\sigma(k+1),2}.$$

We avoid for now completely stating a mathematical model for this problem (see section 4.6.1), but the reader can appreciate the difference between this modeling framework and the classical state equation approach. □

1.3 An overview of management problems in manufacturing systems

In the previous sections we have considered some basic production management problems; in fact there is a huge variety of them. The range of possibilities is wide due to the combination of different factors:

- the hierarchical level at which the decisions must be taken;
- the type of production scenario, both from the point of view of the market interface and the physical structure of the manufacturing system;
- the performance measures of interest.

1.3.1 A hierarchy of management problems

Manufacturing related problems are not necessarily management problems; we may have to deal with a *design* problem. For instance we may have to size the buffers in a multi-stage system like the one considered in the previous section, or we may have to decide in which stage we should add a machine to alleviate bottleneck problems. However, drawing the line separating management and design is difficult when we have a rapidly changing environment. In fact, it is not possible to deal with management ignoring design issues. What is certainly different is the time scale at which management and design problems are solved; we have to solve management problems on a day by day basis, whereas a design problem is solved with a coarser time scale (possibly years). Indeed, there are different decision echelons within the firm in charge of solving decision problems on different time scales.

Within production management problems there is also a hierarchy. We may distinguish at least the following three levels.

Production planning. At this level we plan production without specifying every detail, with a relatively long time horizon and an aggregate view of the manufacturing system. For instance, we might state what we want to produce for each of the next 12 weeks (or months), without specifying exactly when and on which machine.

At the production planning level it is possible to consider the production capacity as a decision variable, whereas at the lower levels the capacity constraints are not negotiable. In some (long term) aggregate planning problems, we do not deal with single items, but with families. The production plans for families must then be disaggregated.

Production scheduling. At this level we take a closer look at the manufacturing facility, with a shorter time horizon. We have a set of tasks to be carried out with a set of resources, and we must specify exactly what happens, where and when.

Production control. This is the lowest level in the hierarchy and it is a task carried out in real time. At this level we must ensure proper implementation of production plans/schedules despite the occurrence of random events. Another task of production control is to monitor the production activities in order to provide the upper levels with statistics about the execution (e.g. about processing times) and up-to-date information necessary to revise the longer term plans or schedules.

Actually, the decision levels can be extended both upward and downward. Going upward we may find facility location problems and manufacturing facility design problems. Going downward we may find issues related to numerical control of single machines. We also have to pay attention to technological issues. For instance, there are classical models aimed at optimizing the cutting parameters for mechanical production. Furthermore, the process plan of a part might not be fixed a priori, and it could be revised in order to ensure a better production plan/schedule or to react to unpredictable events such as machine failures. When there is some flexibility in the process plan or in the assignments of operations to machines, the complexity of the decision problem increases. In such cases we must also solve a **routing** problem.

1.3.2 Production scenarios and manufacturing systems layout

The EOQ model and its variants assume a **make-to-stock** production environment. Given a forecasted demand, we want to manage production in order to meet the demand at minimum cost; what we want to ensure is that all customer orders can be immediately satisfied from stock. This situation is typical of low product variety, high demand systems. A different environment is **make-to-order**: in this case we may have a high variety of demand for expensive items, which makes maintaining inventories too costly. Production is started only after the receipt of an order.

Sometimes, neither a make-to-stock nor a make-to-order strategy is appropriate: consider a family of products which come in a large number of variants, obtained by assembling different combinations of components. Maintaining an inventory for each end item configuration is very expensive. On the other hand, the lead time necessary to produce the necessary components and to assemble them may be too long to adopt a make-to-order strategy. In this case, an **assemble-to-order** strategy may be taken. We maintain an inventory of the components, and we assemble them on order.

Assemble-to-order can be considered a hybrid strategy between make-to-stock and make-to-order. An extreme case of the make-to-order strategy is **engineer-to-order**. Here accepting an order involves both design and production of a highly customized and unique product.

It must be emphasized that, in practice, it may be hard to classify a certain production activity as strictly make-to-order or make-to-stock. For instance, a manufacturing facility may be used both to satisfy the production needs of the firm which owns it (make-to-stock) and to satisfy occasional orders coming from another firm (make-to-order).

Other terms are widely used to denote production strategies:

- mass production
- repetitive production
- batch production
- job shop production
- one-of-a-kind production

Mass production, job shop production and one-of-a-kind production are readily identified with make-to-stock, make-to-order and engineer-to-order strategies. Repetitive production is characterized by a **production mix**, i.e., by a set of production quantities that are periodically repeated; for instance, we might produce 100 items of type A, 150 of type B and 70 of type C per day. This mix is kept stable for a fair amount of time, and it is occasionally changed. Batch production is characterized by **production orders** rather than by a production mix; however such production orders are not necessarily related by a one-to-one correspondence to *customer* orders, unlike a job shop environment.

The various production scenarios call both for different management policies (and mathematical models) and different manufacturing system layouts. In a low variety, high demand environment, it is worthwhile to organize the manufacturing system layout in order to maximize its efficiency: we obtain a **product-oriented** layout, whereby products are allocated to dedicated production lines (see Figure 1.5).

In a job shop environment, characterized by higher variety and lower demand, a larger degree of flexibility is required. It is natural to group machines according to their processing capabilities to obtain a **process-**

An overview of management problems in manufacturing systems 13

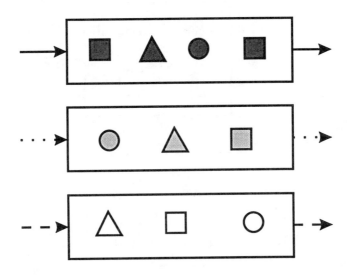

FIGURE 1.5
A product-oriented layout

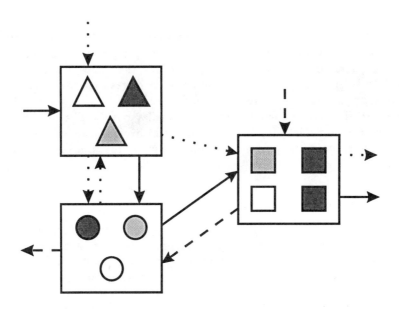

FIGURE 1.6
A process-oriented layout

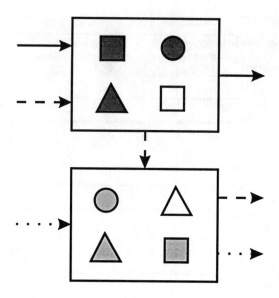

FIGURE 1.7
A group technology layout

oriented layout (see Figure 1.6). A process-oriented layout is more flexible than a product-oriented layout, but material flow is scattered and difficult to manage; apart from material handling problems, there may be difficulties related to large setup times. A compromise is represented by a **group technology** layout (see Figure 1.7): in this case we build a set of cells oriented to the production of a *family* of similar parts. The aim is to rearrange the manufacturing system in order to avoid scattered material flows and to reduce setup problems, without requiring the large investments associated with a product-oriented layout.

1.3.3 Performance measures

Given the diversity in hierarchical levels, production scenarios and manufacturing system layouts, it comes as no surprise that there is a variety of performance measures we may be interested in. It is possible to classify performance measures by relating them to the following basic factors:

Resource-related costs. In production planning problems, the possible need for increasing the production capacity (e.g. through the use of overtime) can be considered. In production scheduling this usually does not make sense, since resource availability is a given constraint. Among the resources we may need to consider, apart from machine availability, are skilled labor, tools, dies, and fixtures.

Inventory-related costs. In a production planning problem within a make-to-stock environment, it is quite natural to consider inventory holding costs. However, even in a production scheduling problem within a make-to-order environment, we must cope with inventory-related issues. For instance, we would like to keep Work in Process (WIP) as low as possible. Furthermore, if due dates are given for completing the orders, we should not complete the corresponding jobs too early.

Lead time. The term *lead time* may take different meanings depending on the context. In an inventory control problem, it is the time elapsing between the release of a replenishment (purchase or production) order and the receipt of the items. In a production scheduling problem, it may be the time elapsed between the release of the raw materials to the shop floor and the completion of the order. Alternatively, it may be the time elapsed between the start of the first and the completion of the last operation of a job. In the two last cases, the lead time is a measure related to the WIP level.

Setup-related costs. Setup issues may play a major role both in production planning and scheduling problems. We must distinguish different cases. First, we may consider setup *costs* and/or setup *times*. Sometimes setup costs are not considered, but setup times heavily influence other performance measures. Second, such costs/times may be *sequence-independent* or *sequence-dependent*. In the first case, cost/time depends only on the next part type to be processed; in the second one, cost/time depends both on the last and the next part type. Clearly, the case of sequence-dependent setups calls for detailed scheduling models and cannot be tackled by production planning models.

Machine utilization. This may be an important measure for capital-intensive equipment. In this case we want the system to be fully utilized. However, the importance of machine utilization should not be overemphasized. Consider, for instance, a manufacturing system consisting of two cascaded machines, as in example 1.4. If the second machine is a bottleneck, there is little point in fully utilizing the first one, since this results in the build-up of WIP. Nevertheless, machine utilization is important in spotting bottlenecks and trying to improve other performance measures.

Throughput. The throughput is defined as the number of parts produced per unit time. Throughput is limited by bottlenecks.

Customer service related costs. In a make-to-stock environment we may require that each incoming order is immediately satisfied from the available stock. Alternatively, we may admit a certain degree of backlog. We will see that, if the clients are patient and accept the delay, this can be modeled by negative inventory levels. In some cases, it is better to consider the **service level**, that is, the percentage of satisfied customers. In a make-to-order environment we must compare the required due date with the completion time of the order; below we show which performance measures can be used to express this requirement.

Technological costs. These costs are often overlooked in the production management literature. However, it might be important to tradeoff the use of more costly process plans for the advantages obtained with respect to other performance measures.

In the following chapters we will see that in production planning problems, inventory and setup costs are usually considered; variants involve overtime and backlog costs.

In scheduling problems, the most important variable is the completion time C_i of each job J_i ($i = 1, \ldots, N$). Given a schedule, we might evaluate it using an *additive* (minsum) function of the form

$$\sum_{i=1}^{N} \gamma_i(C_i),$$

or a minmax function like

$$\max_{i=1,\ldots,N} \gamma_i(C_i).$$

The γ_i are "elementary" functions of the completion time. A common example is the **flow time**: $F_i = C_i - r_i$, where r_i is the **release time** of job J_i, i.e., the time at which the job is released and we may start processing it. The flow time is usually considered synonymous of lead time, and it is a measure related to WIP, since it is the time the job spends on the shop floor. The total flow time, $\sum_{i=1}^{N} F_i$, is related to the average WIP level in the system. A well-known performance measure of the max-type is the **makespan**, defined as $C_{\max} = \max_{i=1,\ldots,N} C_i$. We will see that the makespan can be related to machine utilization issues.

When **due dates** are associated with jobs (as is customary in a make-to-order environment), we may express our ability to meet customer expectations by different functions. We may consider the **lateness** $L_i = C_i - d_i$, or the **tardiness** $T_i = \max\{L_i, 0\}$. Note that the lateness is negative if a job is completed early, whereas the tardiness can only assume positive values. If inventory issues are important, we should consider the **earliness** $E_i = \max\{-L_i, 0\}$. It is natural to consider performance measures such as the **total tardiness** $\sum_{i=1}^{N} T_i$ or the **maximum lateness** $L_{\max} = \max_{i=1,\ldots,N} L_i$. We might also consider the number of tardy jobs.

We may modify the basic performance measures by associating a **weight** w_i to each job. For instance, to obtain the total *weighted* flow time $\sum_{i=1}^{N} w_i C_i$ and the total *weighted* tardiness $\sum_{i=1}^{N} w_i T_i$.

There are other types performance measures that may be considered when dealing with scheduling problems. We will come back to the performance measures for scheduling problems in section 4.2.

Finally, evaluative models are usually concerned with throughput, WIP, utilization and lead time.

1.4 Plan of the book

Including the present chapter, the book consists of six chapters and five appendices.

Chapter 2 *Optimization models and model solving*

Chapter 3 *Discrete-time models*

Chapter 4 *DEDS models for scheduling problems*

Chapter 5 *Evaluative models*

Chapter 6 *Putting things together*

Appendix A *Fundamentals of Mathematical Programming*

Appendix B *Linear Programming and Network Optimization*

Appendix C *Enumerative and Heuristic Methods for Discrete Optimization*

Appendix D *Dynamic Programming*

Appendix E *Stochastic modeling*

Chapters 2, 3 and 4 deal with generative models. Optimization models and methods are introduced in Chapter 2. For the reader's convenience, we have collected background material in the appendices. Appendices A, B, C, D are prerequisites for Chapters 3 and 4. In Chapter 3 we present discrete-time models and discrete-event models are dealt with in Chapter 4. We have preferred a model-based, rather than problem-based, presentation; therefore, Chapter 3 deals both with planning and scheduling problems, whereas Chapter 4 is concerned only with scheduling models. This is due to the fact that DEDS models, being more detailed, are not suitable for high-level problems. Continuous-time models are practically neglected; there is only one example in section 6.2 in Chapter 6.

Evaluative models are described in Chapter 5. The treatment of analytical models requires some background in stochastic modeling, which is given in Appendix E.

Finally, in Chapter 6 we "put things together." On the one hand, we provide some examples of integration between generative and evaluative models; on the other hand, we consider some topics which are fundamental to implement decision support software based on optimization models, i.e. integration with databases, model management and the use of model generators.

Each chapter includes one (or more) supplement. Supplements contain advanced material, which may be safely skipped on the first reading.

After the appendices, a section is provided including some exercises and problems. We have preferred a single section at the end of the book, rather than a separate section at the end of each chapter, because we feel that the reader should tackle these problems having the whole picture in mind, in order to appreciate the tradeoffs involved in the selection of a modeling approach.

For further reading

In this book we deal with mathematical models related to the management of manufacturing systems. This does not imply that mathematical models are the only or the best answer to such problems. There are many commercially available software packages for production management which do not rely on a mathematical approach, such as Manufacturing Resource Planning (MRP) systems. The brief introduction we have given in this first chapter is sufficient for mathematically inclined readers who want to get a glimpse of some possible applications of optimization and performance evaluation models. However, the reader who is truly interested in applying mathematical models to practical problems must first get acquainted with topics in Operations Management which are outside the scope of this book.

A classical reference for MRP systems, Just in Time, and operations scheduling is [203]. Readers interested in a broader introduction to Operations Management may consult [140]. A deeper understanding of some production management issues may be attained by reading [133], which is a most useful introduction to solving business problems by standard MRP packages.

It is not the aim of this book to present *all* the mathematical models pertaining to production management. Such a broad survey can be given in a reasonably sized book either by assuming a solid background on the part of the reader or by keeping things at a superficial level. Our aim is to provide students and practitioners with strong foundations in order to enable them to evaluate and/or implement decision support software based on mathematical models, as well as to gain access to the scattered literature published in scientific journals. As a consequence, we have decided to go into relatively deep detail as far as certain topics are concerned, while others have been deliberately omitted. For instance, we have neglected classical inventory control issues. The interested reader may find a quite accessible introduction to inventory control in [206]; sophisticated models are illustrated e.g. in [101]. Other useful readings are [6] (where the reader may find, for instance, models for layout optimization and Group Technology) and [95] (where exhaustive surveys are given on many advanced topics). Tutorial introductions to some modeling topics may also be found in [34].

2

Optimization models and model solving

In Chapter 1 we have drawn a distinction between generative and evaluative models. Since generative models are basically optimization models, in this chapter we lay down the foundations for writing and solving them on a computer.

Optimization models come in a wide variety of forms, and there is a corresponding variety of solution methods. Optimization models are classified in Section 2.1, and the available methods are associated with each model type in Section 2.2. In this chapter we do not go into any detail about optimization *methods*; after reading Section 2.2, the reader with no optimization background is referred to Appendices A, B, C, D where we deal with the fundamentals of Mathematical Programming, Linear Programming and Network Optimization, Branch and Bound methods, and Dynamic Programming.

The ultimate aim of optimization methods is to find the solution of a *model* by a computer program, in order to help the manager to find the solution of a *problem*. In lucky cases we are able to devise exact optimization methods, ensuring the optimality of the solution of the model (though this does not necessarily imply the optimality of the solution with respect to the problem). Obviously, we cannot take for granted that we are able to devise a solution method for every optimization model; if this is not possible, we must settle for an approximate solution method. A fundamental distinction must be drawn between **exact** and **heuristic** optimization methods, where the latter aim at finding a good solution, without guaranteeing its optimality.

Unfortunately, there are many cases in which finding the optimal solution of the model is trivial from a conceptual point of view, but impossible from a practical one. The typical example is a problem with a finite number of solutions. A brute-force enumeration method is all we need to find the optimal solution, provided we have enough time; in fact, even in moderate size problems, "enough" may mean some billions years. Also in this case, unless we are able to characterize the optimal solution or to enumerate the set of solutions in an efficient way, we must settle for a heuristic optimization method, trading off the solution quality for a reduced computational burden.

In Section 2.3 we clarify the concepts behind efficiency issues by introducing the reader to the theory of **computational complexity**. In particular, we distinguish between the complexity of an algorithm and the complexity of a problem, and we introduce important terms such as **polynomial** and NP-**hard** problem. For the remainder of the book, only an intuitive grasping of such concepts is needed; the interested reader can find more information in Supplement S2.1.

In Section 2.4 we consider the relationship between model building and model solving. At first sight, one could think that they are distinct phases. However, even within the same class of models, the way a model is formulated can have a significant impact on the efficiency of the solution procedure. This is the case when using commercial software packages for solving Mixed-Integer Linear Programming problems. We show some basic modeling principles that may help in improving the solvability of the model.

We have said that, despite all the efforts, optimization models are often too difficult to solve. Indeed, one of the major criticisms of using optimization methods in production management is the computational complexity of the solution procedures; the use of common-sense procedures have been advocated as a practical way of coping with realistic problems. An important point is that mathematical modeling may be a valuable source of excellent heuristic methods able to find near-optimal solutions to practical problems. In Section 2.5 we point out different approaches to obtain approximate strategies by mathematical modeling.

A further criticism of the use of optimization models is that they consider a *single* objective function, whereas many real problems are characterized by multiple, conflicting objectives. A thorough treatment of multiobjective optimization is beyond the scope of this book, but in Supplement S2.2 we show how single objective optimization methods can be used to cope with multiobjective problems.

2.1 Classes of optimization models

We have seen some examples of optimization models in Chapter 1. An optimization model can be cast in the general form:

$$\min_{\mathbf{x} \in S} f(\mathbf{x}). \tag{2.1}$$

We see that there are three elements in an optimization model:

1. **x** is the set of **decision variables**, which encode the different policies that could be adopted to cope with the management problem. Usually **x** is a n-dimensional vector of real variables, i.e. $\mathbf{x} \in \mathbf{R}^n$. The solution of the optimization problem generates a set of decision variables; this is why optimization models are called generative.

2. S is the set of **feasible solutions**. Feasibility is defined with respect to a set of economical and technological restrictions that any solution must satisfy. Usually the set S is expressed by constraints on the decision variables. We have three common forms of constraints:

   Equality constraints are represented by a set of equations

   $$h_i(\mathbf{x}) = 0 \qquad i \in E,$$

   involving the decision variables. We can write equality constraints in vector form:

   $$\mathbf{h}(\mathbf{x}) = \mathbf{0}.$$

   Inequality constraints are expressed as a system of inequalities:

   $$g_i(\mathbf{x}) \leq 0 \qquad i \in I.$$

   In vector form we may write:

   $$\mathbf{g}(\mathbf{x}) \leq \mathbf{0},$$

   having stipulated that a vector inequality is interpreted componentwise. Note that a constraint such as $g_k(\mathbf{x}) \geq 0$ can be trivially rewritten in the above form by considering $-g_k(\mathbf{x}) \leq 0$. A recurring restriction that can be thought of as an inequality constraint is the nonnegativity requirement $x_j \geq 0$, also denoted by $x_j \in \mathbf{R}_+$. We will see that optimization algorithms usually handle nonnegativity constraints in a special way.

   Integrality constraints are enforced when some decision variables may assume only *integer* values, i.e. $\mathbf{x} \in \mathbf{Z}_+^n$, where $\mathbf{Z}_+ = \{0, 1, 2, \ldots\}$ is the set of nonnegative integers (models involving negative integer variables are quite rare).

   Each of these restrictions delimits a subset of $\mathbf{R}^n$; the intersection of these subsets defines the feasible set.
   In some software packages for optimization, other types of restrictions are introduced for the sake of computational efficiency (see Section C.5).

3. $f(\mathbf{x})$ is the **objective function**, which associates a *scalar* value to each solution $\mathbf{x}$, measuring its suitability, e.g. from an economic point of view. More specifically, f is a model of the decisionmaker preferences, in the sense that if $f(\mathbf{x}_1) < f(\mathbf{x}_2)$, then $\mathbf{x}_1$ is a better solution than $\mathbf{x}_2$. In the formulation (2.1), we have assumed that the objective function is to be minimized. Sometimes, for instance when f represents a profit, the objective function must be maximized. Still, (2.1) is a general formulation, since a maximization problem can be readily transformed into a minimization problem:

$$\max_{\mathbf{x} \in S} f(\mathbf{x}) \quad \Rightarrow \quad -\min_{\mathbf{x} \in S}\left(-f(\mathbf{x})\right).$$

The solution of an optimization model yields an optimal solution $\mathbf{x}^*$ and an optimal value $f^* = f(\mathbf{x}^*)$. We also use the notation

$$\arg\min_{\mathbf{x} \in S} f(\mathbf{x})$$

to denote the optimal solution, and

$$\nu\left\{\min_{\mathbf{x} \in S} f(\mathbf{x})\right\}$$

to denote the optimal value.

Depending on the three elements of the optimization model, different classes of optimization problems can be distinguished. In the following we draw some general distinctions; different solution algorithms can be used to solve each type of problem.

2.1.1 Finite/infinite-dimensional problems

In a finite-dimensional problem the solution is represented by an n-dimensional vector of real variables, i.e. $\mathbf{x} \in \mathbf{R}^n$.

Example 2.1
The problem:

$$\begin{aligned} \min \quad & x_1 x_2 + x_3^2 \\ \text{s.t.} \quad & x_1 + x_2 + x_3 = 1 \\ & x_1^2 + x_2^2 \leq 4 \end{aligned}$$

is a finite-dimensional problem with one equality constraint, one inequality constraint, and

$$\mathbf{x} = \begin{bmatrix} x_1 \\ x_2 \\ x_3 \end{bmatrix} \in \mathbf{R}^3. \quad \square$$

In an infinite-dimensional problem, the solution is represented by an infinite collection of decision variables. This is the case when the solution is represented as a function of time over a continuous interval.

Example 2.2
Consider a continuous-time dynamic system represented by a vector state equation
$$\dot{\mathbf{x}}(t) = \mathbf{h}(\mathbf{x}(t), \mathbf{u}(t)),$$
where $\mathbf{x}$ is the vector of state variables and $\mathbf{u}$ is the vector of control inputs. A common problem is to find a suitable control by solving the infinite-dimensional problem:
$$\min \int_0^T f(\mathbf{x}(t), \mathbf{u}(t)) dt$$
$$\text{s.t.} \quad \dot{\mathbf{x}}(t) = \mathbf{h}(\mathbf{x}(t), \mathbf{u}(t)) \quad \forall t \in [0, T]$$
$$\mathbf{x}(0) = \mathbf{x}_0$$
$$\mathbf{u}(t) \in \Omega \quad \forall t \in [0, T],$$
where $[0, T]$ is the time horizon we are interested in, $\mathbf{x}_0$ is the inital state of the system, and Ω is the set of admissible controls. It is also possible to specify the terminal state $\mathbf{x}(T) = \mathbf{x}_T$. Restrictions on the initial and terminal state are referred to as **boundary conditions**. □

Often the terms **Mathematical Programming** and **Optimal Control** or **Dynamic Optimization** are used to distinguish finite- and infinite-dimensional problems. It must be noted that many authors consider Optimal Control problems as particular Mathematical Programming problems [144]. This is why we prefer to use the terms "finite-" and "infinite-dimensional". Furthermore, Mathematical Programming methods may be used to solve Optimal Control problems by discretizing the continuous-time model (see Section 1.2, Example 1.2). The infinite-dimensional problem above can be transformed into the finite-dimensional problem:
$$\min \sum_{k=1}^{K} f(\mathbf{x}_k, \mathbf{u}_k)$$
$$\text{s.t.} \quad \mathbf{x}_k = \mathbf{h}(\mathbf{x}_{k-1}, \mathbf{u}_k) \quad k = 1, \ldots, K$$
$$\mathbf{u}_k \in \Omega \quad k = 1, \ldots, K,$$
with suitable boundary conditions, where $\mathbf{x}_k = \mathbf{x}(k\Delta)$. Note that here $\mathbf{x}_k$ is the state *at the end* of the kth period (i.e., the period between $(k-1)\Delta$ and $k\Delta$), whereas $\mathbf{u}_k$ is the control applied *during* the kth period.

2.1.2 Linear/nonlinear problems

A finite-dimensional problem is called a **Linear Programming** (LP) problem when both the constraints and the objective are expressed by linear functions.[1] The general form of a linear programming problem is

$$\min \sum_{j=1}^{n} c_j x_j$$

$$\text{s.t.} \sum_{j=1}^{n} a_{ij} x_j = b_i \quad \forall i \in E$$

$$\sum_{j=1}^{n} d_{ij} x_j \leq e_i \quad \forall i \in I,$$

which can be written in matrix form

$$\min \mathbf{c}^T \mathbf{x}$$
$$\text{s.t.} \ \mathbf{A}\mathbf{x} = \mathbf{b}$$
$$\mathbf{D}\mathbf{x} \leq \mathbf{e}.$$

Example 2.3
Here is an example of a LP problem:

$$\min \ 2x_1 + 3x_2 + 3x_3$$
$$\text{s.t.} \ x_1 + 2x_2 = 3$$
$$x_1 + x_3 \geq 3$$
$$x_1, x_2, x_3 \geq 0. \quad \square$$

If either condition is not met, i.e., if the objective function or a constraint is expressed by a nonlinear function, we have a nonlinear programming problem.

Example 2.4
The following are examples of nonlinear programming problems:

---

[1] For the sake of precision, it would be better to use the term *affine*, rather than linear: a function of the form $f(x) = ax + b$ is not actually linear.

$$\min\ 2x_1 + 3x_2 + 3x_3$$
$$\text{s.t.}\ x_1 + x_2^2 = 3$$
$$x_1 + x_3 \geq 3$$
$$x_1, x_2, x_3 \geq 0.$$

$$\min\ 2x_1 + 3x_2 x_3$$
$$\text{s.t.}\ x_1 + 2x_2 = 3$$
$$x_1 x_3 \geq 3$$
$$x_1, x_2, x_3 \geq 0.$$

$$\min\ 2x_1^2 + 3x_2^2 + 3x_3^2$$
$$\text{s.t.}\ x_1 + 2x_2 = 3$$
$$x_1 + x_3 \geq 3$$
$$x_1, x_2, x_3 \geq 0.$$

The last problem is characterized by a quadratic objective function and by linear constraints. This kind of problem is called a **quadratic programming** problem. Quadratic programming problems are the simplest nonlinear programming problems (provided that the objective function is convex; see Section A.1). □

2.1.3 Continuous/discrete problems

So far we have assumed that x is a vector of real variables, or, in other words, that the management policy can be expressed by continuous decision variables. Very often, some decision variables must be restricted to integer values. In such a situation we have an **integer programming** or **discrete optimization** problem. To be precise, the term "discrete optimization" has a more general meaning than "integer programming", since it encompasses both integer programming and **combinatorial optimization** problems. Combinatorial optimization problems, such as the scheduling problems discussed in Chapter 4, *may be* represented by integer programming problems, but this is not necessary, nor it is always advisable.

Consider, for instance, a vector x whose components express the production quantity of a set of products. In a chemical production environment a continuous value of x makes sense, whereas in discrete manufacturing it should take only integer values. We speak of **a general integer variable** when we require $x_j \in \mathbf{Z}_+$. A common special case is when we require $x_j \in \{0, 1\}$; in this case we speak of **binary** or **logical** variables. The term **pseudoboolean** optimization is also used to refer to a problem involving only binary variables. Binary variables are used when x is related to a set of discrete choices; for instance, a certain part may be processed using one of M available machines, or we may produce at most one type of product during a certain time period (e.g., a shift).

In practice, general integer variables are much less common than one would expect, whereas binary variables are a most powerful modeling tool. General integer decision variables are often approximated by continuous variables. It turns out that dealing with integer variables makes the problem much more difficult to solve (see Appendix C); therefore, if production quantities are large enough, it is advisable to relax the integrality requirement and then to round the optimal solution. If, for instance, we have $x_1^* = 1000.3$, it is quite acceptable to round it down to 1000; a possibly unacceptable error would be made if we had $x_1^* = 4.3$. It is important to note that rounding a continuous optimal solution to an integer one is not only critical from the optimality point of view. In fact, it may lead to an *unfeasible* solution; recovering a feasible integer solution from the optimal solution of the continuous version of the model is practically impossible for certain models, and relatively easy for others.

It is possible to build nonlinear integer programming models, but linear integer programming models are by far more common. We speak of a **mixed-integer linear programming** (MILP for short) model when only a subset of the decision variables is restricted to integer values. When all the variables are restricted to integer values, we have a **pure integer programming** problem.

Example 2.5
Here is an example of an MILP problem:

$$\min \ \mathbf{c}^T\mathbf{x} + \mathbf{e}^T\mathbf{y}$$
$$\text{s.t.} \ \mathbf{Ax} + \mathbf{Dy} \leq \mathbf{b}$$
$$\mathbf{x} \in \mathbf{R}_+^{n_1} \quad \mathbf{y} \in \mathbf{Z}_+^{n_2},$$

where $\mathbf{R}_+$ and $\mathbf{Z}_+$ are the set of nonnegative real numbers and nonnegative integer numbers respectively.

A pure 0/1 linear problem is:

$$\min \ \mathbf{c}^T\mathbf{x}$$
$$\text{s.t.} \ \mathbf{Ax} \leq \mathbf{b}$$
$$\mathbf{x} \in \{0, 1\}^n. \quad \square$$

Several examples of integer programming models are given in Section C.1.

2.1.4 Deterministic/stochastic problems

So far, we have implicitly assumed that all the relevant data of an optimization problem were known with certainty. In practice this is hardly the case. Common sources of uncertainty in a production management problem are:

- unexpected fluctuations in the demand;
- uncertainties in the supply chain;
- failure-prone machines;
- uncertainty in the processing times (particularly when manual operations are involved).

It is possible to explicitly take into account the stochastic character of a problem; in this case the optimization problem should be written as

$$\min_{\mathbf{x} \in S} E[f(\mathbf{x})],$$

where we are interested in minimizing the *expected value* of the objective function, denoted by $E[\cdot]$. The feasible region S may be random, and the link between $\mathbf{x}$ and $f(\mathbf{x})$ may be random. A thorough treatment of stochastic optimization is definitely beyond the scope of this book, but we outline an example of this approach in Section 6.2, based on the exploitation of Dynamic Programming techniques (see Appendix D). Another possible approach is to use evaluative stochastic models, such as those described in Chapter 5, possibly integrating them with optimization methods (see e.g. Section 6.1.3).

However, the inherent uncertainty in manufacturing systems does not necessarily imply that deterministic models should be dismissed. Uncertainty can be coped with in different ways. It is possible to exploit a hierarchical decision structure. Rough cut decisions are taken at high decisional levels with a relatively long time horizon; the high level problem is used to drive the low level problem solver, which operates with a shorter time horizon. The high level aggregate plan is used to overcome the myopia of the low level problem solver, which in turn can take into account the occurrence of random events. Often a rolling time horizon approach is adopted, i.e. only the first part of the aggregate plan is detailed and implemented. Furthermore, a buffering mechanism such as safety stocks can be adopted (see Section 3.1).

2.1.5 Single/multiobjective problems

We have said that a scalar objective function f is a model of the decision-maker preferences, in the sense that it induces an ordering of the feasible solutions on the basis of their value $f(\mathbf{x})$; assuming a minimization problem, $\mathbf{x}_a$ is better than $\mathbf{x}_b$ if and only if $f(\mathbf{x}_a) < f(\mathbf{x}_b)$. If $f(\mathbf{x}_a) = f(\mathbf{x}_b)$, we assume that the solutions are indifferent; it may happen that an optimization problem has multiple optimal solutions, in which case we take one of them arbitrarily.

Usually, there are different, and *conflicting*, features of a solution that we must evaluate; if such features are expressed by functions $f_1(\mathbf{x}), \ldots, f_m(\mathbf{x})$, we might think of aggregating them by a scalar objective function in order to find a suitable tradeoff. The most natural scalarization is based on a linear combination of the functions f_j through a set of weights w_j:

$$f(\mathbf{x}) = \sum_{j=1}^{m} w_j f_j(\mathbf{x}).$$

Note that a linear combination approach is suitable if certain hypotheses about the preference structure of the decisionmaker(s) hold, but it is not the only possibility. The weights' role is to make the different evaluations "homogeneous". Unfortunately, there are many practical settings in which such a scalar objective function may be hard to assess. The decisionmaker may be unable to identify a reliable set of weights due to the incommensurability among the objectives and to uncertainty. The situation is worse when there are conflicts among different decisionmakers. In such cases a different approach must be taken.

Consider the simple case with $m = 2$. If we assume that we are not able to sensibly combine f_1 and f_2, we have to cope with a **vector optimization** problem:

$$\text{``}\min_{\mathbf{x} \in S}\text{''} \mathbf{f}(\mathbf{x}) = \begin{bmatrix} f_1(\mathbf{x}) \\ f_2(\mathbf{x}) \end{bmatrix},$$

where $\mathbf{f}(\mathbf{x})$ is a *criterion vector*, and the notation "min" means that minimization does not make sense from a strict mathematical point of view. In the vector case we may find two solutions such that $f_1(\mathbf{x}_a) < f_1(\mathbf{x}_b)$ but $f_2(\mathbf{x}_a) > f_2(\mathbf{x}_b)$ (or vice versa). In such a case, unless some way to compare the two objectives is known, it is impossible to establish a complete order among the solutions, but only a partial one. In fact, it is reasonable to assume that $\mathbf{x}_a$ is better than $\mathbf{x}_b$ if $f_1(\mathbf{x}_a) < f_1(\mathbf{x}_b)$ and $f_2(\mathbf{x}_a) \leq f_1(\mathbf{x}_b)$, or $f_1(\mathbf{x}_a) \leq f_1(\mathbf{x}_b)$ and $f_2(\mathbf{x}_a) < f_1(\mathbf{x}_b)$. In this case the criterion vector of $\mathbf{x}_a$ dominates the criterion vector of $\mathbf{x}_b$.

A possible approach to solve the vector optimization problem is to provide the decisionmaker with a set of 'reasonable' solutions, namely, those such that their criterion vector is not dominated by any criterion vector

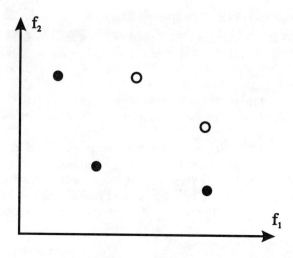

FIGURE 2.1
The efficient set for a two-criterion problem

associated with a feasible solution. These solutions are called **efficient solutions** and form the efficient set (see figure 2.1).

In Supplement S2.2 we outline possible strategies for generating the efficient set by solving a set of auxiliary scalar problems. Unfortunately, there are some drawbacks in such an approach:

- the number of efficient solutions can be too large; screening or interactive strategies may be needed;
- when dealing with integer programming models, determining the efficient set exactly is out of the question, because of excessive computational requirements.

2.2 An overview of optimization methods

We have seen that there is a wide variety of optimization models to which an even wider variety of optimization methods correspond. Optimization methods may be classified according to many different criteria. We do not want to review all of them, but it is fundamental to draw at least the following distinctions.

Optimization methods may be **exact** or **heuristic**. Exact methods ensure the optimality of the solution they find, whereas heuristic methods are only expected to yield a reasonably good solution. Sometimes it is possible to establish a bound on the degree of suboptimality of the solution, i.e. by how much the suboptimal solution value differs, in the worst case, from the optimal value. We would like to know an upper bound ϵ on the relative error, guaranteeing that

$$\frac{\hat{f} - f^*}{f^*} \leq \epsilon,$$

where $\hat{f}$ is the value of the solution obtained by the heuristic method. This bound may be given a priori (i.e., before running the algorithm) or a posteriori. Unfortunately, a priori bounds are mainly of theoretical interest, since they are known for specific and relatively simple problems and are usually rather weak. Heuristic methods are very important in practice because they trade off the solution quality in favor of drastically reduced computation times.

A second major feature of optimization methods is the degree of generality. We distinguish **general purpose** and **special purpose** methods. By general purpose method, we do not mean a method able to cope with a generic optimization problem. We mean a method able to cope with any problem cast in a certain form. For instance, there are commercially available codes able to solve any problem formulated as an MILP. On the opposite side, we have many algorithms able to solve (exactly or approximately) a specific problem. There is a class of heuristic methods lying somewhere in between, in that they are based on a rather general and widely applicable principle, but their application requires a certain degree of "customizing".

At this point we refer the reader to the following appendices.

Appendix A: *Fundamentals of Mathematical Programming.*
There we review some basic concepts such as convexity, local and global optimality, as well as the fundamental concepts behind continuous nonlinear programming problems. Actually, continuous nonlinear programming methods, per se, play a small role in this book. Still, some familiarity with topics such as gradient-based methods and Lagrangian duality is fundamental. We also cover decomposition approaches for large-scale problems and the optimization of nondifferentiable functions.

Appendix B: *Linear Programming and Network Optimization.*
In this appendix we briefly review some basic theory concerning con-

tinuous LP problems. In particular we outline the well-known Simplex method for LP; this algorithm is available on any type of hardware platform, and it is at the basis of software packages for MILP problems. We also review the basics of some Graph and Network Flow Optimization problems which can be formulated as LP problems.

Since there are many excellent books on such topics, we just give an overview of those aspects which are relevant for the remainder of the book. It is understood that once a problem has been cast in LP form, quite efficient algorithms are available for its solution. Greater computational efficiency is attained in the special cases of Network Flow problems.

Appendix C: *Enumerative and Heuristic Methods for Discrete Optimization.*

In this appendix we first give several examples of how a problem can be formulated in MILP form. MILP is the most powerful modeling framework we use in the book. Unfortunately, apart from some specific cases, solving an MILP problem turns out to be awkward. The reason is that spotting the optimal solution of a discrete optimization problem usually calls for a costly enumeration process.

The most successful enumerative method for discrete optimization problems is **Branch and Bound**, which aims at limiting the enumeration process by showing that the optimal solution cannot lie in some subsets of the feasible set. There are both general purpose (commercially available) and special purpose Branch and Bound algorithms. General purpose methods are based on the Simplex method for LP problems and are able to solve MILP problems. Other specific methods do not necessarily need a mathematical model of the problem.

Since many practical discrete optimization problems cannot be solved at optimality, due attention must be paid to effective and efficient heuristic methods. In this book we limit ourselves to general principles that can be adapted to specific problems.

Appendix D: *Dynamic Programming.*

Dynamic Programming (DP) is not a specific optimization algorithm. It is more of a *principle* that can be adapted to a wide variety of problems. In fact, DP is probably the most powerful tool for optimization, since it can cope with discrete optimization problems, but also with infinite-dimensional and stochastic problems. Such a remarkable generality comes at a price. There is no off-the-shelf software package implementing DP. Furthermore, DP may be difficult to exploit from a computational point of view, due to its memory and time requirements. In some cases, it leads to equations that cannot be solved analytically. Still, DP provides us with quite efficient algorithms to solve specific but important problems and may be a valuable source of approximate algorithms. Furthermore, it can be used to spot the *qualitative structure* of the optimal policy, even if determining its numerical parameters may be difficult.

2.3 Complexity of optimization problems

When solving optimization problems by a computer, a fundamental issue is the computational efficiency. A useful characterization of efficient algorithms is obtained by considering the number of elementary steps required as a function of the size of the input problem. As the size of the input problem, we may consider the amount of bits of computer memory required to store the problem data under a suitable encoding. It turns out that it is usually not necessary to be overly precise in the definition of what we mean by "step" and of the encoding scheme.

We must distinguish the concepts of **problem** and **problem instance**. The "problem" can be thought of as the collection of all the instances we can build by specifying the numerical value of the problem data. Let N be the size of the problem instance and $g(N)$ be the number of steps needed by a solution algorithm. We are interested in how fast the computational complexity grows when N grows. Actually there are different measures we could be interested in. We may consider the **average-case** complexity and the **worst-case** complexity, defined over the set of all the possible instances of a problem. Coping with average complexity requires a probabilistic treatment of the possible instances, which is rather difficult; therefore, we deal only with worst-case complexity.

Efficient algorithms are characterized by a complexity which is bounded by a polynomial function of N; inefficient algorithms have an exponential complexity. The class of problems for which a polynomial algorithm is known is denoted by $\mathcal{P}$; this is the class of *easy* problems.

Example 2.6
Consider a sorting problem: given a vector of N numbers, we want to arrange them in nondecreasing order. An approach to this problem is the selection sort algorithm [200]: we first find the smallest element by successive comparisons (this requires $N-1$ comparisons). Then we find the smallest element among the remaining elements (which requires $N-2$ comparisons) and so on. The number of comparisons is easy to compute:

$$\sum_{i=1}^{N}(N-i) = N^2 - \sum_{i=1}^{N} i = N^2 - \frac{N(N+1)}{2} = \frac{N(N-1)}{2} = O(N^2).$$

The notation $O(\cdot)$ has the usual interpretation. We write $f = O(g)$ when there are constants c and K such that $f(n) \leq cg(n)$ for $n \geq K$; g is therefore an *upper bound* on the increase of f as their argument gets large. More information is given when we may find constants c_1, c_2, K such that $c_1 g(n) \leq f(n) \leq c_2 g(n)$ for $n \geq K$. In this case we write $f = \Theta(g)$, and both f and g grow at the same rate. In the insertion sort case, we may write that the complexity is $\Theta(N^2)$.

This complexity of the insertion sort algorithm is independent of the particular vector instance at hand. More sophisticated algorithms may have a better performance for certain problem instances. For example, the quicksort algorithm has an average complexity $O(N \log N)$; its worst-case complexity is $O(N^2)$. □

Sorting a vector of numbers is therefore an "easy" problem. This implies that optimization problems that can be solved by sorting a set of numbers are "easy".

Example 2.7
Consider the following sequencing problem. We are given a set of N jobs to be processed on a single machine. Each job J_i has a processing time p_i and a due date d_i. Given a job sequence, represented by a permutation σ of the job indexes, each job has a completion time C_i. If $\sigma(k)$ is the index of the kth job in the sequence, then

$$C_{\sigma(k)} = \sum_{j=1}^{k} p_{\sigma(j)}.$$

We can measure the violation of the due date by the lateness $L_i = C_i - d_i$. We want to sequence the jobs in order to minimize the maximum lateness $L_{\max} = \max_i L_i$. This problem can be represented as $1//L_{\max}$, i.e. a problem with one machine whereby we want to minimize $L_{\max}$. It can be shown that this problem is easily solved by arranging the jobs in nondecreasing order of their due dates (this sequencing rule is known as the Earliest Due Date, or EDD, rule; see Section 4.3.2). Therefore $1//L_{\max}$ is in $\mathcal{P}$. □

Unfortunately, more often than not it is impossible to find a suitable characterization of the optimal solution of sequencing problems. In such a case, a brute-force enumeration of the feasible solutions would seem a viable solution approach. The following example shows how relatively small problems turn out to be computationally intractable.

Example 2.8
Consider a generic single machine sequencing problem; if we have N jobs, the number of possible permutations is $N!$. In principle, the optimal solution could be found by complete enumeration. Unfortunately, such a brute-force approach is not applicable but for the smallest problems. If $N = 30$, we have to enumerate $30! \simeq 2.65 \cdot 10^{32}$ sequences. If we could generate and evaluate one billion sequences per second (which is not actually the case), it would take about $8.41 \cdot 10^{15}$ years to completely explore the solution set. □

The difficulty is that the factorial $N!$ grows faster than any polynomial in N, resulting in a combinatorial explosion of the complexity for increasing N. A similar combinatorial explosion occurs for binary programming problems.

Example 2.9
Consider the knapsack problem

$$\max \sum_{i=1}^{N} v_i x_i$$

$$\text{s.t.} \sum_{i=1}^{N} w_i x_i \leq W$$

$$x_i \in \{0, 1\} \quad \forall i.$$

A simple enumeration algorithm would solve the problem; unfortunately there are 2^N possible N-dimensional binary vectors to check. Actually, the performance of the enumeration algorithm is not that bad, since not all vectors correspond to a feasible solution: still, the worst-case complexity is of exponential order. □

In Table 2.1 we quantify the effect of combinatorial explosion. The most important thing to note is that a major breakthrough in computer technology has a significant impact on polynomial algorithms, whereas it does not improve the situation for exponential algorithms. We have said that, usually, it is not important to exactly specify the encoding scheme and the type of elementary steps considered when evaluating the complexity as a function of the problem size. This is because changes in the way the complexity is measured imply a polynomial transformation. For instance, the instructions of a powerful computer can be simulated by a simpler computer in polynomial time. Therefore, in the case of a polynomial algorithm,

Worst-case complexity	Number of steps n			
	10	20	100	1000
$O(1000 \cdot n)$	0.01 sec.	0.02 sec.	0.1 sec.	1 sec.
$O(1000 \cdot n^2)$	0.1 sec.	0.4 sec.	10 sec.	16.7 min.
$O(2^n)$	0.001 sec.	1.05 sec.	$4 \cdot 10^{16}$ years	–
$O(n!)$	3.62 sec.	77146 years	–	–

TABLE 2.1
Combinatorial explosion: the worst-case computation times are obtained assuming that we can carry out one million elementary steps per second

such differences may change the order of the polynomial, but the distinction between polynomial and exponential algorithms is not affected.[2]

So far we have dealt with the computational complexity of specific algorithms. However, we should also pay attention to the intrinsic complexity of *problems*. If we know an efficient solution algorithm for a certain problem, we say that the problem is easy. However, if we cannot find an efficient algorithm, does this necessarily imply that the problem is difficult? It might be the case that the problem is easy, but we are just not able to find the "right" algorithm.

Example 2.10
Consider again the $1//L_{\max}$ sequencing problem addressed in Example 2.7. To solve it, we have just to pick up the most urgent job first and so on. This works if all the jobs are simultaneously available at time zero. However, some jobs could be available only after a certain time, called the release time. If job J_i has a release time r_i, we cannot start processing it before r_i. This problem is coded as $1/r_i/L_{\max}$. To solve it, it seems reasonable to adapt the EDD rule. We can select, as the next job to be processed, the most urgent one among those available. Unfortunately, a simple counter-example shows that this rule does not ensure the optimality of the resulting schedule.

Consider a $1/r_i/L_{\max}$ instance characterized by two jobs J_1 and J_2 and the following data:

- processing times $p_1 = 8$, $p_2 = 8$;

---

[2] A notable case where the type of encoding is important is the case of pseudopolynomial algorithms; see Supplement S2.1.

- release times $r_1 = 0$, $r_2 = 2$;
- due dates $d_1 = 20$, $d_2 = 10$.

Any priority rule, including EDD, behaves in the same way in this case. Job J_1 is scheduled first, since it is the only one available at time zero, but this is not the optimal schedule. In fact, it is better to wait until J_2 is available and to schedule J_1 after its completion. □

We have seen that $1/r_i/L_{\max}$ cannot be solved by a simple sorting algorithm; in fact, no polynomial algorithm is known for this problem. Since there seems to be little difference from the easy case, one might well wonder if there are inherent reasons why no polynomial algorithm is known for it. In other words, we would like to say something about the computational complexity of a *problem*, rather than an algorithm. If finding a polynomial algorithm seems impossible, it would be pleasing to show that it is indeed impossible to devise an efficient algorithm for the problem. Unfortunately this seems to be as difficult as finding an efficient algorithm for $1/r_i/L_{\max}$. A way out of the dilemma is to show that the apparently difficult problem $1/r_i/L_{\max}$ is *equivalent*, from the computational complexity point of view, to other problems which have defied the efforts of a large community of researchers. This is what the theory of NP–completeness is all about. Here NP stands for *nondeterministic polynomial*. Roughly speaking, it is a concept related to the possibility of solving a problem by enumeration on an infinitely parallel computer. We define the class $\mathcal{NP}$ as the class of problems which can be solved in polynomial time on a nondeterministic computer. We say that an optimization problem is NP–hard when a polynomial algorithm for its solution would yield a polynomial algorithm for all the problems in $\mathcal{NP}$, and for all other NP–hard problems. This would imply that the classes $\mathcal{P}$ and $\mathcal{NP}$ are the same. Since the class of NP–hard problems includes virtually all difficult discrete optimization problems, for which efficient algorithms are unknown, we may conclude that it is *unlikely* that a polynomial algorithm for a NP–hard problem will ever be discovered.

The interested reader may find further details about NP–completeness in Supplement S2.1, where we also show that $1/r_i/L_{\max}$ is NP–hard. For the remainder of the book it is enough to grasp the intuitive association between NP–hardness and difficult problems. Most discrete optimization problems related to production management are NP–hard; MILP problems are also NP–hard in general. After showing that a problem is NP–hard, two routes can be pursued for its solution: enumerative algorithms, such as Branch and Bound or Dynamic Programming, or heuristic methods.

A caveat is in order about the theory of NP–completeness. It copes with *worst-case complexity*. Though all NP–hard problems are equivalent, in practice, some of them are easier than the others. Furthermore, it may well be the case that, on the average, an exponential algorithm turns out to be better than a polynomial one. The most striking example is the Simplex algorithm for LP problems. It is actually an exponential algorithm, since pathological problems can be devised requiring an exponential number of steps [154, Chapter 8]. Still, the Simplex method is, in practice, a most efficient tool to solve LP problems. On the other hand, polynomial algorithms have been developed for LP, but the early ones were not competitive. The most recent ones do outperform the Simplex method in some cases, but it is not at all clear if they will actually replace it in common practice.

2.4 Good and bad model formulations

Several software packages are available for the solution of continuous and mixed-integer linear programming problems. One is tempted to think that, once a problem has been cast into one of these forms, the job is done (assuming there is no uncertainty about model parameters, and there is no need for a sensitivity analysis). This is often the case with continuous linear programming, since years of experience have shown that the Simplex method is sufficiently efficient and reliable for solving large scale problems. But when a mixed-integer formulation has to be coped with, things are very different. We have seen in the previous section that such problems are usually NP–hard. Although commercial software packages are available for solving general MILP problems, it is often necessary to exploit the problem structure to come up with a special purpose exact algorithm or to settle for a heuristic algorithm. This, however, does not imply that the general purpose approach of Branch and Bound with continuous relaxation should be dismissed. Implementing a special purpose software can be costly.

If one wants to adopt an off-the-shelf Branch and Bound approach, special care must be taken in developing the "right" model. In fact, for a given problem, different correct models may be written, but they are not always equivalent from the computational point of view.

Example 2.11
Consider a part type selection problem in a simple Flexible Manufacturing System (FMS). The FMS consists of a single machine, which has a tool magazine of capacity C (expressed in tool slots). The FMS must process a set of I part types indexed by i; each part type has a priority w_i related to the urgency of the corresponding order. Each part type i needs a tool set T_i; the number of tools in the tool set is denoted by $|T_i|$. Since the tool magazine size is not large enough to contain all the tools required to produce all the part types simultaneously, we must select the next subset

of parts to be processed. We assume here that tools do not wear out, that all of them require exactly one slot on the tool magazine, and that there is no Tool Logistic System able to support a change in the loaded tools while the machine is operating.

To build a mathematical model, we express the selection of part type i by the binary variable x_i, set to 1 if the part is selected, and the decision of loading tool t by the binary variable y_t, set to 1 if the tool is loaded. A possible model is:

$$\max \sum_{i=1}^{I} w_i x_i$$

$$\text{s.t.} \sum_{t=1}^{T} y_t \leq C$$

$$x_i \leq \frac{\sum_{t \in T_i} y_t}{|T_i|} \quad \forall i \tag{2.2}$$

$$x_i, y_t \in \{0, 1\} \quad \forall i, t.$$

Eq. (2.2) implies that the part type selection variable cannot be set to 1, unless all the required tools are loaded. From a theoretical point of view, this constraint is correct; in practice there are two problems with it. First, there may be numerical problems, whose consequence are that some part types can never be selected. Consider a part type requiring three tools (say 1, 2, 3). Eq. (2.2) is in this case

$$x_i \leq \frac{1}{3} y_1 + \frac{1}{3} y_2 + \frac{1}{3} y_3.$$

If the whole tool set is loaded, we have

$$x_i \leq \frac{1}{3} + \frac{1}{3} + \frac{1}{3} = 3\frac{1}{3}.$$

Unfortunately, on a computer with finite precision arithmetic, $3\frac{1}{3} \neq 1$, unless some correction mechanism is used. For instance, if we have $\frac{1}{3} = 0.33333$, then the constraint turns out to be

$$x_i \leq 0.99999,$$

and part type i is never selected. Therefore (2.2) should be rewritten as

$$|T_i| \cdot x_i \leq \sum_{t \in T_i} y_t \quad \forall i.$$

An alternative formulation of the same constraint is:

$$x_i \leq y_t \quad \forall i, \forall t \in T_i. \tag{2.3}$$

According to this second formulation, we have more constraints. Still, the second formulation is preferable, since (2.2) is a surrogate constraint for (2.3). Actually (2.2) is obtained by summing up (2.3) over each tool set. Although the set of discrete solutions does not change, the two formulations have different feasible regions in the continuous relaxation. Since surrogating constraints relaxes the feasible region (see Section C.4.4), the aggregate constraint results in weaker lower bounds. □

The point is the relationship between the convex hull $[S]$ of the feasible region S of an MILP problem and its continuous (or linear) relaxation $\overline{S}$. For instance, given a problem of the form:

$$\min \ \mathbf{c}^T \mathbf{x}$$
$$\text{s.t.} \ \mathbf{A}\mathbf{x} \leq \mathbf{b}$$
$$\mathbf{x} \in \{0,1\}^n,$$

we have:

$$S = \{\mathbf{x} \in \mathbf{R}^n \mid \mathbf{A}\mathbf{x} \leq \mathbf{b}; \ x_j \in \{0,1\}\}$$
$$\overline{S} = \{\mathbf{x} \in \mathbf{R}^n \mid \mathbf{A}\mathbf{x} \leq \mathbf{b}; \ x_j \in [0,1]\}$$
$$[S] = \left\{ \mathbf{x} \in \mathbf{R}^n \mid \mathbf{x} = \sum_i \lambda_i \mathbf{z}_i; \ \sum_i \lambda_i = 1, \ \lambda_i \geq 0, \ \mathbf{z}_i \in S \right\}.$$

An illustration of these sets is shown in Figure 2.2. Clearly we have $S \subset [S] \subseteq \overline{S}$.

In the lucky cases whereby $[S] = \overline{S}$, the MILP problem can be easily solved as an LP problem, since its optimal solution has integer coordinates; this happens in the case of a totally unimodular matrix $\mathbf{A}$ (see Section B.7). In general, however, solving the optimization problem over $\overline{S}$ just yields a lower bound for the optimal solution the discrete problem, the larger $\overline{S}$, the weaker (i.e. the smaller) the lower bound. Now, the representation of S by $\mathbf{A}$ and $\mathbf{b}$ is not unique. We may find $\hat{\mathbf{A}}$ and $\hat{\mathbf{b}}$ such that

$$T = \left\{ \mathbf{x} \in \mathbf{R}^n \mid \hat{\mathbf{A}}\mathbf{x} \leq \hat{\mathbf{b}}; \ x_j \in \{0,1\} \right\} \equiv S.$$

Even if $T \equiv S$, we have, in general, $\overline{T} \neq \overline{S}$ (see Figure 2.3). The efficiency of a standard MILP code based on continuous relaxation may be strongly enhanced by a representation such that the linear relaxation is as close as possible to the convex hull of the feasible region of the discrete problem.

This is exactly what happens in the previous example. We give another example of how a formulation can be improved in Supplement S3.1. Such model-dependent reformulations require some art and experience, but there are other simpler manipulations involving single constraints.

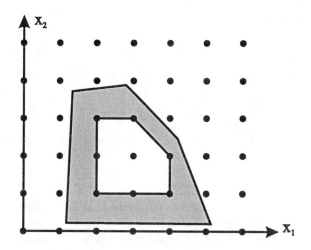

FIGURE 2.2
Relationship between the continuous relaxation and the convex hull of the feasible region of an MILP problem

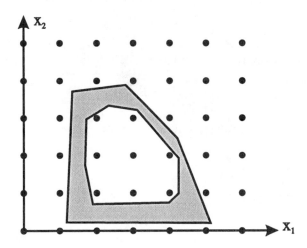

FIGURE 2.3
Different continuous relaxations

Example 2.12
Consider the constraint

$$2x_1 + 8x_2 + 6x_3 \geq 15,$$

where all the variables are restricted to integer values. Dividing by 2, we get:

$$x_1 + 4x_2 + 3x_3 \geq 7.5,$$

which, considering the integrality restriction, may be transformed into the tighter constraint:

$$x_1 + 4x_2 + 3x_3 \geq 8.$$

The second constraint is tighter in the sense that any point with integer coordinates which satisfies it, also satisfies the first constraint and vice-versa. However there are points with noninteger coordinates satisying the first constraint but not the second one. $\square$

Apart from transforming constraints into tighter ones, we may add constraints which are redundant from the discrete point of view, but may improve the continuous relaxation. They are called **valid inequalities**. Sometimes, it is also possible to eliminate constraints and to fix variables.

Example 2.13
Consider a knapsack type constraint:

$$\sum_{i \in I} w_i x_i \leq W,$$

where x_i are binary variables, and assume that

$$\sum_{i \in I} w_i > W, \qquad w_i, W > 0.$$

A set $C \subseteq I$ is called a *cover* if

$$\sum_{i \in C} w_i > W.$$

Given a cover C, we can write the valid inequality:

$$\sum_{i \in C} x_i \leq |C| - 1. \quad \square$$

Example 2.14
Consider the constraint

$$5x_1 + 2x_2 \geq 3,$$

where both variables are binary. It is easy to see that such a constraint is satisfied if and only if $x_1 = 1$. Therefore we may fix $x_1 = 1$ and eliminate the constraint. ☐

The constraint manipulation methods we have considered are relatively simple and can be automated. Indeed, automatic preprocessing of constraints is included in some software tools for MILP, with the aim of strengthening the formulation and to improve the branching process.

2.5 Developing heuristics from mathematical models

In the book we present a wide variety of mathematical models for production management. By far, the most common modeling approach is based on MILP. Yet, we have pointed out the difficulty of solving MILP models at optimality and the consequent need for heuristic procedures. Heuristic procedures are often based on commonsense. Therefore, one may well wonder why we should bother building mathematical models. One of the main point of the book is that mathematical modeling may be a valuable source of good heuristics. In the following chapters we will show examples of the following strategies.

Heuristics from continuous relaxation. Given an MILP problem, its continuous relaxation can be solved by the Simplex method. In some cases, the resulting solution is feasible, though not necessarily optimal, for the discrete problem. Usually, the continuous solution is not feasible. For general MILP models, there is no easy way to generate a feasible discrete solution from the continuous one. Nevertheless, it is possible in some cases to obtain a good solution by clever rounding strategies. These strategies must exploit the problem structure and are advantageous when the number of fractional variables in the continuous solution is not too high. Examples of this approach are given in Sections 3.2.5 and 3.2.7.

Heuristics from Lagrangian relaxation. In Lagrangian relaxation we relax a set of constraints by dualization, obtaining an easier problem. By maximizing the corresponding dual function, we obtain a lower bound on the optimal value of the original problem, but we may also obtain good feasible solutions. Again, a problem-specific mechanism must be devised to generate feasible solutions from the solutions of the relaxed problem. Examples of this approach are given in Sections 3.2.3, 3.3.2, 4.7.4, and 4.8.2.

Heuristics from decomposition in subproblems. Sometimes, a complex problem can be hierarchically decomposed in subproblems to be solved in sequence. For instance, consider a joint problem of routing and scheduling, i.e., a problem whereby we must assign operations to machines and schedule them. In Section 4.8.6 we tackle such a problem by first solving the routing subproblem and then the scheduling problem. Such hierarchical decomposition schemes are called *one-way*, when the solution of the upper level subproblems is not revised given the solution of the lower level subproblems; we have a *two-way* scheme when there are iterations among different levels. Clearly, this requires the ability of devising sensible objective functions for the subproblems, since it is generally not possible to use the original objective function for all of them.

Heuristics from approximate models. The continuous relaxation of an MILP problem can be regarded as an approximate model of the problem at hand. It may be worthwhile to deliberately build approximate but easier problems, whose optimal solutions can be exploited in order to generate a good solution for the original problem. In Sections 3.4 and 6.2, we show the use of continuous flow models, based on the approximation of the discrete material flow by a continuous flow.

Another argument in favor of the study of optimization models and algorithms, even within a heuristic perspective, is that approximate strategies may be obtained by "truncating" exact enumerative procedures, such as Branch and Bound and Dynamic Programming. Moreover, easy optimization problems may be used as subroutines in approximate strategies for solving more complex problems. We will also see that sometimes we do not build a complete mathematical model of the problem, since part of it is left implicit. This may happen when applying Lagrangian relaxation. The "easy" part of the model needs not to be made explicit (see Section 4.7.4). Finally, we may build a mathematical model just to find the optimal combination of "elementary recipes", without building a complex model for obtaining such recipes. This happens with modal formulations (see Sections 3.2.5 and C.2.3).

S2.1 The theory of NP–completeness

In this supplement we provide the reader with a basic understanding of the theory of computational complexity. For a thorough and more rigorous treatment, we suggest the classical book by Garey and Johnson [82].

Branch and Bound and enumerative algorithms have exponential worst-case complexity. They are, in this sense, "bad" algorithms. We have already defined the class $\mathcal{P}$ of problems for which a polynomial algorithm is known. It turns out to be convenient to restrict the attention to *decision* problems rather than optimization problems. By decision problem, we mean finding if a given question has a *yes* or *no* answer.

Example 2.15
This is a decision version of the knapsack problem. Given a set of positive integer numbers $a_1, a_2, \ldots, a_n$ and b, is it possible to find an assignment of the binary variables $x_1, \ldots, x_n$ such that $\sum_{i=1}^{n} a_i x_i = b$? □

For any optimization problem

$$\min_{\mathbf{x} \in S} f(\mathbf{x}),$$

it is quite natural to devise a related decision version. Given a number k, is it possible to find out an $\mathbf{x} \in S$ such that $f(\mathbf{x}) \leq k$? Clearly, if we are able to solve the optimization problem, we are also able to solve its decision version. In particular, if there is a polynomial algorithm for the optimization problem, then the corresponding decision problem is in $\mathcal{P}$. On the other hand, if we are able to show that the decision problem is difficult, then the optimization problem must be difficult too (otherwise, the efficient algorithm for the optimization problem would yield an efficient algorithm for the decision problem). Since our aim is exactly to show that some optimization problems are difficult (in a sense to be specified), we may restrict our attention to decision problems.

Consider again the decision problem of Example 2.15. A possible way to solve it is simply to enumerate all possible assignments of the 0/1 variables x_i as shown in Figure 2.4. Clearly, such an enumeration approach has an exponential worst-case complexity. Nevertheless, for each assignment (i.e. for each leaf of the tree), it is easy to check if it is a solution of the problem or not. Technically speaking, given a *solution guess*, it is possible to *check* in polynomial time if it is an acceptable solution. If we had a *nondeterministic computer*, i.e., a computer able to spawn an infinite number of computation processes, we would be able to solve the decision problem in polynomial time. We could assign each node of the enumeration tree to a process

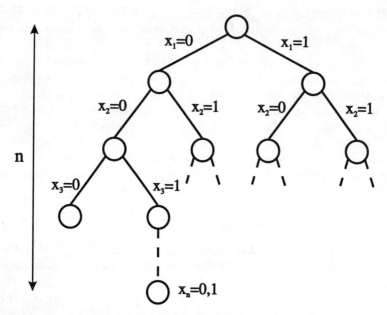

FIGURE 2.4
Solving a decision problem by enumeration

and, since the depth of the tree is n and each leaf requires a polynomial time check, the decision version of the knapsack problem could be solved in polynomial time on this nondeterministic computer. The class of decision problems solvable in polynomial time on a nondeterministic computer is denoted by $\mathcal{NP}$. Note that, by definition, $\mathcal{P} \subseteq \mathcal{NP}$.

DEFINITION 2.16 *If we have two decision problems P and Q such that any instance I_P of P can be transformed in polynomial time to an instance I_Q of Q such that*

I_P *has a yes answer if and only if I_Q has a yes answer,*

then, we say that P is **polynomially reducible** *to Q. This is denoted by $P \prec Q$.*

The notation $P \prec Q$ is due to the fact that the complexity of P is not larger than the complexity of transforming P to Q and, then, solving Q,

$$\mathrm{compl}(P) \leq \mathrm{compl}(Q) + \mathrm{compl}(P \rightarrow Q).$$

If the transformation has a negligible complexity, reducing P to Q shows

The theory of NP–completeness

that Q is not easier than P. If P is difficult and $P \prec Q$, Q cannot be easy. Otherwise, we could transform P to Q and, then, use the efficient solution algorithm for Q.

DEFINITION 2.17 *A problem P is said* **NP–hard** *if any problem in the class $\mathcal{NP}$ is reducible to P.*

DEFINITION 2.18 *A decision problem P is said* **NP–complete** *if it is in $\mathcal{NP}$ and is NP–hard.*

The practical implications of these definitions are:

1. an NP–hard problem is not easier than any problem in $\mathcal{NP}$;
2. the class of the NP–complete problems is the class of the most difficult problems in $\mathcal{NP}$.

Note that, by definition, the class of NP–complete problems contains only decision problems; an optimization problem is NP–hard if its decision version is NP–complete. If we were able to show that a decision problem Q is NP–complete, it would be relatively easy to prove that another problem P is NP–complete by showing

1. that P is in $\mathcal{NP}$;
2. that the NP–complete problem Q can be reduced to P.

Usually, the first part of the proof is trivial. Note that, since Q is NP–complete, $P \prec Q$ and, if we neglect the complexity of the transformation,

$$\text{compl}(P) \leq \text{compl}(Q).$$

But, if $Q \prec P$, we also have

$$\text{compl}(Q) \leq \text{compl}(P);$$

Putting it all together we have

$$\text{compl}(Q) = \text{compl}(P).$$

In other words, NP–complete problems are equivalent from a computational complexity point of view. A polynomial algorithm for one of them yields a polynomial algorithm able to solve *all* NP–complete problems.

At first sight, it is difficult to think of an NP–complete problem. In fact, it can be shown that the following problem is NP–complete: deciding if a boolean formula in canonical conjunctive form is satisfiable or not. For instance, the formula

$$(A \text{ or } B) \text{ and } (\text{not}(A) \text{ or } C)$$

is satisfied if B and C are both true. But

$$(A \text{ or } B) \text{ and } (\text{not}(A) \text{ or } B) \text{ and } (\text{not}(B))$$

cannot be satisfied by any assignment of truth values to the variables. This problem is called **satisfiability** and has been the first problem to be shown NP–complete. By reducing the satisfiability problem to other decision problems it is possible to enlarge the class of NP–complete problems; in particular, by a chain of polynomial reductions, it is possible to show that the decision version of knapsack is NP–complete (see e.g. [107, pp. 30-35]).

Consider now a decision version of $1/r_i/L_{\max}$. Given release times r_i, due dates d_i, and processing times p_i, find if there exists a schedule in which no job is late. This problem is clearly in $\mathcal{NP}$, since it is easy to verify if a given schedule is admissible. We can reduce the knapsack problem to this scheduling problem as follows. We set:

$$\begin{array}{lll} r_0 = b & r_i = 0 & i = 1, \ldots, n \\ p_0 = 1 & p_i = a_i & i = 1, \ldots, n \\ d_0 = b + 1 & d_i = 1 + \sum_{j=1}^n a_j & i = 1, \ldots, n. \end{array}$$

A feasible schedule has necessarily no idle time, since the due dates of the jobs $i = 1, \ldots, n$ are exactly the sum of the processing times. Job 0 must start at time b to be completed in time. But these two conditions can be met only if there exists a subset $\mathcal{I}$ of indices such that

$$\sum_{i \in \mathcal{I}} a_i = b.$$

We have shown that the decision version of $1/r_i/L_{\max}$ is NP–complete; this in turn shows that the optimization version is NP–hard. The implication is that a polynomial algorithm for $1/r_i/L_{\max}$ would yield a polynomial algorithm for all the NP–complete problems. Furthermore, this would prove that $\mathcal{P} \equiv \mathcal{NP}$. However, since $\mathcal{NP}$ includes thorny problems which have defied the efforts of thousands of researchers, this seems unlikely.

A last remark is in order. We have said that the way we encode a problem is irrelevant for the determination of the complexity of a problem. In fact, this is not always true. Consider for instance the optimization version of the knapsack problem with N items and a knapsack capacity W; it can be shown that this problem can be shown by an algorithm with complexity $O(NW)$ (see the dynamic programming algorithm described in Example D.2). One would be tempted to say that this problem is polynomial; in fact, this is not true if we adopt a binary encoding, i.e., if we use the customary representation of numbers on a computer, based on symbols 0 and 1, only $\lceil \log_2 W \rceil$ bits are needed to encode the capacity W, and the computational complexity is exponential with respect to the size of the encoding. But if we use a unary encoding, i.e. we use only the symbol 1 and a separator, W bits are needed to represent W; with respect to this encoding, the problem is polynomial. In summary, we cannot say that knapsack is polynomial, but we can expect that in practice it is relatively easy with respect to other NP–hard problems. In fact, relatively large knapsack problem can be solved with a reasonable computational effort by special purpose Branch and Bound or Dynamic Programming algorithms. We say, in such a case, that the $O(NW)$ algorithm is **pseudopolynomial**.

S2.2 An outlook on multiobjective optimization

Multiobjective optimization is a broad topic, which cannot be thoroughly treated in a book such as this. Nevertheless, we would like to give some hints on how scalar optimization methods can be used to cope with vector problems. Consider a vector optimization problem with two objectives, f_1 and f_2, to be minimized with respect to $x \in S$. Such a problem, can be tackled by the following approaches:

- the *scalarization* method, whereby the objectives are combined into a scalar objective function and a set of auxiliary scalar problems is solved;
- the *constraints* method, based on the idea of transforming all objectives but one into constraints.

Probably the most intuitive approach to the solution of a multiobjective problem is to scalarize it by forming a convex combination of the objectives and obtaining the problem

$$\min_{\mathbf{x} \in S}\{\lambda f_1(\mathbf{x}) + (1-\lambda)f_2(\mathbf{x})\} \qquad \lambda \in [0,1]. \tag{2.4}$$

This problem is solved with different values of λ. This method would in some cases identify the set of efficient solutions (e.g. in a multiobjective Linear Programming problem, each efficient solution is the optimal solution for a certain scalarized objective function). Unfortunately, the convex combination scalarization is theoretically wrong for discrete optimization problems, since, in such a case, the set of criterion vectors is not convex, and there is no guarantee that the whole set of efficient solutions will be generated. The issue is best illustrated by figure 2.5.

A theoretically sound approach was proposed in [29] and adopted in [26] for scheduling problems. The idea is to find solutions such that their criterion vectors minimize the distance from the ideal criterion vector f^*

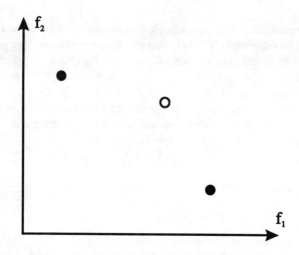

FIGURE 2.5
A convex combination approach does not always yield the whole efficient set

(i.e., the vector whose coordinates are the optimal values for the single objectives):

$$\min_{\mathbf{x} \in S} \| f(\mathbf{x}) - f^* \|,$$

where the distance is measured by a modified Tchebycheff norm. We obtain the problem:

$$\min_{\mathbf{x} \in S} \{ \max[\lambda(f_1(\mathbf{x}) - f_1^*), (1-\lambda)(f_2(\mathbf{x}) - f_2^*)] \} \qquad 0 < \lambda < 1 \qquad (2.5)$$

where f_i^* is the optimal value with respect to objective i.

As to the constraints approach, in our case it would require the solution of a set of subproblems of the form:

$$\min f_1(\mathbf{x})$$
$$\text{s.t.} \begin{cases} \mathbf{x} \in S \\ f_2(\mathbf{x}) \leq \bar{f}_2, \end{cases}$$

for different values of $\bar{f}_2$. If e is an efficient solution, and f_2 is integer-valued, a suitable value of $\bar{f}_2$ is $f_2(e)-1$; solving the corresponding problem yields another efficient solution.

For further reading

At the end of appendices we have listed many references concerning optimization methods. Here we just recall [144], which is a broad book dealing with continuous, discrete and infinite-dimensional problems, and [150] which is a very comprehensive reference for discrete optimization.

The topic of *model building* in mathematical programming is covered in many texts on Operations Research; a notable reference, rich in good examples, is [210]. In particular [210, Chapter 10] deals with the fundamental issue of good and bad model formulations, which is also touched on in [150, Chapter I.1]. Different classes of tightening constraints are given in [54] and [199]. Applications of these principles to production planning problems are described in [109] and [199]. Computational applications of preprocessing and supernode processing for Branch and Bound codes based on continuous relaxation are described in [177]; see also [166] for probing and preprocessing techniques.

We recall again [82] for computational complexity issues. Less thorough, but easier treatments can be found in books on discrete optimization such as [107], [154], and [155]. For multiobjective problems, a good starting point is [176].

3

Discrete-time models

In this chapter we describe a range of models based on the segmentation of the time horizon in equal time intervals, known in the production management literature as **time buckets**. Basically, two classes of problems will be addressed here: production planning problems at an aggregate level, whereby the exact timing of events is not specified, and production scheduling problems, whereby decisions are taken at a much more detailed level.

We consider lot-sizing problems in Sections 3.1, 3.2 and 3.3. By lot-sizing problems we broadly intend those in which production quantities for a set of items must be determined in order to meet a time-varying demand in a production-inventory environment. There is actually a wide variety of problems of this kind, differing in their strategic or operational character: indeed, we consider both aggregate planning and scheduling problems.

In Section 3.1 we deal with Aggregate Production Planning problems. Such problems are usually considered at the strategic decision level, involve product families rather than single items, and can be modeled by continuous linear programming. When we go down the decision hierarchy, we consider more detailed models, involving, for instance, setup costs and/or times. This is the subject of Section 3.2, where MILP models are obtained for the Capacitated Lot-Sizing Problem (CLSP). Another issue we consider is how models can be properly reformulated in order to improve their solvability by a standard Branch and Bound code. A simple reformulation of CLSP is proposed in Section 3.2.2 and a stronger, but counter-intuitive one in Supplement S3.1. In Section 3.3, the most detailed models are covered, namely, the Discrete Lot-Sizing and Scheduling models, whereby the exact timing of resource usage is specified. For this class of models, we outline a solution approach based on Lagrangian decomposition; the solution of the single item problem by dynamic programming is relatively involved and is dealt with in Supplement S3.2.

In Section 3.4 we consider approximate models for scheduling problems. Scheduling problems are extensively treated in Chapter 4, but we anticipate here an approximate approach based on continuous flows, due to its conceptual relationship with discrete-time models based on state equations. Unlike lot-sizing problems, in a scheduling problem the size of the jobs is specified. This happens in a make-to-order environment, or in a make-to-stock environment when lot-sizing decisions are taken at a higher decision level and a lower decision level must cope with detailed sequencing and scheduling.

The emphasis throughout the chapter is on developing heuristics *from* mathematical models. An important source of such heuristics is Lagrangian decomposition, which is applied in a similar manner to all lot-sizing problems considered here. We also consider the use of continuous relaxation for this purpose.

The segmentation of the time horizon implies a modeling error, which can be sometimes critical (as shown in Section 3.2.7). Nevertheless discrete-time modeling is a very flexible and powerful tool. In Section 3.5 we show how such flexibility can be exploited to develop models for complex problems.

3.1 Aggregate Production Planning

The aim of Aggregate Production Planning (APP) is to determine production, inventory, and workforce levels in order to meet a time-varying demand pattern. The APP problem pertains to strategic decision levels; therefore, APP models represent suitable aggregations of manufacturing systems. Aggregation has different facets:

- similar products are often grouped in families; in this way the uncertainty in demand is reduced, since it is easier to forecast the demand for a family of items than for the single item;
- time is discretized in relatively large time-buckets; this means that more than one product/family is produced during each time bucket, but no sequencing decision is taken;
- resources are grouped in workcenters; since we are not concerned with short-term scheduling at this level, it does not make sense to specify which machine in a workcenter will carry out a certain job.

Such aggregations imply that what we obtain is not a detailed production schedule, but only a rough-cut indication of what should be done. A further difference between APP and more detailed model is that resource availability needs not to be given a priori, but it can be a decision variable.

In the literature it is possible to find much more complex models than those presented here; we will omit such models for two reasons. First, APP models have raised some controversy (see e.g. [38]). The Simplex method is able to solve large scale APP problems formulated in LP form, but the use of complex models requiring great amount of data is questionable, since such data may be difficult to gather. Second, our aim is didactic, since we present simple APP models in order to introduce and interpret decomposition strategies suitable for difficult MILP models.

We consider first a basic APP problem with fixed resource availability. Then, we introduce overtime and variable workforce.

3.1.1 The basic APP model

A plant produces a set of N products/families, indexed by i; from the formal point of view we do not distinguish products from families. The time horizon is divided in T time buckets indexed by t. For each item i, we have the forecasted demand d_{it} during time bucket t. The problem is capacity-constrained: let M resource types be available, indexed by $m = 1, \ldots, M$. The resource availability during each time bucket is R_{mt}; resource availability may be time-varying, e.g. due to maintenance periods or workforce hiring/firing. Producing one unit of item i requires an amount r_{im} of resource m.

The planning problem consists of choosing the production quantity x_{it} of each item i *during* each time bucket t, subject to resource availability constraints, in order to minimize inventory costs. Note that we do not want to specify the sequence in which the items will be processed on the machines; this kind of decision pertains to lower levels of the decision hierarchy. To quantify inventory costs, we introduce a set of decision variables I_{it} measuring the inventory level of product i *at the end* of time bucket t; let c_i be the unit inventory cost of product i and I_{i0} the starting inventory level. A further requirements might be that the inventory level cannot fall below a safety threshold $\bar{I}_{it}$, that depends on the demand uncertainty. A further role of the constants $\bar{I}_{it}$ is to specify a final inventory level $\bar{I}_{iT}$ at the end of the time horizon. If a rolling horizon approach is pursued, it may not make sense to let the final state be free (since we are minimizing inventory costs, this would imply empty inventories at the end of the horizon).

The basic APP problem can be modeled by the following LP problem:

$$\min \sum_{t=1}^{T} \sum_{i=1}^{N} c_i I_{it} \tag{3.1}$$

$$\text{s.t.} \quad I_{it} = I_{i,t-1} + x_{it} - d_{it} \qquad \forall i, t \tag{3.2}$$

$$\sum_{i=1}^{N} r_{im} x_{it} \leq R_{mt} \qquad \forall t, m \tag{3.3}$$

$$x_{it} \geq 0; \qquad I_{it} \geq \bar{I}_{it} \qquad \forall i, t. \tag{3.4}$$

The objective function (3.1) sums the inventory levels over the planning horizon. Eq. (3.2) models inventory balance, linking the production and inventory decision variables. This equation simply states that the inventory level at the end of time bucket t equals the inventory level at the end of the previous time bucket, plus what is produced during t, minus the demand. Note that these constraints can be interpreted as state equations for a discrete-time dynamic system, where the inventory levels I_{it} are the state variables, the production quantities x_{it} are the control variables, and the demand d_{it} is an exogenous input. Finally, (3.3) is the capacity constraint. Production quantities are restricted to nonnegative values; unlike the production levels, negative inventory levels could make sense in order to model demand backlogging. This is done in the next Section. Note that if the resources were not constrained, i.e. if Eq. (3.3) were omitted, the obvious optimal solution would be to produce "just-in-time". Neglecting safety stock issues, we would have $x_{it}^* = d_{it}$.

3.1.2 Extensions of the basic APP model

We consider here various extensions of the basic APP model of the previous section, taking into account demand backlogging and variable capacity.

A first observation is that the basic APP formulation has a poor diagnostic capability. Consider the situation in which the available capacity is not sufficient to meet the demand. In this case the feasible set is empty, and this would be reported by any LP software package. Nevertheless, it could be more appropriate to relax the capacity constraints by a suitable penalty in order to help the decisionmaker in the task of spotting what resources are critical and when. This diagnostic task can be accomplished by rewriting (3.3) as:

$$\sum_{i=1}^{N} r_{im} x_{it} \leq R_{mt} + O_{mt} \quad \forall t, m,$$

where $O_{mt} \geq 0$ is the overuse of resource m during time bucket t, and by modifying the objective function accordingly:

$$\min \sum_{t=1}^{T} \left(\sum_{i=1}^{N} c_i I_{it} + P_o \sum_{m=1}^{M} O_{mt} \right),$$

where P_o is a large penalty coefficient. If the penalty coefficient is large enough, the overuse variables will be activated only if no feasible solution exists for the original problem. We have translated a hard constraint into a soft one; now it is possible to spot what resources are critical and when, and to manage the problem properly.

This relaxation of the capacity constraints is also required when the possibility of using overtime work or subcontracting must be modeled. It is also possible to develop a variable workforce model. We now formulate an LP model for a more complex APP problem, including variable workforce and demand backlogging. If we consider workforce as the only capacitated resource, we obtain:

$$\min \sum_i \sum_t \left(c_{it} I_{it}^+ + b_{it} I_{it}^-\right) + \sum_t (w_t W_t + o_t O_t + h_t H_t + f_t F_t)$$

s.t.
$$I_{it}^+ - I_{it}^- = I_{i,t-1}^+ - I_{i,t-1}^- + x_{it} - d_{it} \qquad \forall i,t \qquad (3.5)$$

$$\sum_i r_i x_{it} \leq W_t + O_t \qquad \forall t \qquad (3.6)$$

$$W_t = W_{t-1} + H_t - F_t \qquad \forall t \qquad (3.7)$$

$$x_{it}, I_{it}^+, I_{it}^- \geq 0 \qquad \forall i,t$$

$$W_t, O_t, H_t, F_t \geq 0 \qquad \forall t.$$

Backlog is modeled as a negative inventory level. The inventory level is decomposed in two parts such that $I_{it} = I_{it}^+ - I_{it}^-$, with $I_{it}^+, I_{it}^- \geq 0$. This is the usual way to express a nonrestricted variable to two nonnegative ones. Note that $I^+ \cdot I^- = 0$ in the optimal solution, since both variables are penalized in the objective function. W_t is the workforce level during period t; O_t is the amount of overtime used; H_t and F_t are, respectively, the increase and decrease of W_t, i.e., hiring and firing. The objective function includes two terms: the first one takes inventory and backlogging costs; the second one is related to the workforce level. Each decision variable has a (possibly time-varying) cost coefficient. Eq. (3.5) relates the inventory levels to production x_{it} and demand d_{it}. Eq. (3.6) is a capacity constraint allowing for overtime. Eq. (3.7) states that the current workforce level is the previous one plus the hired workforce minus the fired workforce.

3.1.3 A decomposition approach for the basic APP model

We have shown in Section A.6 how a large scale optimization problem can be decomposed by dualizing the interaction constraints coupling subproblems. We apply this strategy to a basic APP problem. The aim here is not to propose a good computational strategy for large scale LP problems, but to introduce a decomposition framework that will be used in the next sections for MILP problems. We will show how a suitable decomposition yields a set of simpler subproblems, as well as an intuitive economic interpretation.

Consider again the LP formulation of the basic APP problem:

$$\min \sum_{i=1}^{N} \sum_{t=1}^{T} c_i I_{it}$$

$$\text{s.t.} \sum_{i=1}^{N} r_{im} x_{it} \leq R_{mt} \qquad (3.8)$$

$$I_{it} = I_{i,t-1} + x_{it} - d_{it}$$

$$I_{it}, x_{it} \geq 0.$$

It is easy to see that if we could get rid of the capacity constraints (3.8), the problem could be decomposed in N single-item subproblems. There is no interaction among the products through the inventory state equations. Products interact since they compete for the available capacity.

This can be obtained by dualization of the coupling constraints. Let us introduce $T \times M$ nonnegative Lagrange multipliers μ_{tm}; by dualizing (3.8) we obtain the relaxed problem:

$$w(\mu) = \min \sum_{t=1}^{T} \sum_{i=1}^{N} c_i I_{it} + \sum_{t=1}^{T} \sum_{m=1}^{M} \mu_{tm} (\sum_{i=1}^{N} r_{im} x_{it} - R_{mt}) \qquad (3.9)$$

$$\text{s.t.} \quad I_{it} = I_{i,t-1} + x_{it} - d_{it}$$

$$I_{it}, x_{it} \geq 0.$$

The optimal value $w(\mu)$ of the relaxed problem is the dual function. Since we are dealing with a convex problem, we can solve the original (primal) problem, by solving the dual problem:

$$\max_{\mu \geq 0} w(\mu).$$

We can rewrite the objective function (3.9) as

$$\sum_{i=1}^{N} \left(\sum_{t=1}^{T} c_i I_{it} + \sum_{t=1}^{T} \sum_{m=1}^{M} \mu_{tm} r_{im} x_{it} \right) - \sum_{t=1}^{T} \sum_{m=1}^{M} \mu_{tm} R_{mt}.$$

The last term is, for a fixed set of multipliers, a constant; the first one is a decomposable function with respect to products. We obtain N separate subproblems:

$$w_i(\mu) = \min \sum_{t=1}^{T} c_i I_{it} + \sum_{t=1}^{T} \sum_{m=1}^{M} \mu_{tm} r_{im} x_{it}$$

$$\text{s.t.} \quad I_{it} = I_{i,t-1} + x_{it} - d_{it}$$

$$I_{it}, x_{it} \geq 0.$$

The coordination problem is:

$$\max_{\mu \geq 0} \left(\sum_{i=1}^{N} w_i(\mu) - \sum_{t=1}^{T} \sum_{m=1}^{M} \mu_{tm} R_{mt} \right).$$

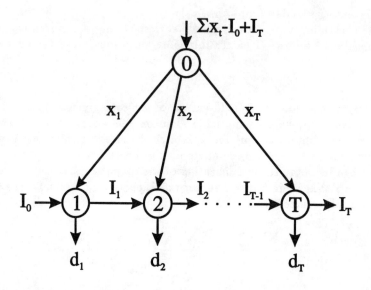

FIGURE 3.1
The network for the single-item problem

An important thing to notice is that the capacity constraint has been replaced by production costs $\mu_{tm}r_{im}$.

Now we must (1) solve the single item problems and (2) maximize the dual function by a subgradient method. In the next sections we show how the single-item problem can be solved as a minimum cost network flow problem, and how the procedure to update the multipliers can be interpreted from an economic point of view.

Reformulation of the single-item problem as network flow problem

We have decomposed the LP model of the APP problem into a set of smaller LP problems. Now, given the efficiency of the Simplex method, the reader could wonder if this decomposition yields some computational advantage.

Consider the following inventory state equations (where we have omitted the item index for clarity):

$$I_t - I_{t-1} - x_{t-1} + d_{t-1} = 0$$
$$I_{t+1} - I_t - x_t + d_t = 0.$$

Aggregate Production Planning

The inventory variable I_t appears in the two equations, with opposite sign. In the constraint matrix, each column corresponding to an inventory variable has a $+1$ and a -1 entry; this looks suspiciously like the incidence matrix of a network optimization problem (see Example B.17). Indeed, the single-item problem is a minimum cost network flow problem. The equivalent network is shown in Figure 3.1. There is a node (node 0) in the upper part of the network, with an entering flow representing the overall production $\sum_t d_t - I_0$. Each other node from 1 to T corresponds to a production period. On each arc from node 0 to the other ones, the amount of flow is the production level during the corresponding period, which is penalized by the production cost; on each horizontal arc, there is a flow corresponding to the inventory left at the end of the period, which is penalized by the inventory cost. For each node there are two in-flows, corresponding to inventory carried over from the previous period and to production during the period, and two out-flows, corresponding to inventory carried over to the next period and to demand. The inventory balance equation is simply the flow equilibrium condition in each node.

The basic APP problem can be interpreted as a multicommodity network flow problem, which is decomposed into a set of single-commodity network flow problems. Since the minimum cost flow problem is much easier to solve than the general LP problem, there may be some computational advantages with such a procedure. It should be noted that some problems may occur from the point of view of the convergence of the dual variables, and that alternative methods should be used if a strictly feasible solution is needed.

Economic interpretation of the coordination problem

We have seen in Section A.4.5 that dual variables can often be interpreted as resource costs. Indeed, in our case, the dual variables μ_{tm} multiplied by the resource absorption r_{im} define the production cost of item i on resource m during period t.

The multipliers are adjusted according to a subgradient procedure. Given an optimal solution $(\hat{x}, \hat{I})$ of the relaxed problem, a subgradient is readily obtained by evaluating the dualized constraints in $(\hat{x}, \hat{I})$ (see Section A.7). Each multiplier is adjusted as:

$$\mu_{tm}^{(k+1)} = \max\left\{0, \mu_{tm}^{(k)} + \alpha^{(k)}\left(\sum_{i=1}^{N} r_{im}\hat{x}_{it} - R_{mt}\right)\right\},$$

for a suitable step length $\alpha^{(k)}$.

The coordination module establishes the cost of each resource during each time period, following a demand-offer rule: cost are increased for scarce resources, and decreased for the partially unused ones.

3.2 The Capacitated Lot-Sizing Problem

In this section we consider a more operational problem than APP. Given a demand pattern for a set of products, we want to decide the production quantities during each time bucket in order to minimize the sum of inventory and setup costs. The setup cost makes few large lots interesting, whereas the inventory cost requires many small lots. We generalize the EOQ model we have seen in Section 1.1. The EOQ model is characterized by uncapacitated resources and stationary demand. We consider here the capacitated case with time-varying demand over a discrete-time horizon. The size of the time buckets is large enough to accommodate the production of more than one product. We speak of a **large bucket** model. Such a problem is referred to as a Capacitated Lot-Sizing Problem (CLSP). The case of small time buckets, whereby at most one product can be produced during each time bucket, is dealt with in Section 3.3.

We first formulate CLSP as a MILP problem in Section 3.2.1. We discuss how to solve it with standard MILP codes in Subsection 3.2.2 and with Lagrangian relaxation in Subsection 3.2.3. Dual approaches to CLSP exploit efficient algorithms for the single-item uncapacitated problem. We present the basics of such algorithms in Subsection 3.2.4. The results for the single-item uncapacitated problem are also exploited in Section 3.2.5 to develop an LP approach to multi-item lot-sizing. In Subsection 3.2.6 we deal with hierarchical approaches for CLSP, which are partially able to cope with lot-sizing problems involving assembly operations. The issues raised by multilevel lot-sizing problems, including assembly systems, are also discussed in Subsection 3.2.7.

3.2.1 Formulation of the CLSP

The CLSP problem can be formulated as the following MILP problem:

$$\min \sum_{t=1}^{T} \sum_{i=1}^{N} (c_i I_{it} + s_i \nu_{it})$$

$$\text{s.t. } I_{it} = I_{i,t-1} + x_{it} - d_{it} \quad \forall i, t \tag{3.10}$$

$$\sum_{i=1}^{N} (r_{im} x_{it} + r'_{im} \nu_{it}) \leq R_{mt} \quad \forall t, m \tag{3.11}$$

$$x_{it} \leq M \nu_{it} \quad \forall i, t \tag{3.12}$$

$$I_{it}, x_{it} \geq 0 \quad \forall i, t$$
$$\nu_{it} \in \{0, 1\} \quad \forall i, t.$$

The notation is essentially the same as in the basic APP problem. The decision variables are the inventory levels I_{it}, the production quantities x_{it}, and a set of binary variables modeling setup decisions:

$$\nu_{it} = \begin{cases} 1 & \text{if item } i \text{ is produced during time bucket } t \\ 0 & \text{otherwise.} \end{cases}$$

The objective function includes inventory and setup costs. Note that setup costs do not depend on the amount produced, since they are *fixed* costs. Eq. (3.10) is the inventory state equation, and Eq. (3.11) is the capacity constraint. In this formulation we have considered both setup costs s_i and setup times; r'_{im} is the time lost during a setup on resource m for item i. Setup times and costs are assumed sequence-independent; coping with sequence-dependent setups requires a scheduling model. It is clearly possible to consider models dealing only with setup times or costs. Eq. (3.12) links the setup and the production variables. We have already met setup variables in Section 1.2 (Example 1.3), where we used an expression of the type $\nu = \delta(x)$; such a formulation may not be computationally convenient, because it is nonlinear. Here we have a typical application of the "big M" to express fixed costs in a linear format. If ν_{it} is 0, then x_{it} must be 0; if ν_{it} is 1, then we have $x_{it} \leq M$. If M is a suitably large constant, this is a nonbinding constraint. Note that the production and inventory variables are not restricted to integer values; this is a reasonable approximation if such variables take relatively large values.

Unlike the APP problem, CLSP is not easy to solve, and it can be shown that it is a NP–hard optimization problem, calling for enumerative solution procedures such as Branch and Bound. If setup times are included, even finding a feasible solution is a difficult problem. Deciding if a given instance is feasible is itself an NP–complete decision problem [136]. On the contrary, if setup times are negligible, the continuous relaxation of the above model readily yields a feasible, though not necessarily optimal, solution, if any exists.

3.2.2 Solving the CLSP with standard MILP codes

The CLSP can be solved in principle by any commercial code implementing a Branch and Bound procedure based on continuous relaxation (see Section C.4.2). Unfortunately, this is not a viable approach for realistic size problems. This difficulty is not due to the number of binary variables: it is due to the particular model structure.

Eq. (3.12), when continuously relaxed, yields a rather weak lower bound. This is geometrically illustrated in Figure 3.2. Here we show the feasible

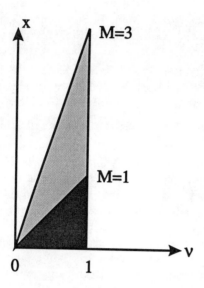

FIGURE 3.2
A geometrical illustration of a fixed-charge constraint

region on the plane (x, ν), corresponding to a constraint $x \leq M\nu$. The continuous feasible region is the triangle, whereas the discrete feasible region consists of the origin and the vertical cathetus. M is the slope of the hypotenuse of the triangle; the larger M, the larger the continuous feasible region. This means that for large values of M, the lower bound computed by continuous relaxation is weak. This is a typical behavior of fixed-charge constraints.

To improve the performance we can try to tighten the big M as much as possible, by introducing a set of constants

$$M_{it} = \sum_{\tau=t}^{T} d_{i\tau}$$

that are obvious upper bounds on the production quantities, and then requiring $x_{it} \leq M_{it}\nu_{it}$. A possible alternative is to use constants related to the capacity constraints: $r_{im}x_{it} \leq R_{mt}\nu_{it}$. Unfortunately, this is not enough to solve large problems. To increase the size of CLSP problems we may solve exactly, we must

- either use a stronger formulation of the problem,
- or adopt a Lagrangian decomposition approach.

Lagrangian relaxation, as shown in the next subsection, yields stronger lower bounds than continuous relaxation. In Supplement S3.1 we describe a strong, but counter-intuitive, formulation of the CLSP problem. Here we describe a simpler reformulation based on a disaggregation concept.

To reduce the big M, we can introduce a set of decision variables y_{itp}, which represent the amount of item i produced during time bucket t to satisfy the demand during time bucket p. This new variable represents a disaggregation of the x variables:

$$x_{it} = \sum_{p=t}^{T} y_{itp}.$$

Now we can rewrite (3.12) as:

$$y_{itp} \leq d_{ip}\nu_{it} \qquad \forall i, t, p \geq t.$$

The advantage of this disaggregated formulation is that we use smaller big Ms, yielding stronger lower bounds.

3.2.3 A dual approach to the CLSP

In Section 3.1.3 we have applied dual decomposition to the basic APP problem. The same procedure can be applied to the CLSP problem. We have to dualize the capacity constraint (3.11) with a set of Lagrangian multipliers μ_{mt}. We do not go into the details, which are quite similar to the APP case, but we write the single-item subproblem we have to solve, where the item index i has been omitted:

$$\min \sum_{t=1}^{T} \left[cI_t + \left(s + \sum_m r'_m \mu_{mt}\right)\nu_t + \left(\sum_m r_m \mu_{mt}\right) x_t \right]$$

$$\text{s.t.} \quad I_t = I_{t-1} + x_t - d_t \quad \forall t$$

$$x_t \leq M\nu_t \quad \forall t$$

$$I_t, x_t \geq 0 \quad \forall t$$

$$\nu_t \in \{0, 1\} \quad \forall t.$$

What we obtain is a single-item, uncapacitated lot-sizing problem, with inventory costs and time-varying setup and production costs. In the next subsection we show that this problem is easily solved by a polynomial algorithm.

By maximizing the dual function for the CLSP problem, we *do not* find an optimal solution of the primal problem; in this case the feasible region is not convex, and only weak duality applies. Therefore we can use the dual decomposition approach in two ways:

1. to compute strong lower bounds within a Branch and Bound procedure;
2. to devise dual heuristics.

We will see in the next section that the single-item problem is polynomially solvable, but it does not enjoy the integrality property. This implies that the Lagrangian lower bound is stronger than the lower bound obtained by the continuous relaxation of the basic model.

Applying dual heuristics requires the ability to recover a primal feasible solution from the dual solution; note that, since we have relaxed capacity constraints, the solution of the relaxed problem is not necessarily primal feasible in general. A possible strategy to obtain a primal feasible solution from a dual solution is to set the setup variables according to the dual solution, and to re-optimize with respect to the continuous variables. The solution of the re-optimization problem can be streamlined by noting that the resulting problem, if we use the disaggregated variables of Section 3.2.2, can be recast as a transportation problem with T source nodes and $T \times N$ demand nodes. It is important to apply the re-optimization procedure to the solution of the relaxed problem at each subgradient iteration, and not only to the optimal dual solution. At each step we obtain a lower bound and possibly an upper bound on the optimal value; their gap can be used in order to stop the procedure. When setup times are included, recovering a primal feasible solution gets more difficult and some perturbation must be applied to the setup pattern.

3.2.4 The single-item uncapacitated lot-sizing problem

The application of Lagrangian decomposition to CLSP requires the ability to solve an uncapacitated single-item lot-sizing problem. Here we show that this problem can be reformulated as a shortest path problem and efficiently solved; actually, we do not deal with the latest algorithms for this problem, because this is beyond the scope of this book. This relationship with the shortest path problem is also exploited in Supplement S3.1 to obtain a strong formulation for CLSP.

For the sake of simplicity we do not consider production costs here, but only setup and inventory costs; what follows can be easily adapted to the more general case. This lot-sizing problem can be written as

$$\min \sum_{t=1}^{T} [s_t \delta(x_t) + c_t I_t]$$
$$\text{s.t.} \quad I_t = I_{t-1} + x_t - d_t \quad \forall t$$
$$I_t, x_t \geq 0 \quad \forall t,$$

where:

$$\delta(x_t) = \begin{cases} 0 & \text{if } x_t = 0 \\ 1 & \text{if } x_t > 0. \end{cases}$$

We have not used the big M to model the link between production and setup variables, since this is necessary only for a MILP solution approach.

One way to cope with the problem is to use dynamic programming (see Section D.2). A backward recursive equation is readily obtained:

$$f_t(I_{t-1}) = \min\left[s_t \delta(x_t) + c_t(x_t + I_{t-1} - d_t) + f_{t+1}(x_t + I_{t-1} - d_t)\right],$$

where we minimize with respect to nonnegative values of x_t, such that $x_t + I_{t-1} \geq d_t$, and $f_t(I_{t-1})$ is the optimal cost for a problem starting at time bucket t with an initial inventory I_{t-1}. The final equation is:

$$f_T(I_{T-1}) = \min\left[s_T \delta(x_T)\right],$$

where we minimize with respect to nonnegative values of x_T, such that $x_T + I_{T-1} = d_T$.

We can solve the functional equation numerically by building tables for the cost-to-go $f_t(I_{t-1})$, corresponding to integer values of the argument (unlike the MILP approach, here it is convenient to consider integer valued production and inventory variables). Unfortunately, the computational complexity of the resulting algorithm is not polynomial. This can be seen by considering the lot-sizing problem as a shortest path problem in a multi-layer graph representing the dynamic state evolution of the system (see Figure 3.3). The graph consists of a starting and final node, and a vertical array of nodes, one for each time bucket; each node represents a possible value of the inventory variable, that plays the role of a state variable. There are polynomial (and extremely efficient) algorithms for solving the shortest path problem, but here we have a large graph, with respect to the amount of data for the lot-sizing problem; therefore, the resulting algorithm is not polynomial with respect to the size of the original problem.

However, the computation can be streamlined by noting that not all the possible values of the state variable must be considered. In fact the following properties hold for the optimal solution:

1. there is always an optimal policy such that:

$$I_{t-1} x_t = 0, \qquad t = 1, \ldots, T;$$

this property is known as the **Wagner-Whitin** property;

2. we can consider only optimal policies such that:

$$x_t = 0, \text{ or } x_t = \sum_{\tau=t}^{t+k} d_\tau,$$

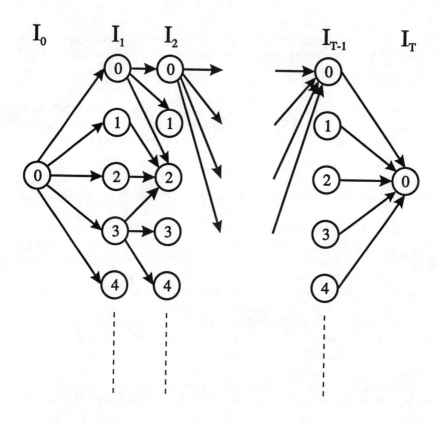

FIGURE 3.3
A shortest path representation of the single-item lot-sizing problem

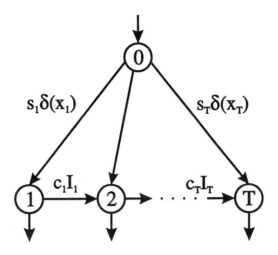

FIGURE 3.4
A fixed-charge network flow representation of the uncapacitated single-item lot-sizing problem

for some k to be determined.

The first property states that it is not economical to produce unless inventory is empty. To see this intuitively, consider the network flow representation shown in Figure 3.4, which is strictly related to the network representation of Figure 3.1. The only difference is that here we have fixed charges associated to production arcs. Now consider a generic node corresponding to time bucket t, and assume $I_{t-1} \neq 0$ and $x_t \neq 0$; it is easy to see that by redirecting the horizontal flow I_{t-1} along the production arc we improve the overall cost, since we decrease the inventory cost without increasing the setup cost.

The second property is a direct consequence of the first, and it states that if $x_t \neq 0$, we produce a quantity corresponding to the demand during the next k periods, with $k \geq 0$ unknown. A consequence of this property is that we can reformulate the single-item problem as a shortest path on the much smaller graph shown in Figure 3.5 for the case of four time buckets. There are an initial node 0, representing the initial state, and an array of nodes corresponding to each time bucket; for each time bucket t we have a set of arcs linking it to time buckets $t+1, t+2, \ldots$. We must go from the start node to the last one along the minimum cost path; the arcs are selected depending on the number of time buckets we cover with a production run, and the cost is computed accordingly. For instance, a path

$$0 \to 2 \to 3 \to 4$$

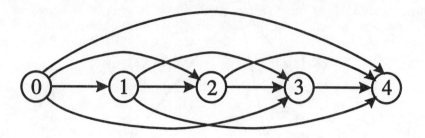

FIGURE 3.5
A shortest path representation of the single-item lot-sizing problem exploiting the Wagner-Whitin condition

corresponds to a production policy

$$x_1 = d_1 + d_2$$
$$x_2 = 0$$
$$x_3 = d_3$$
$$x_4 = d_4.$$

The cost of this path is the sum of three setup costs and the cost of carrying d_2 units in inventory for one period. Due to limited number of nodes of the graph, computing the minimum-cost path is accomplished by a polynomial time algorithm.

3.2.5 A modal formulation for the CLSP problem

In the last section we have shown an important property of the solution of single-item uncapacitated lot-sizing problems: the optimal production quantities cover the demand of a number of periods to be determined. For instance, if we have three periods, we may consider the following schedules:

1. $x_1 = d_1 + d_2 + d_3$; $x_2 = 0$; $x_3 = 0$;
2. $x_1 = d_1 + d_2$; $x_2 = 0$; $x_3 = d_3$;
3. $x_1 = d_1$; $x_2 = d_2$; $x_3 = d_3$;
4. $x_1 = d_1$; $x_2 = d_2 + d_3$; $x_3 = 0$.

Such schedules are referred to as **dominant**. This property can be exploited to build a heuristic strategy for the multi-item capacitated problem.

The basic idea is to use the dominant schedules as basic recipes for a modal approach (see Section C.2.3). Let x_{ijt} the amount of item i produced during time bucket t according to the dominant schedule j. It is easy to compute the cost c_{ij} of the jth schedule for item i, including setup and inventory costs. We may also compute the resource requirements for each dominant schedule; assuming a single capacitated resource:

$$l_{ijt} = r'_i \delta(x_{ijt}) + r_i x_{ijt}.$$

Let $J = 2^{T-1}$ be the number of dominant schedules for each item; the problem of combining the recipes can be formulated as:

$$\min \sum_{i=1}^{N} \sum_{j=1}^{J} c_{ij} \theta_{ij} \qquad (3.13)$$

$$\text{s.t.} \sum_{i=1}^{N} \sum_{j=1}^{J} l_{ijt} \theta_{ij} \le R_t \quad \forall t \qquad (3.14)$$

$$\sum_{j=1}^{J} \theta_{ij} = 1 \quad \forall i \qquad (3.15)$$

$$\theta_{ij} \ge 0 \quad \forall i, \forall j. \qquad (3.16)$$

Here R_t is the resource availability during time bucket t; θ_{ij} is the "proportion" of recipe j for item i that contributes to the overall solution. Note that the problem is formulated as a continuous LP problem; restricting the θ variables to integer values yields a large-scale MILP problem, which would be difficult to solve exactly. This implies that this modal formulation is only an approximation of the original problem; a fractional value of a θ means that only a portion of the setup cost is charged, and the optimal continuous solution is not in general an optimal solution for the original problem. A worse problem occurs when setup times are significant; the solution of the continuous problem may be unfeasible. Nevertheless, this modal formulation works well when the number of products is large with respect to the number of time buckets. Since there are $N+T$ constraints, at most $N+T$ variables will be basic (and therefore positive); if N is large, the number of products for which more than one θ is positive will be small.

A further issue is the number of recipes we should consider; since there are 2^{T-1} dominant schedules for each item, the size of the problem may grow considerably. In practice, column generation procedures must be exploited in order to keep the number of recipes at a reasonable level. It has also been argued that good columns should be generated by applying heuristic methods for the single-item lot-sizing problem, rather than considering all the dominant schedules.

3.2.6 Hierarchical approaches for CLSP

Hierarchical Production Planning (HPP) is an approach aiming at reducing both the computational burden of solving a CLSP and the difficulties due to demand uncertainty. The basic idea is common to APP. Similar items are aggregated in families, and a high-level problem is solved for the family lot-sizing problem; then the problem is solved at the item level, with some constraints enforced by the solution of the high-level problem. A further consideration is that demand for families is affected by smaller uncertainties than the demand for items.

Here we describe a mathematical model for lot-sizing with major family setups and minor item setups (as a typical example, consider a chemical product sold in different packages); therefore, unlike APP, we have to solve MILP problems at both levels. The model was proposed in [186] and is also aimed at lot-sizing in assembly systems. We will see in the next section an alternative approach to such a problem; here the idea is to use time-phased load profiles to compute the capacity absorption for producing a certain item. We first state the monolithic model, then we show the HPP version.

$$\min \sum_{t=1}^{T} \left[\sum_{k=1}^{K} S_k \xi_{kt} + \sum_{i=1}^{N} (s_i \nu_{it} + c_i I_{it}) \right] \quad (3.17)$$

$$\text{s.t.} \quad I_{it} = I_{i,t-1} + x_{it} - d_{it} \quad \forall i, t \quad (3.18)$$

$$x_{it} \leq M \nu_{it} \quad \forall i, t \quad (3.19)$$

$$\nu_{it} \leq \xi_{kt} \quad \forall k, \forall i \in F_k \quad (3.20)$$

$$\sum_{\tau=0}^{T-t} \left[\sum_{k=1}^{K} R_{km\tau} \xi_{k,t+\tau} + \sum_{i=1}^{N} (r'_{im\tau} \nu_{i,t+\tau} + r_{im\tau} x_{i,t+\tau}) \right] \leq Q_{mt}$$
$$\forall m, t \quad (3.21)$$

$$\xi_{kt}, \nu_{it} \in \{0, 1\} \quad \forall i, k, t$$

$$x_{it}, I_{it} \geq 0 \quad \forall i, t.$$

The model is a variation of the CLSP model, and part of the notation is similar; here $k = 1, \ldots, K$ is the family index and F_k is the set of items belonging to family k. The objective function (3.17) includes a family setup cost S_k which is paid whenever the family setup variable ξ_{kt} is activated. Eqs. (3.18) and (3.19) are identical to those we have already met in the CLSP model and require no explanation. Eq. (3.20) simply states that an item cannot be produced without a setup for its family. The capacity constraint (3.21) includes family and item setup times, as well as item processing times; $r_{im\tau}$ is the time-phased load profile, i.e., the amount of resource m required during a generic time bucket t to produce one unit of item i which is completed at time bucket $t+\tau$. A similar interpretation holds for $r'_{im\tau}$ and $R_{km\tau}$. This model aims at capturing the essence of multilevel lot-sizing within a single-level modeling framework. The idea is that the production of a certain item has different requirements in different time buckets; these requirements can be determined by considering the production and the assembly of the components on different workcenters. Obviously, the model has some limitations, since the load profile is postulated a priori.

This formulation can be decomposed into two subproblems: an aggregate family level problem and a disaggregation problem at the item level. The family level problem is obtained by aggregating the decision variables by family and by neglecting item setup times. Introducing family inventory levels and production quantities

$$I^F_{kt} = \sum_{i \in F_k} I_{it} \qquad x^F_{kt} = \sum_{i \in F_k} x_{it},$$

we obtain the following model:

$$\min \sum_{t=1}^{T}\sum_{k=1}^{K}\left(S_k \xi_{kt} + c^F_k I^F_{kt}\right)$$

s.t. $I^F_{kt} = I^F_{k,t-1} + x^F_{kt} - d^F_{kt} \qquad \forall k,t$

$x^F_{kt} \leq M\xi_{kt} \qquad \forall k,t$

$$\sum_{\tau=0}^{T-t}\sum_{k=1}^{K}\left(R_{km\tau}\xi_{k,t+\tau} + r^F_{km\tau}x^F_{k,t+\tau}\right) \leq Q_{mt} \qquad \forall m,t$$

$\xi_{kt} \in \{0,1\} \qquad \forall k,t$

$x^F_{kt}, I^F_{kt} \geq 0 \qquad \forall k,t,$

where the family parameters, c^F_k and $r^F_{km\tau}$, are obtained by averaging over the corresponding item parameters.

Let $\hat{x}^F_{kt}$ and $\hat{\xi}_{kt}$ be the optimal solution of the aggregate problem. The disaggregation problem can be formulated as:

$$\min \sum_{t=1}^{T}\sum_{i=1}^{N}(s_i \nu_{it} + c_i I_{it})$$

s.t. $I_{it} = I_{i,t-1} + x_{it} - d_{it} \qquad \forall i,t$

$x_{it} \leq M\nu_{it} \qquad \forall i,t$

$$\sum_{i \in F_k} x_{it} = \hat{x}^F_{kt} \qquad \forall k,t \qquad (3.22)$$

$$\sum_{\tau=0}^{T-t}\sum_{i=1}^{N}(r'_{im\tau}\nu_{i,t+\tau}+r_{im\tau}x_{i,t+\tau}) \leq \hat{Q}_{mt} \quad \forall m,t$$

$$\nu_{it} \in \{0,1\} \quad \forall i,t$$

$$x_{it}, I_{it} \geq 0 \quad \forall i,t,$$

where

$$\hat{Q}_{mt} = Q_{mt} - \sum_{\tau=0}^{T-t}\sum_{k=1}^{K} R_{km\tau}\xi_{k,t+\tau}$$

is the remaining capacity, obtained by subtracting the family setup times from the available capacity. Note how (3.22) enforces the consistency of the items and families plans. Now, instead of one large CLSP problem, we have two simpler problems: if they are still too difficult to solve exactly, the techniques we have described in the previous sections can be applied.

Sometimes the disaggregation process is tackled by heuristic methods. Consider a rolling horizon approach to the overall problem: the family level problem is solved, and only the first portion of it is disaggregated at the item level. A simple principle to allocate the family production quantity among the items of the family is the equalization of the runout times: the idea is that the inventory for all the items of a family should get empty at the same time. The literature on HPP is vast, and the interested reader is referred to the few references given at the end of the chapter. Here we just point out one thorny issue: feasibility. It may happen that, given a solution of the family level problem, there is no way to disaggregate it into feasible item level solutions. Ways to overcome such difficulties are described in the literature.

3.2.7 Multilevel lot-sizing

In multilevel lot-sizing problems, we consider assembly operations and/or multi-stage production-inventory systems. Here we present a model that, unlike the previous HPP formulation, represents explicitly the link between the independent demand for end items and the dependent demand for components. We denote by α_{ij} the number of items of type i necessary for assembling one unit of type j, and by $\mathcal{P}_i$ the set of items which are "parents" of i in the sense that i is one of their components. The following model has been proposed in [136]:

$$\min \sum_{t=1}^{T}\sum_{i=1}^{N}(c_i I_{it} + s_i \nu_{it})$$

$$\text{s.t.} \quad I_{it} = I_{i,t-1} + x_{it} - d_{it} - \sum_{j \in \mathcal{P}_i} \alpha_{ij} x_{jt} \quad \forall i, t \quad (3.23)$$

$$\sum_{i=1}^{N} (r_{im} x_{it} + r'_{im} \nu_{it}) \leq R_{mt} \quad \forall t, m$$

$$x_{it} \leq M \nu_{it} \quad \forall i, t$$

$$I_{it}, x_{it} \geq 0 \quad \forall i, t$$

$$\nu_{it} \in \{0, 1\} \quad \forall i, t.$$

The key difference from the CLSP formulation is clearly Eq. (3.23), where we subtract from the inventory level both the independent demand and the dependent demand induced by the production of parent items. Usually $d_{it} = 0$ if i is not an end item; however, there may be cases in which there is an independent demand for components sold as spare replacement parts.

The model is more difficult to deal with than the single-level CLSP: dualizing the capacity constraints does not yield a set of easy problems. Another way to develop heuristic methods based on MILP models is to round the optimal solution of the continuous relaxation. Although this idea does not work for general MILP models, it may be a good approach for specific models. We outline here some LP-based heuristics originally proposed in [136]:

1. The simplest idea is to solve the continuous relaxation and to fix to 1 the largest noninteger setup variable. Then the continuous problem is solved again until the resulting solution is integer.

2. Another idea is to fix different setup variables at once to 1 or 0. It has been observed that fractional setup variables can often be grouped in sets such that their sum is 1, and that in the optimal solution they are substituted by a single setup. Therefore, it is possible to obtain a good integer solution not only by fixing setup variables to 1 but also to 0. In the selection of the variables, knowledge of the specific problem may be exploited, paying attention to the tightest capacity constraints and highest setup costs.

3. Finally, a third possibility is to solve the continuous relaxation, to fix all integer setup variables to their values, and to perform a limited Branch and Bound search on the remaining setup variables.

The above multilevel formulation is rather simple, but it has a potential drawback. It assumes that it is possible to produce a component and use it to assemble a parent item during the same time bucket. If the bill of materials is deep (i.e., there are many different levels and long chains of components/parents), it may happen that the solution of the lot-sizing problem cannot be translated to feasible schedules. To avoid this problem, some form of delay must be introduced by modifying Eq. (3.23). One possibility is to associate a minimal lead time L_i with item i [184]:

$$I_{it} = I_{i,t-1} + x_{i,t-L_i} - d_{it} - \sum_{j \in \mathcal{P}_i} \alpha_{ij} x_{jt}.$$

Here the lead time L_i is not due to capacity constraints, but e.g. to material handling. An alternative idea is to stipulate that the production of a component during time bucket t is available for assembling the parent item only at time bucket $t+1$. Then the inventory can be modeled as an excess of production [161]:

$$I_{it} = \sum_{\tau=1}^{t} x_{i\tau} - \sum_{\tau=1}^{t} d_{i\tau} - \sum_{j \in \mathcal{P}_i} \sum_{\tau=1}^{t} \alpha_{ij} x_{j,\tau+1}.$$

However, even this approach may not be completely satisfactory. If there are many levels, high inventory costs may be incurred, since the total lead time from raw materials to end items may be large. Actually the difficulty is due to adoption of a discrete-time representation; we verify the inventory balance only at discrete instants, disregarding what happens during the time bucket. In Eq. (3.23) the material flow assumed by the model may be too fast; in the last formulation it may be too slow. We discuss issues related to time discretization in Section 3.5.

3.3 The Discrete Lot-Sizing and Scheduling Problem

The CLSP formulation considers setup costs/times with a large time bucket; during each time bucket more than one product is processed, but no sequencing decision is considered. Consider two consecutive time buckets, and assume that a product is processed during both time buckets. According to the CLSP formulation, we pay the setup charge twice, which could be avoided by clever sequencing. If the number of products processed during a time bucket is large enough, the error is negligible. If we need a model encompassing both lot-sizing and scheduling, we can reduce the size of the time buckets, so that *at most one product* is processed during each period. We obtain a **small bucket** model known as Discrete Lot-Sizing and Scheduling Problem (DLSP). We assume that all the available capacity is used during each time bucket. This may be reasonable or not, but the resulting problem is much easier to solve than the case in which the only a portion of the time bucket is used. Such a detailed model allows carrying over the setup between consecutive time buckets.

In Section 3.3.1 we formulate different versions for the DLSP. Then, we outline a dual decomposition strategy in Section 3.3.2. The dual decomposition is based on a dynamic programming approach for the single item problem, described in Supplement S3.2.

3.3.1 The DLSP formulation

There are many different variations of small time bucket lot-sizing problems. Here we present three single machine versions of DLSP: in the first one setup times are neglected; in the second one there are sequence-independent setup times smaller than the time buckets; and in the third one, the setup times are multiple of the time bucket size.

Note that, given the detail of the model, the number of resources must be explicitly considered. The models discussed in the following deal with single-machine problems.

The basic DLSP model

The single machine DLSP model with setup costs but no setup times is the following:

$$\min \sum_{i=1}^{N} \sum_{t=1}^{T} (s_i \max\{0, x_{it} - x_{i,t-1}\} + c_i I_{it})$$

$$\text{s.t.} \sum_{i=1}^{N} x_{it} \leq 1 \quad \forall t \tag{3.24}$$

$$I_{it} = I_{i,t-1} + p_i x_{it} - d_{it} \quad \forall i, t \tag{3.25}$$

$$I_{it} \geq 0 \quad \forall i, t$$

$$x_{it} \in \{0, 1\} \quad \forall i, t.$$

Here I_{it} is, as usual, the inventory level of item i at the end of time bucket t, and d_{it} is the demand. The inventory and setup costs are c_i and s_i. Note that we are assuming *sequence-independent* setup costs. The decision of producing item i during period t is modeled by the binary variable x_{it}; p_i is the number of items of type i produced during one period; recall that we consider only full capacity production runs during each period. Eq. (3.24) is the interaction constraint, stating that at most one product is processed during a period; this is the constraint linking the products. Note that (3.24) is not a strict analogue of the capacity constraints in the CLSP. The capacity constraints are expressed partly by the interaction constraints, and partly by the $p_i x_{it}$ term in (3.25); in the DLSP case the production quantity per time bucket is either p_i or 0. The setup cost is charged for item i during period t if and only if $x_{i,t-1} = 0$ and $x_{it} = 1$. To express this in linear form, we may introduce setup variables ν_{it} subject to the constraints

$$\nu_{it} \geq x_{it} - x_{i,t-1} \quad \forall i, t$$
$$\nu_{it} \geq 0 \quad \forall i, t.$$

The objective function is, then,

$$\min \sum_{i=1}^{N} \sum_{t=1}^{T} (s_i \nu_{it} + c_i I_{it}).$$

It is important to note that the ν_{it} variables are not an exact translation of the expression $\max\{0, x_{it} - x_{i,t-1}\}$; as far as the constraints are concerned, we could have $\nu_{it} = 1$ even when $x_{it} - x_{i,t-1} = 0$. However, the minimization of the objective function rules out this possibility. Furthermore, there is no need to require that ν_{it} is binary: this requirement is obviously satisfied by the optimal solution of the problem.

DLSP with sequence-independent setup times

There are two main cases to consider:

1. the case of setup time smaller than the time bucket;
2. the case of setup time larger than the time bucket.

In the first case, we have only to modify Eq. (3.25) as follows:

$$I_{it} = I_{i,t-1} + p_i x_{it} - d_{it} - r_i \nu_{it} \quad \forall i, t,$$

where r_i is the number of parts that has not been produced due to the time lost during setup.

The second case is more involved. Let a_i denote the setup time for item i, expressed as a number of time buckets. Whenever $x_{i,t-1} = 0$ and $x_{it} = 1$, we must forbid production for time buckets $t - a_i, t - a_i + 1, \ldots, t - 1$. This is obtained by making sure that the right number of consecutive setup variables are set to 1:

$$\nu_{i, t-a_i+\tau} \geq x_{it} - x_{i,t-1} \quad t = a_i + 1, \ldots, T, \ \tau = 0, \ldots, a_i - 1,$$

and by modifying Eq. (3.24);

$$\sum_{i=1}^{N} (x_{it} + \nu_{it}) \leq 1 \quad \forall t.$$

Note that the production and setup variables are disjunctive.

3.3.2 A dual decomposition approach for the multi-item DLSP problem

Consider the DLSP problem without setup times. By dualizing the capacity constraints (3.24) with multipliers $\mu_t \geq 0$, we can decompose the problem into a set of single-item subproblems. The subproblem for item i is:

$$\min \sum_{t=1}^{T} (s_i \max\{0, x_{it} - x_{i,t-1}\} + c_i I_{it} + \mu_t x_{it})$$

$$\text{s.t.} \quad I_{it} = I_{i,t-1} + p_i x_{it} - d_{it} \quad \forall t$$

$$I_{it} \geq 0 \quad \forall t$$

$$x_{it} \in \{0, 1\} \quad \forall t.$$

As customary, the multipliers define time-varying production costs. The single-item problem can be solved by a dynamic programming approach described in Supplement S3.2. Note that we cannot say that the single-item problem is uncapacitated, since we cannot produce more than p_i parts of type i per time period.

Let $w_i(\mu)$ be the optimal value for each single-item problem; the dual problem is

$$\max_{\mu \geq 0} \left(\sum_{i=1}^{N} w_i(\mu) - \sum_{t=1}^{T} \mu_t \right),$$

which can be tackled by subgradient optimization along the lines illustrated for the CLSP case. The dual problem can be exploited within Branch and Bound procedures and for developing dual heuristics.

3.4 Continuous-flow models for production scheduling

In this section we deal with discrete-time models for production scheduling. DEDS models for the same problem are described in Chapter 4. DEDS models are detail models whereby the timing of each event is specified. The models we present here are approximate, in the sense that the discrete material flow is approximated by a continuous material flow. This type of approximation is also exploited in Section 6.2, where a continuous-time model is presented. The basic motivation behind the approach is that a detail scheduling model may be impractical for two reasons:

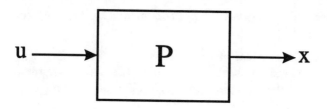

FIGURE 3.6
A model P represents the relationship between controls u and states x

1. computational complexity: scheduling problems are NP–hard as a rule, and we will see in Chapter 4 why obtaining optimal solutions is usually out of the question;
2. uncertainty and need for real-time re-scheduling: in a manufacturing system affected by random phenomena, a schedule specifying the timing of all the events may be of little use, since it is too vulnerable to disruptions.

These difficulties have led many practitioners to the use of myopic priority rules for scheduling (Section 4.8.1). The approach we describe here is a possible way to overcome the above difficulties without sacrificing the quality of the solution. We use an approximate discrete-time dynamic model to generate an "ideal" trajectory for certain state variables, which is tracked by a detail scheduler. This idea is described in Section 3.4.1. Then we describe the mathematical model in Section 3.4.2.

3.4.1 Discrete-time dynamic models and trajectory tracking

In Chapter 1 we have introduced discrete-time dynamic models of the form:

$$\mathbf{x}_t = \mathbf{h}(\mathbf{x}_{t-1}, \mathbf{u}_t) \qquad t = 1, 2, \ldots, T. \tag{3.26}$$

The vector of state variables x evolves under the influence of an external input u, which is the **control vector**, in the sense that it represents the decision variables by which we influence the system. The relationship between controls and states can be depicted as in Figure 3.6, where the 'black box' P represents a way for solving the state equations (let P be the model represented by the state equations).

Controlling the system requires finding a suitable control u in order to

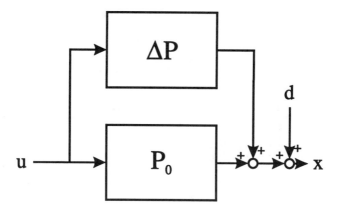

FIGURE 3.7
The actual trajectory is affected by modeling errors and uncertainty

obtain a satisfactory state trajectory. We want to solve

$$\min_{\mathbf{u}} f(\mathbf{x}, \mathbf{u}),$$

subject to (3.26). Further constraints restrict the set of applicable controls and enforce initial and/or final values of state variables. Given a way to solve this optimization problem, we can obtain an optimal control $\mathbf{u}^*$, and an optimal state trajectory $\mathbf{x}^*$.

When the state equations of model P are complicated, the solution of the above optimization problem can be impractical. The optimization problem can be solved only for a simplified representation, P_0, of the system, called the **nominal** model. The nominal model is affected by a modeling error ΔP. Furthermore, the state trajectory may be affected by uncontrollable inputs, called **disturbances**, d. For instance, we have a modeling error if we assume a continuous material flow, which is only an approximation of the discrete material flow. A typical disturbance is a variation of the demand. Another type of disturbance may be a variation of the production capacity due to a machine failure. We will see an example of such a *parametric* disturbance in Section 6.2.

The application of the optimal control $\mathbf{u}^*$, computed according to the nominal model, P_0, does not result in the optimal trajectory $\mathbf{x}^*$ (the situation is depicted in figure 3.7). In control theory, the approach to overcoming these difficulties is the adoption of closed-loop or feedback control (see figure 3.8). Instead of computing the optimal control $\mathbf{u}^*$ and applying it in open loop, i.e., neglecting real-time information on system performance, the optimal trajectory $\mathbf{x}^*$ (computed according to the nominal model, P_0)

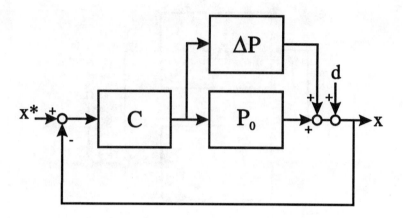

FIGURE 3.8
A closed-loop control system

is used as a **reference trajectory** to be tracked. When a tracking error $x^* - x$ is detected, a corrective control is computed by a suitable black box C. In our case the tracking problem can be stated as

$$\min \sum_{t=1}^{T} \| x_t^* - x_t \|,$$

where $\| \cdot \|$ is, for instance, the usual euclidean norm. A simple example will clarify the idea.

Example 3.1
Consider an assembly line for a set of n end items. We are in a repetitive production environment, and each end item must be produced according to a given mix. Let m_i be the mix for end item i $(i = 1, \ldots, n)$. The total number of items to assemble during each cycle is $T = \sum_{i=1}^{n} m_i$.

Each end item is assembled using a set of q components. Let j be the index for components $(j = 1, \ldots, q)$. Assume a flat assembly structure, i.e., item i requires a known quantity b_{ij} of component j, but there is no subassembly. For example, consider the case of a line assembling two part types, P_1 and P_2, where the production mix is three units of the first type and five of the second one. Three components, C_1, C_2, and C_3, are required for the assembly, and the b_{ij} coefficients are as follows:

$$b_{11} = 1, \, b_{12} = 0, \, b_{13} = 2,$$

$$b_{21} = 0, \ b_{22} = 2, \ b_{23} = 1.$$

According to just-in-time principles, the end items should be sequenced in such a way as to assure the smoothest absorption of components. A sequence like

$$P_1, P_1, P_1, P_1, P_1, P_2, P_2, P_2$$

would not be a good sequence, since it implies a lumpy demand for components C_1 and C_2.

The sequencing process can be thought of as a discrete-time decision process, whereby at each step t an end item must be selected for assembly. Let u_{it} be a binary variable set to 1 if item i is assembled at step t, 0 otherwise; the set of these variables is our set of controls. Since one item can be assembled at a time, the following constraint must be enforced:

$$\sum_{i=1}^{n} u_{it} = 1 \qquad (t = 1, \ldots, T).$$

Let z_{it} be the number of end items i assembled at the end of step t:

$$z_{it} = z_{i,t-1} + u_{it} = \sum_{k=1}^{t} u_{ik}.$$

The state variables z_{it} must satisfy initial and final constraints:

$$z_{i0} = 0; \qquad z_{iT} = m_i.$$

We must now relate the controls u_{it} to state variables expressing the trajectory of component absorption. The amount of component j required during step t is $b_{ij} u_{it}$, and the amount consumed at the end of step t is expressed by state variables x_{it} whose dynamic equation is

$$x_{jt} = x_{j,t-1} + \sum_{i=1}^{n} b_{ij} u_{it}.$$

The aim is to obtain a smooth trajectory of x_{jt} for all j.

Apparently, obtaining the smoothest trajectory is not a trivial task, but it is easily computed in an idealized case. If a continuous, rather than discrete, material flow is assumed, the smoothest trajectory for each component is simply a straight line whose slope is

$$\frac{Q_j}{T},$$

where Q_j is the total amount of components of type j required for each

cycle,
$$Q_j = \sum_{i=1}^{n} m_i b_{ij}.$$
This means that the optimal trajectory for this ideal case would be
$$x_{jt}^* = \frac{Q_j}{T} t;$$
this trajectory is ideal because in general it does not consist of integer numbers, whereas the real trajectory does.

This would be the optimal trajectory if the production were continuous; in our approach it is the optimal solution for a model affected by a modeling error. The ideal solution can be used as a reference to be tracked, i.e., one would like to minimize the overall tracking error
$$\sum_{t=1}^{T} \sum_{j=1}^{q} (x_{jt}^* - x_{jt})^2.$$
A simple approach to minimize the tracking error is to adopt a *greedy* rule (see Section C.8). At each step we select the end item yielding the locally optimal value, i.e., the one minimizing the tracking error up to the current step, neglecting the future ones. Let $\hat{x}_{j,k}$ be the (already determined) trajectory for time buckets 1 throught $t-1$; we determine the next item i to be assembled by solving
$$\min_{i} \sum_{j=1}^{q} \left(x_{jt}^* - \hat{x}_{j,t-1} - b_{ij} \right)^2$$
This greedy rule is known as Toyota Goal Chasing method. □

The idea of tracking a reference trajectory is illustrated in Figure 3.9. In the next section we apply this idea to scheduling in a Flexible Flow Line.

3.4.2 A discrete-time continuous-flow model for batch production

Consider a flexible flow line (FFL), i.e., a manufacturing system composed of a series of stages. Each stage (indexed by $k = 1, \ldots, K$) consists of a parallel arrangement of identical machines (see Figure 3.10). This structure is used for illustrative purposes, but the applicability of the approach is not restricted to this case. A set of batches must be manufactured on the FFL

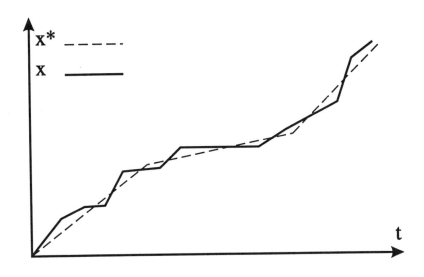

FIGURE 3.9
Tracking a reference trajectory

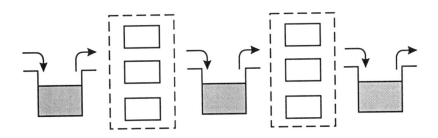

FIGURE 3.10
A flexible flow line

over a scheduling horizon of known length. For each batch $i = 1, ..., N$, a batch size l_i is given (i.e., the number of parts that must be produced); release times r_i and due dates d_i may be given. Let t_i^s and t_i^c denote the start and the completion time for each batch i; the set of batches must be scheduled in such a way as to minimize a given objective function of the form $f(\mathbf{t}^s, \mathbf{t}^c)$, where $\mathbf{t}^s, \mathbf{t}^c$ are the vectors of the start and completion times of all batches. We have seen in Section 1.3.3 that such objective functions may be of the minsum or minmax type.

We could try tackling the problem using detail DEDS models (see Chapter 4). This may be impractical if the problem is complicated by issues such as the following. A first problem is that there might be limits on the interstage work in process levels, due to buffer capacity constraints. Another complicating factor could be tooling: this means that some machines must be equipped with a set of tools, which are stored in a limited capacity tool magazine on each machine. If enough tools are available, a batch may be processed on more than one machine of the same stage; if enough tools may be loaded on the same machine, more than one batch may be processed on it at the same time. We assume that the setup time is negligible; this assumption is justified when appropriate tool changing technology is available. A last problem is that here we have *batches* of parts. We may consider each part as a single job, but this would result in a huge problem; on the contrary, we could consider each batch as a single job, but we would loose the opportunity of using *transfer batches* to reduce the work in process and the lead time. When using transfer batches, the processing of the same batch on different stages may overlap (see figure 3.11); the first parts of a batch which have been processed on a stage are immediately transferred to the next stage, without waiting for the completion of the whole batch on that stage. This practice is encouraged by production scheduling approaches such as OPT (Optimized Production Technology).

Due to such difficulties, we may attempt a solution approach based on a nominal model, as in Example 3.1. The nominal model here is also based on a continuous flow approximation: the parts are considered as infinitely divisible. A consequence is that the production capacity of each machine can be time-shared. Unlike DEDS scheduling models, we do not require that each machine processes one part at a time. The state variables of the dynamic model of the FFL are the **cumulative production levels** x_{ikt} for batch i, on stage k, at the end of time bucket t. Due to the continuous flow approximation, they are continuous variables. The control variables are the production quantities u_{ikt} for each part type i, on each stage k, during time bucket t; again, these are continuous variables. State and control variables are related by equations of the form

$$x_{ikt} = x_{ik,t-1} + u_{ikt}.$$

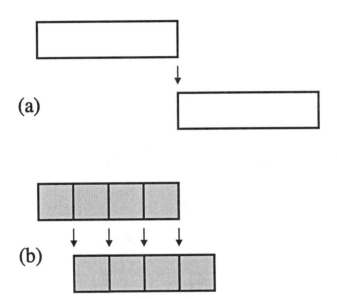

FIGURE 3.11
The use of transfer batches

If we consider an objective function of the form

$$\sum_{i=1}^{N} f_i(t_i^c),$$

i.e., an additive function of the completion times for each batch i, we need to express t_i^c as a function of the state variables. This can be accomplished by introducing a set of binary variables z_{it} set to 1 if the batch has not yet been completed at the beginning of time bucket t. We have, then, $t_i^c = \sum_{t=1}^{T} z_{it}$. Now the continuous-flow problem formulated in MILP form[1] is

$$\min \sum_{i=1}^{N} \sum_{t=1}^{T} w_{it} z_{it} \tag{3.27}$$

$$\text{s.t.} \quad z_{it} \geq 1 - x_{iK,(t-1)}/l_i \quad \forall i, t \tag{3.28}$$

$$x_{ikt} = x_{ik,t-1} + u_{ikt} \quad \forall i, k, t \tag{3.29}$$

$$x_{i10} = 0, \; x_{iKT} = l_i \quad \forall i \tag{3.30}$$

$$x_{ikt} \leq x_{i,(k-1),(t-1)} \quad \forall i, k, t \tag{3.31}$$

$$\sum_{i=1}^{N} \tau_{ik} u_{ikt} \leq M_k \quad \forall k, t \tag{3.32}$$

$$u_{ikt} \in [0, \bar{u}_{ik}] \quad \forall i, k, t \tag{3.33}$$

$$z_{it} \in \{0, 1\} \quad \forall i, t.$$

$$\tag{3.34}$$

The objective function (3.27) has a quite general form. It can be used to express many objective functions, depending on how the weights w_{it} are set:

- total completion time: $w_{it} = 1 \quad \forall i, t$,
- total tardiness: $w_{it} = 1 \quad \forall i, t > d_i$; 0 otherwise, and
- number of late jobs: $w_{it} = 1 \quad \forall i, t = d_i + 1$; 0 otherwise.

We have assumed that the due date d_i is set at the end of the corresponding time bucket. Weights can be easily associated with each job.

Eq. (3.28) sets the correct values for the variables z_{it}. Since z_{it} is binary, it is forced to 1 whenever $x_{iK,t-1} < l_i$; z_{it} is set to 0 only when the cumulative production on the last stage has reached the batch size.

---

[1] The model presented here is not suitable from the computational point of view; we will present a better formulation in Section 6.3.

Eq. (3.29) relates the evolution of cumulative production to the production during each time bucket. Eq. (3.30) sets initial and final conditions on the state variables; we start from 0 at the first stage, and all the required material is processed on the last stage within the allowed time horizon T. Eq. (3.31) relates cumulative production between adjacent stages; the cumulative production on stage k at the end of time bucket t cannot be larger than the cumulative production at stage $k-1$ at the end of time bucket $t-1$.

Eqs. (3.32) and (3.33) are related to capacity constraints and require some explanation. M_k is the number of (identical) machines within stage k; τ_{ik} is processing time of part type i on stage k (expressed in time buckets). Therefore, $\sum_{i=1}^{N} \tau_{ik} u_{ikt}$ is the capacity required for production (expressed in time buckets) during time bucket t on stage k; this cannot be larger than the number of available machines, as expressed by (3.32). Eq. (3.33) enforces an upper bound $\bar{u}_{ik}$ on the production of part type i on stage k during one time bucket. This is a further limitation, due to the fact that it may be impossible to produce a certain part type i on *all* the machines within a stage. For instance, if stage k consists of four machines, but only two of them can be setup for batch i, e.g. due to tooling limitations, then a limit $\bar{u}_{ik}$ on the production of that part during one time bucket must be enforced:

$$u_{ikt} \leq \bar{u}_{ik} = \frac{2}{\tau_{ik}};$$

in such a case the capacity of the stage must be shared, or left partially unused.

A rather remarkable characteristic of this discrete-time model is its flexibility, that is its ability to adhere to several different features of a real situation just by specifying some parameters or introducing minor modifications within the same general structure.

Minmax objective functions. The formulation can be easily accommodated to the case of a maximum type objective function

$$\max_{i=1,\ldots,N} f_i(t_i^c).$$

For instance, the model for the minimization of maximum lateness

$$\min L_{\max} = \max_{i}\{t_i^c - d_i\} \qquad (3.35)$$

simply requires the addition of the constraints,

$$\sum_{t=1}^{T} z_{it} - d_i \leq L_{\max} \qquad \forall i.$$

The minimum makespan (maximum completion time $C_{\max}$) problem is obtained by setting $d_i = 0$.

Release times. If a release time r_i is given, then we have simply to enforce $u_{ikt} = 0$ for $t < r_i$.

Nonregular objective functions. Earliness costs can be considered by introducing a set of binary variables y_{it}, whose value is 1 if batch i has been started at time bucket t, 0 otherwise. The correct value is obtained by enforcing the following constraint:

$$y_{it} \geq \frac{x_{i1t}}{l_i} \qquad \forall i, t.$$

Then the start time can be expressed as $t_i^s = T - \sum_{t=1}^{T} y_{it}$. In order to penalize early start times, objective function (3.27) can be modified as follows:

$$\min \sum_{i=1}^{N} \sum_{t=1}^{T} (w_{it} z_{it} + c_{it} y_{it}),$$

having introduced suitable cost coefficients c_{it}.

Assembly systems. The model can also be accommodated to assembly systems. For example, if one unit of the part type of batch α and two units of the part type of batch β are manufactured at stage k and assembled in one unit of batch γ in stage $k + 1$, then the material consistency constraints, expressed by Eq. (3.31), should be rewritten as

$$x_{\gamma,(k+1),t} \leq x_{\alpha,k,(t-\Delta)} \qquad \forall t$$

$$x_{\gamma,(k+1),t} \leq \frac{1}{2} x_{\beta,k,(t-\Delta)} \qquad \forall t.$$

Shop structure different than a flexible flow line can be considered, as well as reentrant flows.

Limited capacity buffers. For example, if the buffer capacity among stages $k - 1$ and k is B_k, and each unit of the part type of batch i consumes a space b_{ik} of the buffer, then the following constraint can be enforced:

$$\sum_{i=1}^{N} b_{ik} (x_{i,(k-1),(t-\Delta)} - x_{ikt}) \leq B_k \qquad \forall k \geq 2, t.$$

Once the continuous-flow problem has been solved, we can track the ideal trajectory by interpolating it and by adopting a greedy chasing strategy; clearly, better approaches can be devised. However, we should wonder if the continuous-flow model is easily solvable or not, since it is a potentially difficult MILP problem. A first thing to note is that if we fix z_{it}, we obtain a feasibility problem; we should try to obtain a feasible schedule with fixed deadlines. The feasibility problem is an LP problem, which is polynomial. If we have N batches and T time buckets, then we could solve the feasibility problem for each possible assignment of the z_{it}; since there are $N \times T$ possible assignments, we must solve $N \times T$ polynomial problems. One would be tempted to say that the overall problem is polynomial; however, the time horizon T may depend on the problem data, such as the due dates. Therefore, we can only say that the strategy based on the feasibility problems yields a pseudopolynomial algorithm (see Supplement S2.1); in fact, it can be shown that the problem with the minsum objective function is NP–hard. Still, this does not imply that the problem cannot be solved in practice. Using the feasibility approach, it is easy to see that for certain objective functions, such as the maximum lateness $L_{\max}$, we can devise a polynomial algorithm, based on a binary search of the possible values of the objective. Given an upper bound $\hat{L}_{\max}$, we must verify if a lateness $\lceil \hat{L}_{\max}/2 \rceil$ is feasible; this is easily accomplished, since a value of the maximum lateness yields the values of the z_{it}. If it is feasible, we try with the midpoint of the interval

$$[0, \lceil \hat{L}_{\max}/2 \rceil];$$

otherwise we try with the midpoint of the interval

$$[\lceil \hat{L}_{\max}/2 \rceil, \hat{L}_{\max}]$$

and so on. It is worth noting that the minimization of $L_{\max}$ remains polynomial even if complicating features such as release times and limited capacity buffers are added.

3.5 Discussion: the flexibility of discrete-time models

At the end of the previous section we have pointed out the flexibility of the discrete-time continuous-flow model for FFL scheduling. Flexibility is a remarkable peculiarity of discrete-time models. The aim of this section is to discuss discrete-time models from the point of view of *modeling power*, without considering computational issues.

In Section 3.5.1 we give an example of the modeling power of discrete-time models. We consider the problem of periodic scheduling in a press shop consisting of parallel (nonidentical) lines, with several complicating features. The problem is essentially modeled within a DLSP framework.

Here we want to point out the pitfalls of this modeling approach. The first, and most obvious, is that time discretization induces a modeling error. For instance, a scheduling model tackled within a DLSP framework is an approximation of the "true" DEDS model. We assume that, during each time bucket, we produce at full capacity (hence, more than necessary) and that we can switch production only at certain time instants. The first assumption could be relaxed (with a considerable increase in computational difficulties), but in such a case we would keep the machine idle until the end of the time bucket.

We have also pointed out some difficulties in modeling the synchronization of material flow within multilevel lot-sizing problems (Section 3.2.7): if we adopt one model, components can be released to assembly only at the end of a time bucket, with an increase in the overall lead time; with the other model, infeasibilities may arise at the scheduling level. The same problem occurs in the continuous-flow scheduling model (Section 3.4): this model has only a limited capability of dealing with transfer batches, since parts may be released to the next stage only at the and of the time bucket.

Such difficulties do not imply that discrete-time models cannot be successfully exploited; however, some care is needed in selecting the length of the time bucket and in validating the model. In practice, it is possible to correct part of the modeling error by post-processing the solution of the discrete-time model.

3.5.1 Periodic scheduling in a press shop

In order to illustrate the advantages of a discrete-time modeling approach, we outline here a rather complex model for production scheduling in a press shop consisting of parallel lines. Each line consists of a series arrangement of machines, carrying out the sequential steps needed to obtain the final part; from the planning point of view each line can be considered as a single machine. The lines are not identical, since some of them are more recent than others; therefore, we must consider the difference in production costs for each (item,line) pair. We want to schedule production on the lines in a steady-state situation characterized by constant demand; the solution has a periodic character. The aim is to obtain "reference" inventory targets for operational scheduling with time-varying demand; the initial inventory values for the periodic problem are the target levels that the operational scheduling should try to reach at the end of the scheduling horizon.

The planning problem consists of a set of interrelated subproblems:

- choose the period at which the cycle is repeated (larger periods imply higher inventory costs, but better capability of facing unpredictable events);
- lot-sizing: after choosing the period, the quantity produced in each cycle is obviously fixed, but we must consider the possibility of splitting lots for items with high demand in order to minimize inventory costs;
- assign lots to the lines, taking into account costs and load balancing

issues;
- sequence the lots on each line.

The objective is to minimize the sum of the inventory, production and setup costs, taking into account a set of complicating features which constrain the solutions of the problem.

The characteristics of the problem can be summarized as follows:

- We must consider both setup costs and times. Setup costs are also due to a production loss, since the first parts produced after a setup are defective. The setup times are significant (actually we must change the dies on a set of machines to switch production on a line), but sequence-independent.
- There is a limit on the number of setups which can be carried out simultaneously, due to labor limitations; in the model we consider the number of setup teams as a given constant, but it could be treated as a decision variable.
- The load on the lines should be balanced in order to minimize the period while leaving a certain idle time to recover from machine breakdowns and to carry out maintenance.
- Some pairs of items should be scheduled concurrently on different lines; this happens when the raw materials for the two items are obtained by cutting the same coil, and one wants to avoid the buildup of work in process for one of them.
- Some pairs of items should be scheduled as far as possible from each other; this happens e.g. when the two items are stored in the same container, and we want to minimize the number of containers. As a practical example, consider the right and the left doors of a car: if the corresponding lots are scheduled simultaneously, a peak in the joint inventory level occurs, calling for a high number of containers.

The resulting problem is obviously rather complex. It is practically impossible to build a manageable mathematical model encompassing all the features of the problem. Here we aim at obtaining good solutions by an approach dealing with all the aforementioned issues simultaneously. To make this possible, a discrete-time modeling framework can be adopted; in particular, we build here a DLSP-like model. What we obtain is a relatively large MILP model. In order to be able to solve it with a reasonable effort, due attention must be paid to the formulation. The following assumptions have been made with this in mind.

- It is assumed that during each time bucket the production is of the *all-or-nothing* type, i.e. the full capacity during a time bucket is exploited; clearly, this is a modeling error, but it is justified by the improved solvability of the model.

Indexes	
$i, j = 1, \ldots, N$	item index
$t = 1, \ldots, T$	time bucket index (T is the cycle period)
$k = 1, \ldots, K$	line index
Problem data	
P_{ik}	productivity for item i on line k
R_{ik}	production loss due to the setup of item i on line k (including setup time and defective parts)
S_{ik}	setup cost for item i on line k
H_i	unit inventory cost for item i
C_{ik}	unit production cost for item i on line k
D_i	demand for item i during one time bucket
G	number of setup teams
$\mathcal{O}$	set of item pairs (ij) for which a minimal overlap is required
$O_{(ij)}$	minimal overlap for items i and j
$\mathcal{S}$	set of item pairs (ij) for which a maximum joint inventory level is allowed
$\tilde{I}_{(ij)}$	maximum joint inventory level for items i and j
α_T	maximum number of time buckets any line can work
Problem variables	
x_{ikt}	binary decision variable for production of item i on line k during time bucket t
$z_{(ij)t}$	binary variable indicating simultaneous production for items i and j during time bucket t
s_{ikt}	binary decision variable for setup of item i on line k during time bucket t
I_{it}	inventory level for item i at the end of time bucket t

TABLE 3.1
Notation for the mathematical model

- It is assumed that the complicating features of the problem are treated as *constraints*, rather than as *objectives*. This has a beneficial effect, if a commercial MILP code is used for solving the model, and is more compatible with managerial practice. For instance, the cycle period is assumed known in the formulation, and it is up the user to experiment with different values; apart from computational issues, it is reasonable to assume that only a few values are meaningful. The overall idea is that the mathematical model should be the "engine" of an interactive Decision Support System.

The mathematical model can be stated as follows (the notation is summarized in Table 3.1):

$$\min \sum_{i=1}^{N} \sum_{t=1}^{T} H_i I_{it} + \sum_{i=1}^{N} \sum_{t=1}^{T} \sum_{k=1}^{M} [(S_{ik} - R_{ik} C_{ik}) s_{ikt} + C_{ik} x_{ikt}] \quad (3.36)$$

$$\text{s.t.} \quad \sum_{i=1}^{N} x_{ikt} \leq 1 \quad \forall k, \forall t \quad (3.37)$$

$$\sum_{k=1}^{M} x_{ikt} \leq 1 \quad \forall i, \forall t \quad (3.38)$$

$$I_{it} = I_{i,t-1} - d_i + \sum_{k=1}^{M} P_{ik} x_{ikt} - \sum_{k=1}^{M} R_{ik} s_{ikt}$$
$$\forall i, \, t = 2, \ldots, T \quad (3.39)$$

$$I_{i1} = I_{iT} - d_i + \sum_{k=1}^{M} P_{ik} x_{ik1} - \sum_{k=1}^{M} R_{ik} s_{ik1} \quad \forall i \quad (3.40)$$

$$s_{ikt} \geq x_{ikt} - x_{ik,t-1} \quad \forall i, \forall k, \, t = 2, \ldots, T \quad (3.41)$$

$$s_{ik1} \geq x_{ik1} - x_{ikT} \quad \forall i, \forall k \quad (3.42)$$

$$\sum_{k=1}^{M} (s_{ik,t-1} + s_{ikt}) \leq 1 \quad \forall i, \, t = 2, \ldots, T \quad (3.43)$$

$$\sum_{k=1}^{M} (s_{ik1} + s_{ikT}) \leq 1 \quad \forall i \quad (3.44)$$

$$\sum_{i=1}^{N} \sum_{k=1}^{M} s_{ikt} \leq G \quad \forall t \quad (3.45)$$

$$\sum_{t=1}^{T} \sum_{i=1}^{N} x_{ikt} \leq \alpha_T \quad \forall k \quad (3.46)$$

$$I_{it} + I_{jt} \leq \tilde{I}_{(ij)} \qquad \forall t, \ \forall (ij) \in \mathcal{S} \tag{3.47}$$

$$z_{(ij)t} \leq \sum_{k=1}^{M} x_{ikt} \qquad \forall (ij) \in \mathcal{O}, \ \forall t \tag{3.48}$$

$$z_{(ij)t} \leq \sum_{k=1}^{M} x_{jkt} \qquad \forall (ij) \in \mathcal{O}, \ \forall t \tag{3.49}$$

$$\sum_{t=1}^{T} z_{(ij)t} \geq O_{(ij)} \qquad \forall (ij) \in \mathcal{O} \tag{3.50}$$

$$x_{ikt}, z_{(ij)t} \in \{0,1\}; \ I_{it}, s_{ikt} \geq 0 \qquad \forall i, \ \forall (ij) \in \mathcal{O}, \ \forall k, \ \forall t. \tag{3.51}$$

We now comment on the objective function and the constraints of the model; the decision variables are restricted in (3.51).

The first term of the objective function (3.36) takes into account the inventory costs; the second term accounts for setup and production costs. In computing the production costs, we consider that the setup time implies a production loss; hence the term $(-R_{ik}C_{ik}s_{ikt})$, which simply states that the production cost must not be charged for parts *not* produced during the setup.

Eq. (3.37) states that, at most, one item can be produced during a time bucket on each line.

Eq. (3.38) states that it is not possible to produce the same item on more than one line concurrently, due to limitations on the available dies.

Eq. (3.39) and (3.40) express the evolution of the inventory levels: (3.39) simply states that the inventory level of item i at the end of time bucket i is the previous level minus the demand, plus what has been produced on all the lines, minus the parts not produced during the setup time; (3.40) is required to take into account the periodic character of the problem.

Eq. (3.41) is such that the setup variable s_{ikt} is set to 1 whenever item i on line k is produced during time bucket t, but not during time bucket $t-1$; (3.42) is the periodic equivalent. Note that since $s_{ikt} \geq 0$, there is no need to require that these variables are binary.

Eq. (3.43) and (3.44) ensure that it is forbidden to perform two setups on the same line in consecutive time buckets; if this constraint is consistent with the problem data, it helps in strengthening the formulation.

Eq. (3.45) limits the number of simultaneous setups, where G is the number of setup teams. This constraint assumes that the setups are always started at the beginning of the time bucket. This may appear a limitation, since if the setup time is smaller than the time buckets, two setups could be carried out by the same team during the same time bucket. Nevertheless, it is advisable to avoid this possibility, since this would complicate the model and the resulting schedule would be vulnerable to disruptions.

Eq. (3.46) requires that no line used at full capacity; each line must be idle at least $T - \alpha_T$ time buckets.

Eq. (3.47) limits the joint inventory levels of some pairs of items; this is one way to require that the corresponding lots cannot overlap. The maximum joint inventory level $\tilde{I}_{(ij)}$ is related to the number of containers available.

Finally, Eqs. (3.48), (3.49) and (3.50) are used to enforce overlapping of pairs of items sharing raw materials. The requirement of synchronizing the production of some items can be expressed in different ways:

- we can model the amount of work in process and penalize it in the objective function;
- we can require that the setups are simultaneous;
- we can enforce a minimum amount of overlap between the corresponding lots.

We have chosen the last alternative, since it yields a better model than the first one from the point of view of model solvability, and the second alternative is too rigid (and in contrast with the need to avoid too many simultaneous setups). For each pair of items (ij) that must overlap, we introduce a binary variable $z_{(ij)t}$ set to 1 if both items are processed during time bucket t; this is accomplished by (3.48) and (3.49). Then, in Eq. (3.50), we require that a minimum overlap $O_{(ij)}$ is obtained.

S3.1 Strong formulations for lot-sizing problems

In this supplement we outline a strong formulation for the CLSP problem, without setup times, but with a single capacitated resource, whose basic formulation is

$$\min \sum_{t=1}^{T} \sum_{i=1}^{N} (c_i I_{it} + s_i \nu_{it})$$

$$\text{s.t.} \quad I_{it} = I_{i,t-1} + x_{it} - d_{it} \quad \forall i, t$$

$$\sum_{i=1}^{N} r_i x_{it} \leq R_t \quad \forall t$$

$$x_{it} \leq M \nu_{it} \quad \forall i, t$$

$$I_{it}, x_{it} \geq 0 \quad \forall i, t$$

$$\nu_{it} \in \{0, 1\} \quad \forall i, t.$$

The formulation is strong in the sense that its continuous relaxation yields larger lower bounds than those obtained with the basic formulation. The formulation, proposed in [68], exploits the special structure of the single-item uncapacitated problems, which can be reformulated as shortest path problems. Here we will skip many technical proofs about the equivalence of the original and the reformulated model; the reader is referred to [117] for details concerning the technique of **variable redefinition**, which is the basis of the approach.

Consider an MILP problem,

$$\min \ \mathbf{c}^T \mathbf{x}$$

$$\text{s.t.} \quad \mathbf{Ax} \geq \mathbf{a} \tag{3.52}$$

$$\mathbf{Bx} \geq \mathbf{b} \tag{3.53}$$

$$\mathbf{x} \geq 0; \quad x_j \text{ integer } \forall j \in \mathcal{J}; \tag{3.54}$$

where $\mathcal{J}$ is the set of integer variables. Assume that (3.53) are the "easy" constraints, in the sense that they are associated with a special structure, whereas (3.52) are the "difficult" constraints. In our case, the difficult constraints are the capacity constraints; if we drop them, we get a set of shortest path problems (see Section 3.2.4). Denoting by X the set of points satisfying both (3.53) and (3.54) we can reformulate problem

$$(P) \quad \min \ \mathbf{c}^T \mathbf{x}$$
$$\text{s.t.} \quad \mathbf{Ax} \geq \mathbf{a}; \quad \mathbf{x} \in X$$

as problem

$$(\hat{P}) \quad \min \ \mathbf{c}^T \mathbf{T} \mathbf{z}$$
$$\text{s.t.} \quad \mathbf{AT} \mathbf{z} \geq \mathbf{a}; \quad \mathbf{z} \in \overline{Z}$$
$$\mathbf{T}_j \mathbf{z} = x_j, \ x_j \geq 0, \ x_j \text{ integer } \forall j \in \mathcal{J},$$

where $\mathbf{T}_j$ is the jth column of $\mathbf{T}$ (the one pertaining to x_j). The basis of the reformulation is the linear transformation $\mathbf{x} = \mathbf{T}\mathbf{z}$, where $\mathbf{T}$ is a matrix of suitable dimension (not necessarily a square one). Z is, in the new variable space, the feasible region which is in some sense "equivalent" to X; the equivalence basically requires that no solution is lost or created by the transformation, and can be expressed technically by the conditions:

1. every vertex of $[X]$ is contained in the image of Z under $\mathbf{T}$;
2. $\mathbf{T}(Z) \subseteq X$.

A further condition for the validity of the above reformulation is

$$\mathbf{T}(\overline{Z}) \subseteq \overline{X}$$

(recall that $\overline{X}$ and $[X]$ denote the continuous relaxation and the convex hull of X respectively). The transformation is correct in the sense that the optimal solution $(\mathbf{z}^*, x_j^*)$ of $(\hat{P})$ yields an optimal solution for (P), with $\mathbf{x}^* = \mathbf{T}\mathbf{z}^*$. This transformation is computationally advantageous if we are able to recognize a special structure X and a valid transformation such that $[Z] = \overline{Z}$. In this case it can be shown that optimal value of the continuous relaxation of $(\hat{P})$ is equal to the optimal value of

$$\min \mathbf{c}^T \mathbf{x}$$
$$\text{s.t.} \quad \mathbf{A}\mathbf{x} \geq \mathbf{a}; \quad \mathbf{x} \in [X].$$

We have seen in Section C.4.5 (Theorem C.1) that this is the value of the lower bound obtained by Lagrangian relaxation; the advantage here is that we can use standard MILP codes.

We apply, now, these ideas to the CLSP problem, omitting the proofs of the validity of the transformation. We have seen that the structure underlying the single item lot-sizing problem is that of a shortest path (Figure 3.5). To model the shortest path problem, we introduce a decision variable z_{itk}, which is set to 1 if in period t we produce a quantity of item i in order to meet the demand for periods t through k. The shortest route equations are (see Section B.6.4)

$$\sum_{k=1}^{T} z_{i1k} = 1; \quad \sum_{k=1}^{T} z_{itk} = \sum_{k=1}^{t-1} z_{ik,t-1} \quad \forall i,\ t = 2,\ldots,T. \quad (3.55)$$

The setup variables are linked to the shortest route variables by:

$$\sum_{k=1}^{T} z_{itk} \leq \nu_{it} \quad \forall i,\ \forall t. \quad (3.56)$$

It is worth noting how no big M is required for these constraints. Furthermore, we require

$$z_{itk}, \nu_{it} \in \{0,1\} \quad \forall i,\ \forall t,\ k = t,\ldots,T. \quad (3.57)$$

Eqs. (3.55), (3.56), and (3.57) define the feasible region Z. It can be shown that $[Z] = \overline{Z}$; note that to prove this result it is not sufficient to exploit the total unimodularity of network flow problems, since we have to cope with setup variables.

The transformation $\mathbf{T}$ is defined by

$$x_{it} = \sum_{k=t}^{T} d_{itk} z_{itk},$$

where d_{itk} is the demand for item i during periods t through k; the ν_{it} is left unchanged. Now, requiring that the z_{itk} are integral is correct for the single-item uncapacitated problem. However, when we consider CLSP, it may happen that the demand for a period is satisfied by the production in different periods. This can be coped with by continuously relaxing the z_{itk}, as required in the definition of $(\hat{P})$. Problem $(\hat{P})$, is in this case,

$$\min \sum_{i=1}^{N}\sum_{t=1}^{T}\sum_{k=t}^{T} c_{itk} z_{itk} + \sum_{i=1}^{N}\sum_{t=1}^{T} s_i \nu_{it}$$

$$\text{s.t.} \quad \sum_{i=1}^{N}\sum_{k=1}^{T}(r_i d_{itk}) z_{itk} \leq R_t \quad t=1,\ldots,T \quad (3.58)$$

$$\sum_{k=1}^{T} z_{i1k} = 1 \quad \forall i$$

$$\sum_{k=1}^{T} z_{itk} = \sum_{k=1}^{t-1} z_{ik,t-1} \quad \forall i,\ t=2,\ldots,T$$

$$\sum_{k=1}^{T} z_{itk} \leq \nu_{it} \quad \forall i,\ \forall t$$

$$z_{itk} \geq 0 \quad \forall i,\ \forall t,\ k=t,\ldots,T$$

$$y_{it} \in \{0,1\} \quad \forall i,\ \forall t,$$

where c_{itk} is the cost of producing during t a quantity of i to meet the demand from t to k and carrying the implied inventory. Here (3.58) corresponds to $\mathbf{ATz} \geq \mathbf{a}$, the remaining constraints correspond to $\mathbf{z} \in \overline{Z}$, and $\mathbf{T}_j \mathbf{z} = x_j$ is not required, since, in this case, the integer variables ν_{it} are unchanged in the reformulation.

S3.2 The single item DLSP problem: a dynamic programming approach

We outline here a dynamic programming algorithm proposed in [76] for the single item DLSP problem with time-varying production costs.

The problem, as formulated in Section 3.3.1, is a mixed-integer programming problem; however, it is convenient to transform it to a pure binary problem by getting rid of the inventory variables. We introduce a *normalized cumulative* demand D_{it}, corresponding to the number of time buckets necessary to satisfy the demand up to period t,

$$D_{it} = \max\left\{0, \left\lceil \frac{\sum_{\tau=1}^{t} d_{i\tau} - I_{i0}}{p_i} \right\rceil \right\},$$

where $\lceil z \rceil$ denotes the smallest integer larger than z. We obtain

$$\min \sum_{i=1}^{N}\sum_{t=1}^{T}(s_i \max\{0, x_{it} - x_{i,t-1}\} + h_{it}x_{it}) - C \qquad (3.59)$$

$$\text{s.t.} \sum_{i=1}^{N} y_{it} \leq 1 \qquad \forall t \qquad (3.60)$$

$$\sum_{\tau=1}^{t} y_{i\tau} \geq D_{it} \qquad \forall i, 1 \leq t \leq T-1 \qquad (3.61)$$

$$\sum_{\tau=1}^{T} y_{i\tau} = D_{iT} \qquad \forall i \qquad (3.62)$$

$$y_{it} \in \{0,1\} \qquad \forall i, t, \qquad (3.63)$$

where we have defined:

$$h_{it} = c_i p_i (T - t + 1) + \mu_t$$
$$C = \sum_i \sum_t c_i (T - t + 1) d_{it} - T \sum_i c_i I_{i0}.$$

Clearly, the constant C is irrelevant for the optimization problem. The problem is purely binary; we must allocate D_{iT} time buckets to item i, satisfying the cumulative demand profile and optimizing setup and production costs. The demand constraints can be strengthened on the basis of the two following observations.

1. If in a certain period there is a demand peak, such that the production during one time bucket would not be able to meet it, we must anticipate production. Therefore we can use the cumulative normalized demands:

$$\tilde{D}_{iT} = D_{iT}$$
$$\tilde{D}_{it} = \max\{\tilde{D}_{i,t+1} - 1, D_{it}\} \qquad t = T-1, \ldots, 1.$$

2. The second consideration is that, over a certain time horizon, there is an upper bound on the time buckets that we can allocate to item i, since we must also produce the other items. Therefore, we obtain the following constraints:

$$\tilde{D}_{it} \leq \sum_{\tau=1}^{t} x_{i\tau} \leq t - \sum_{j \neq i} \tilde{D}_{jt} \qquad t = 1, \ldots, T-1.$$

The main feature of this DP formulation is that the state variable is t_k, defined as the period during which the kth production of item i takes place;

the functional equation defines the value function, $f_k(t)$, the minimal cost for k periods of production, where the last period is t.

From the strengthened demand constraints, the following bounds on the state variable are derived:

$$t_k^{\min} = \min\left\{t \mid \tau - \sum_{j \neq i} \tilde{D}_{j\tau} \quad (\tau = 1, \ldots, T)\right\}$$

$$t_k^{\max} = \min\left\{t \mid \tilde{D}_{it} \geq k\right\}.$$

The starting point of the recursion is, for $t = t_1^{\min}, \ldots, t_1^{\max}$,

$$f_1(t) = \begin{cases} h_{i1} & \text{if } t = 1 \text{ and } x_{i0} = 1 \\ h_{i1} + s_i & \text{otherwise.} \end{cases}$$

To find the general recursion for $k > 1$, we note that the state t in step k is reachable from a state τ in step $k-1$ if

$$t_{k-1}^{\min} \leq \tau \leq \min\{t - 1, t_{k-1}^{\max}\}.$$

Now we must evaluate the cost of going from state τ to state t. If $\tau = t-1$, we have two consecutive periods of production, and the cost is $h_i t$; if $\tau \leq t - 2$, we must add a setup cost s_i. The recursive equation is, then,

$$f_k(t) = \begin{cases} h_{it} + f_{k-1}(t-1) \\ \qquad \text{if } t = t_{k-1}^{\min} + 1 \\ h_{it} + \min\left(f_{k-1}(t-1), s_i + \min\left\{f_{k-1}(\tau) \mid t_{k-1}^{\min} \leq \tau \leq t - 2\right\}\right) \\ \qquad \text{if } t_{k-1}^{\min} \leq \tau \leq t_{k-1}^{\max} + 1 \\ h_{it} + s_i + \min\left\{f_{k-1}(\tau) \mid t_{k-1}^{\min} \leq \tau \leq t_{k-1}^{\max}\right\} \\ \qquad \text{if } t \geq t_{k-1}^{\max} + 2 \end{cases}$$

for $k = 2, \ldots, \tilde{D}_{iT}$ and $t = t_k^{\min}, \ldots, t_k^{\max}$.

For further reading

We have considered many production/inventory problems, but we have neglected classical inventory control models based on generalizations of the well-known Economic Order Quantity; the interested reader is referred e.g. to [101, Chapter 4] or [206].

We have said that the decomposition approach to the APP problem is not the best way to solve it as a large-scale LP problem. See [125] for alternative approaches to the solution of large scale LP problems. However, decomposition approaches for multicommodity network flow problems may be useful; the reader is referred e.g. to [4, Chapter 17] or [16].

A review of lot-sizing problems is given in [10]; see also [184]. Dual heuristics for CLSP were proposed in [183]. The generation of a primal feasible solution from the solution of the relaxed problem is not trivial if setup times are considered; this issue is discussed in more detail in [60] and [189]. The Lagrangian relaxation for obtaining lower bounds for CLSP is discussed in [50].

The modal formulation for CLSP was proposed in [137]; computational improvements were suggested in [126] and [132]; see also [101, pp. 94-99] for column generation. Column generation based heuristics for CLSP are described in [9] and [44]. The single item uncapacitated problem is treated extensively in [204, Chapter 9].

We have outlined a hierarchical approach for a lot-sizing problem, along the lines proposed in [186]. A complete architecture based on these ideas is described in [98]. The interested reader is referred to [101, Chapter 6] for another HPP methodology. Feasibility issues for HPP are discussed e.g. in [25]. The equalization of runout times to disaggregate family plans is dealt with in [113]. See e.g. [3] and [136] (and references therein) for multilevel lot-sizing.

A good overview of DLSP problems is given in [165]; lot-sizing in general is considered in [164], where some sophisticated algorithms for DLSP are also described. DLSP with setup times is tackled by column generation and dual ascent in [45]; DLSP with sequence-dependent setup costs is considered in [77]. In [68] strong formulations are proposed for DLSP; see also [56] for an application of Lagrangian relaxation to a parallel machine DLSP problem.

The Toyota Goal Chasing method is illustrated in [145]; see [143] for more sophisticated models and algorithms. Continuous-flow models were originally proposed in [116]. Alternative scheduling approaches for Flexible Flow Lines are described in [213]. Proofs of the computational complexity of the continuous-flow model are given [33]. We have hinted at the possibility of using transfer lots; see [80] for an overview and a discussion of how to size transfer lots.

4

DEDS models for scheduling problems

In Chapter 3 we have dealt with discrete-time models; by discretizing the time horizon we enforce a priori some constraints on the occurrence of events. Here we assume that time is continuous, so that the events may occur at any instant. Still, the system state changes only in correspondence to such events; we have to cope with a Discrete-Event Dynamic System (DEDS). In this chapter we cover generative DEDS models; evaluative DEDS models are treated in the next chapter. The generative DEDS models we consider are machine scheduling models. With respect to large bucket lot-sizing models, they are more detailed models, since the exact timing of resource usage is considered; the events we are interested in are the start of an operation of a job on a machine and its completion.

First we deal with the statement of classical scheduling problems in terms of objective functions and technological constraints (Section 4.1). A classification scheme for machine scheduling problems is reviewed in Section 4.2, where we also describe some nonstandard problems. Then we consider solution procedures and the related complexity issues in Section 4.3, where we list the few lucky cases in which a polynomial time algorithm yields the optimal solution of a scheduling problem; in fact, more often than not, the exact solution of scheduling problems calls for enumerative methods, such as Dynamic Programming and Branch and Bound. In order to apply Dynamic Programming, we need to define the state of a decision process. This is quite natural for discrete-time models, but it is trickier in the DEDS case; in Section 4.4 we show how this can be accomplished. Unfortunately a Dynamic Programming approach is severely limited by its memory and computation requirements; still, it may prove a valuable tool to solve some nonstandard single processor problems, as shown in Supplement S4.1. A different modeling framework, based on the assignment of potentials to nodes of a graph is introduced in Section 4.5. This approach is most useful both for developing exact and heuristic methods and can be extended to cope with periodic problems, as shown in Supplement S4.2.

Job	Process Plan
J_1	$(M_1,10), (M_2,5), (M_3,6)$
J_2	$(M_2,5), (M_1,8)$
J_3	$(M_1,2), (M_3,10), (M_2,4)$

TABLE 4.1
A sample job shop scheduling problem

Different approaches to building MILP models for scheduling problems are reviewed in Section 4.6. Branch and Bound algorithms are then discussed in Section 4.7, where we consider special purpose, integer programming and Lagrangian relaxation methods. In practice, however, the solution of machine scheduling problems usually calls for the application of heuristic algorithms; many different classes of general and special purpose heuristic methods are covered in Section 4.8. We deal with list scheduling rules, greedy and dual heuristics, local search, heuristics for parallel machines and job shop scheduling, and hierarchical decomposition approaches for FMS scheduling. All the methods we consider aim at solving a single-objective problem: a quick glimpse of how a multiobjective problem can be tackled by solving a set of auxiliary single-objective problems is given in Supplement S4.3, where we describe approaches to cope with a machine loading problem involving load balancing and cost issues.

4.1 Classical machine scheduling theory

Scheduling problems involve a set of jobs and a set of processors. Each job consists of a set of operations. For each operation we know the set of machines able to perform it and its processing time. Note that the processing time is known because lots, unlike in lot-sizing problems, are already sized. Operations are related by precedence constraints.

Example 4.1
In Table 4.1 we list the data of a typical job shop scheduling problem. We have three machines M_1, M_2, M_3 and three jobs J_1, J_2, J_3; for each job we have a process plan consisting of a sequence of operations, and for each operation we have the machine on which it must be executed and its processing time. A possible solution, represented by a Gantt chart, is depicted in Figure 4.1. ☐

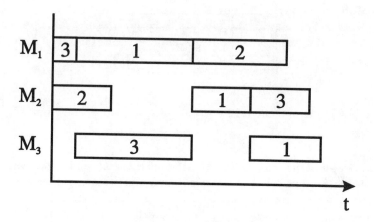

FIGURE 4.1
A Gantt chart for the sample job shop problem

Classical machine scheduling models rely on a set of restrictive assumptions; here we list them and we point out when they prove inadequate to represent a realistic scheduling problem:

- **Single parts and batches of parts are always treated as a single job.**
  This assumption may be appropriate in many situations, but in the case of large lots it may produce poor quality schedules. Consider, for instance, a lot of parts that must undergo two operations in two machines arranged in series; the classical machine scheduling formulation forbids starting processing on the second machine until *all* the parts in the lot are processed on the first one. However, a better schedule may be obtained by transferring a sub-batch of a part from machine 1 to machine 2 before the completion of the whole batch; such sub-batches are called **transfer batches** (see Figure 3.11) and sometimes the term **lot streaming** is used. Another difficulty may occur when equivalent machines are available; a large batch could be split and processed on two machines simultaneously, but this is impossible if the batch is treated as a whole.
- **Preemption is not allowed.**
  Preemption occurs when an operation is stopped and resumed at a later time. This is clearly unacceptable from the technological point of view in most manufacturing environments. This is not the case when scheduling tasks on computer systems.

- **Job cancellation is not allowed.**
  This means that all the jobs are to be processed eventually. This is not always true; for instance, it may be useful to draw a distinction between high-priority and low-priority jobs; low-priority jobs may be scheduled only if the execution of the high-priority ones leaves enough idle time on the machines. In other situations the time horizon is limited and not all jobs can be processed within its limits; for instance, we may have to select, among a list of waiting jobs, those processed during the next shift. Note that all jobs will eventually be processed during the next shifts, but, from the modeling point of view, a selection must be done; this is typical of **part type selection** problems.

- **Processing times are independent of the schedule.**
  Essentially, this assumption implies that setup times are sequence-independent and can be included in the processing time. This is not the case when sequence-dependent setup times are to be dealt with. A classical case occurs in paint production, where we may have chromatic and/or chemical compatibility issues; sequencing a lot of white paint after the production of black paint is a bad idea, due to the large setup time needed to clean the machine.

- **Work-in-process is allowed.**
  This means that jobs may wait in a queue until the next machine required for processing is free. In some metallurgical processes this is not possible. When a part finishes processing on one station, it must be immediately processed on the next one. An intermediate case occurs when buffers are of limited capacity; parts may wait, but blocking occurs when too many parts are queued.

- **Machines are able to process one job at a time.**
  This is a correct assumption in most mechanical operations. The assumption is violated by batch processors, such as those used for thermal treatment of metals or burn-in of electronic circuits; in this case a whole batch of parts is processed together.

- **Each job visits all machines exactly once.**
  In practical problems, some machines may not be required for a certain job. In other cases reentrant flows may occur, i.e., a machine is visited more than once by a job. A typical example is semiconductor manufacturing. Reentrant flows may be also due to reworking of defective parts.

- **Machines are always available.**
  In practice, machines may be not available because of controllable events (e.g., preventive maintenance) or uncontrollable events (e.g., failures). The machine availability itself may be part of the decision

problem: for instance, we may have to decide when to shut the machines down to save energy or for a preventive maintenance period.

- **Machines are the only resources modeled.**
  In practice, it may be necessary to model additional resources such as transportation devices (e.g., AGVs), fixtures, tools, dies, or skilled labor.

- **Jobs are all known in advance.**
  This is the characteristic that distinguishes **static** and **dynamic** scheduling problems. In the last case new jobs arrive at unpredictable times.

- **The problem is purely deterministic.**
  In practice, scheduling problems are stochastic in nature: a certain degree of uncertainty is due to machine failures. Furthermore, processing times may be difficult to predict; typically this happens when manual operations are involved; another issue could be the dependence of the process on environmental factors such as temperature or humidity.

4.2 Classification of scheduling problems

There is a huge variety of scheduling problems; therefore, a quick way of specifying a scheduling problem is useful. We review here a classification scheme due to Graham, which is based on a three field coding $\alpha \mid \beta \mid \gamma$. The three fields specify respectively the machine environment, the job characteristics, the objective function[1].

4.2.1 Machine environment

The field α is a string consisting of two symbols α_1 and α_2. The symbol α_1 specifies the type of machine arrangement. The symbol α_2 specifies the number of processors. Two cases occur: single-operation and multiple-operation jobs.

The simplest single-operation case occurs when only one machine is available: this situation is known as **single machine**, and is denoted by $\alpha_1 = \emptyset$ and $\alpha_2 = 1$. The case of single-machine but multiple-operation jobs is handled by the introduction of precedence constraints among operations. Another single-operation case is that of **parallel machines** (see figure 4.2.a). Three parallel machine subcases are distinguished; let p_{ij} be the processing

---

[1] A different classification scheme, with four fields, is used in some textbooks such as [79].

Classification of scheduling problems 107

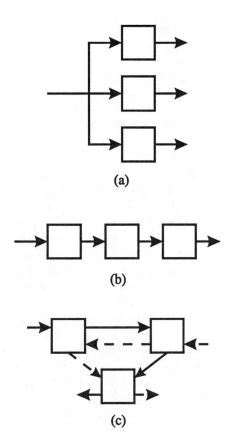

FIGURE 4.2
Part flow in different shop structures: parallel machines (a), flow shop (b), and job shop (c)

time of job J_i on machine M_j. In the **identical** machines case (denoted by $\alpha_1 = P$) we have $p_{ij} = p_i$, i.e., for each job there is the same processing time on any machine. We speak of **unrelated** machines ($\alpha_1 = R$) when processing times differ on the machines. An intermediate case occurs when processing times are given by $p_{ij} = p_i/s_j$, i.e., they depend on machine "speed"; in this case we speak of **uniform** machines ($\alpha_1 = Q$). Note that the distinction between single- and multi-operation jobs is drawn from the scheduling point of view, and not from the technological point of view. In many Flexible Manufacturing Systems, each part must undergo a set of operations, but if all of them are performed during a single *machine visit*, the scheduling problem can be considered as a single-operation one. Indeed, many FMS scheduling problem are essentially dealt with as parallel machine scheduling problems.

We consider now multi-operation jobs. Different cases are distinguished, depending on the structure of the operations precedence constraints. When there is no precedence relation, we have an **open shop** problem ($\alpha_1 = O$); the operations of each job can be carried out in any order. The opposite case, in a sense, is the **flow shop** ($\alpha_1 = F$); all the jobs must follow the same routing and the machines are arranged in series (see figure 4.2.b). We speak of a **permutation flow shop** when it is also assumed that job sequences are maintained on all the machines. This restriction may be imposed by the material handling system; sometimes it is an assumption made for the sake of simplicity; in a few cases it can be shown that we can find an optimal solution among the set of permutation schedules (see Section 4.3.3). In the scheduling literature, the flow shop problem is usually assumed to be of the permutation type[2]. A flow shop arrangement is typical of product-oriented layouts. Finally, we speak of a **job shop** problem ($\alpha_1 = J$) when the jobs have different routings (see figure 4.2.c). A job shop arrangement is typical of process-oriented layouts. Note that in the classical job shop problem there is just one machine able to execute a certain operation, and that each operation has at most one predecessor and at most one successor (i.e., the operation precedence graph for each job is linear, as shown in Figure 4.3.a).

It is worth noting that the aforementioned scheduling problems are just simple prototypes; in practice, one can find hybrid situations, such as flow shop arrangements of stages, each one consisting of a parallel arrangement of machines (see the flexible flow Line of Section 3.4.2). Another hybrid situation occurs when there are some degrees of freedom in the process plan of a job; we might have operations not related by precedence constraints, and the resulting precedence graph is nonlinear (see Figure 4.3.b).

---

[2] In the four field classification scheme, different codes are assigned to general and permutation flow shops.

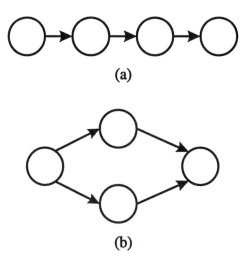

FIGURE 4.3
Different operations precedence graphs: linear (a), and nonlinear (b)

4.2.2 Job characteristics

The field β is actually a string of symbols; when given, such symbols specify characteristics of the jobs or of the resources:

pmtn means that preemption is allowed;

res denotes the presence of additional resources;

prec is used to specify that jobs are related by precedence constraints;

r_i means that nonzero release times are considered;

$\bar{d}_i$ is used when **deadlines** must be complied with; note that a due date represents a soft constraint, in the sense that it can be violated at a certain price, whereas a deadline is a hard constraint;

s_{ij} denotes sequence-dependent setup times;

no − wait means that jobs cannot wait in a queue, and work in process is forbidden.

This list of fields is not exhaustive; new fields are often introduced when dealing with a nonstandard problem.

4.2.3 Objective functions

In Section 1.3.3 we have outlined the different performance measures that can be considered when solving a scheduling problem. Here we formalize this topic a little better. The classical objective functions are either of the 'minsum' or the 'minmax' type. They are mostly built by combining "elementary" functions $\gamma_i(C_i)$ of the completion time of job J_i. When $\gamma_i(C_i)$ is a nondecreasing function of its argument, we speak of a **regular** objective function. Given N jobs along with their elementary functions, a minsum objective function is

$$\sum_{i=1}^{N} \gamma_i(C_i),$$

and a minmax objective function is

$$\max_{i=1,\ldots,N} \gamma_i(C_i).$$

The classical objective functions are built by considering the following elementary functions:

flow time: $F_i = C_i - r_i$;
lateness: $L_i = C_i - d_i$;
tardiness: $T_i = \max\{L_i, 0\}$;
earliness: $E_i = \max\{-L_i, 0\}$;
unit penalty: $U_i = \begin{cases} 1 & \text{if } C_i > d_i, \\ 0 & \text{oherwise.} \end{cases}$

A comment is in order about the definition of the flow time. Here we have used the release time assuming it is given a priori in the problem statement; sometimes it is better to consider the time elapsed between the start of the first operation of a job and the completion of the last operation of that job. In this case, the release time is a decision variable.

The most significant minsum objective functions are

$\sum_i C_i$: total completion time;
$\sum_i T_i$: total tardiness;
$\sum_i U_i$: number of tardy jobs;
$\sum_i w_i C_i$: total weighted completion time;
$\sum_i w_i T_i$: total weighted tardiness;
$\sum_i w_i U_i$: weighted number of tardy jobs.

The most significant minmax objective functions are

$L_{\max} = \max_i L_i$: maximum lateness;
$C_{\max} = \max_i C_i$: makespan.

Example 4.2
In Chapter 2 we have already met problems $1//L_{\max}$ and $1/r_i/L_{\max}$, i.e., single-machine problems characterized by the objective of minimizing maximum lateness; in the second case some release times are not zero. By $1/s_{ij}/\sum_i w_i C_i$ we denote a single-machine problem with sequence-dependent setup times in which we want to minimize the total weighted completion times. $F2//C_{\max}$ is a two machine flow shop problem in which we want to minimize makespan. The most classical job shop scheduling problem is $J//C_{\max}$. □

Equivalent objective functions

The reader may have noticed that in the last section we have not listed all the possible combinations of elementary functions; for instance, total lateness has not been mentioned. This is due to the fact that some objective functions are equivalent to others, in the sense that they are optimized by the same schedule.

It is easy to see that total lateness and total flow time are both equivalent to total completion time, since they just differ by a constant.

$$\sum_{i=1}^{N} F_i = \sum_{i=1}^{N} (C_i - r_i) = \sum_{i=1}^{N} C_i - \sum_{i=1}^{N} r_i,$$

$$\sum_{i=1}^{N} L_i = \sum_{i=1}^{N} (C_i - d_i) = \sum_{i=1}^{N} C_i - \sum_{i=1}^{N} d_i.$$

Note that total tardiness is *not* equivalent to total completion time. In some textbooks mean performance measures are considered, such as the mean tardiness $\bar{T} = (1/N) \sum_{i=1}^{N} T_i$; since mean and total performance measures differ by the multiplicative constant $1/N$, they are equivalent.

As to minmax objective functions, there is a partial equivalence between $L_{\max}$ and $T_{\max}$, in the sense that an optimal solution with respect to $L_{\max}$ is also optimal for $T_{\max}$, but not vice versa. To see this, note that:

$$\begin{aligned} T_{\max} &= \max\{T_1, T_2, \ldots, T_N\} \\ &= \max\{\max\{0, L_1\}, \max\{0, L_2\}, \ldots, \max\{0, L_N\}\} \\ &= \max\{0, \max\{L_1, L_2, \ldots, L_N\}\} = \max\{0, L_{\max}\}. \end{aligned}$$

If the optimal $L^*_{\max}$ is negative, we obviously have $T^*_{\max} = 0$; if it is positive, then $L^*_{\max} = T^*_{\max}$.

Interpreting objective functions

It is important to link the above objective functions to practical performance measures.

- Total flow time is related to Work in Process and lead time, since it is linked to the time a job spends on the shop floor.
- Makespan is related to machine utilization for a static scheduling problem; minimizing makespan implies minimizing machine idle time. Note that in a dynamic scheduling problem things may be different; to tackle such a problem, we could adopt a rolling horizon strategy, whereby static scheduling problems are solved in sequence, taking into account the newly arrived jobs and the state of the old ones. In such a case minimizing makespan does not seem a good idea. Indeed, it is more sensible to consider a lead time related performance measure such as the total flow time; we could also consider $F_{\max}$, which is not equivalent to makespan if release times are nonzero.
- Finally, $L_{\max}$, $T_{\max}$, and $\sum_i T_i$ are obviously related to customer service. We adopt a minmax function when we aim at minimizing the worst-case lateness/tardiness; a minsum function is adopted when we are willing to sacrifice a customer to insure better service to the others. Consider for instance a problem involving four jobs, and compare the following situations:

    a: $T_1 = 1, T_2 = 1, T_3 = 1, T_4 = 1$;
    b: $T_1 = 0, T_2 = 0, T_3 = 0, T_4 = 4$.

    The two cases are equivalent from the point of view of total tardiness, but the second one is worse from the point of view of max tardiness.

Other objective functions can be built, related to inventory and setup costs; note that in this case the objective function is no longer a regular one.

It is natural to expect that some objective functions are negatively correlated to others. For instance, increasing WIP generally increases machine utilization; this implies that to improve makespan we may have to worsen total flow time. A clear example of this is the Johnson rule for the $J2//C_{\max}$ problem (see Section 4.3.3). Another example of conflict may arise in the case of sequence-dependent setup times. One would like to minimize total setup time (or cost) by batching, i.e., by trying to form subsequences of similar jobs; this, however, may have a detrimental effect on due date performance, since urgent lots could be delayed because of their setup requirements.

It turns out that scheduling problems should be considered as multiobjective problems: the inherently single-objective nature of most scheduling literature is one of the major criticisms of the application of such techniques to production management. Commercially available software packages for scheduling, which are usually based on list scheduling strategies (see Section 4.8.1), may evaluate a schedule with respect to multiple performance measures; they may also combine different scheduling rules aimed at different performance measures. This however, is *not* a truly multiobjective approach, since it does not allow evaluating tradeoffs among efficient solutions (see Supplement S2.2). An example of multiobjective approach to a scheduling problem can be found in Supplement S4.3.

4.2.4 Semiactive, active and nondelay schedules

Consider a single-machine problem with a regular objective function. An obvious feature of the optimal schedule is that there will be no idle time on the machine, unless it is enforced by nonzero release times or technological constraints. There is no point in deliberately keeping the machine idle. In such a case, the solution can be given in the form of a permutation, and the scheduling problem boils down to a *sequencing* problem. Also in a multi-machine problem the solution can be represented, in the case of a regular objective function, by the job sequence on each machine. On the contrary, if the objective function is nonregular, there may be some advantage in keeping the machine idle, and we need to specify the timing of each event. In this case the problem is harder since there is no one-to-one correspondence between sequencing and scheduling. Here we make this idea more precise in a general multi-machine environment.

Consider the first schedule in Figure 4.4; the schedule can be made more compact by *left-shifting* the first operation of job 3, resulting in the second Gantt chart. Note that the left-shift is local, in the sense that the sequence on the machines is not changed. A schedule such that no local left-shift is possible is called **semiactive**; in a semiactive schedule there is no operation that could be scheduled earlier without changing the sequence, or violating the technological constraints or the release times.

A local left-shift does not change the sequence on a machine. It may happen that in the Gantt chart there is a 'hole', such that an operation can be moved earlier without delaying any other operation. We say that a schedule is **active** if no operation could be started earlier without delaying another operation or violating technological or release time constraints. It may happen that a semiactive schedule is not active; this issue is illustrated in Figure 4.5. Actually, active schedules are a subset of semiactive schedules. The importance of active schedules is due to the following theorem.

THEOREM 4.1
There is always an active schedule which is optimal for a regular objective function.

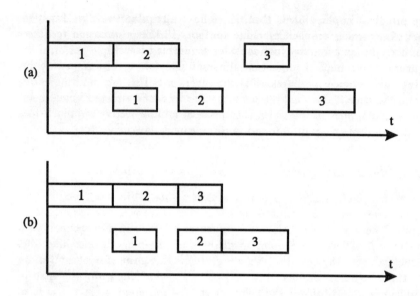

FIGURE 4.4
A non semiactive schedule (a), and a semiactive schedule (b)

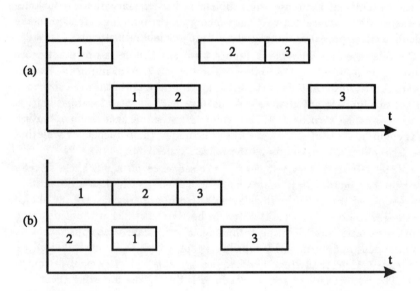

FIGURE 4.5
A semiactive but not active schedule (a) and the corresponding active schedule (b)

The practical implication is that, if we have a regular objective function, a set of sequences on each machine contains all the information to derive the schedule, and we may just consider sequencing issues.

There is still another important class of schedules, such that no machine is kept idle when it could start processing a job; this type of schedule is called **nondelay**. Nondelay schedules are not necessarily optimal; to see why, go back to Example 2.10. The interest in nondelay schedules lies in the fact that they are quite easily obtained by dispatching procedures and that they are a subset of active schedules. We consider the generation of active and nondelay schedules in Section 4.8.1.

4.2.5 Nonstandard scheduling problems

In Section 4.1 we have reviewed the typical assumptions of standard machine scheduling problems. Here we consider in more detail practical situations in which such assumptions are not valid.

Scheduling with batch processors

Classical machine scheduling models assume that each machine is able to process one job at a time. Notable exceptions to this assumption are the machines used for thermal treatment; such machines are common both in the mechanical industry (e.g., annealing processes) and the electronic industries (e.g., ovens for the burn-in of the cards). Such machines are called batch processors, since a whole batch of jobs is processed together.

Scheduling batch processors is rather difficult, since it is not easy to evaluate when a batch should be started. For instance, if we have a set of waiting jobs amounting to 90% of machine capacity, should we start the batch, or should we wait for another job? The problem is that a full-batch policy, i.e. a policy, whereby the processor starts a batch only when it has reached full capacity, is not optimal in general. Further complicating factors might be the different processing requirements of the jobs.

To cope with batch processors, it would be possible, at least in principle, to build a MILP model such as those that we will describe in Section 4.6; unfortunately, the derived model would be hard to solve in practice. A possible method for such a problem is to isolate the batch processor from the other unit-capacity machines (if any) by associating a lead time with it. In other words, we schedule the whole shop considering an infinite capacity batch processor, taken into account by a standard and load-independent delay associated with the batch processor; this results in release times and deadlines for the local scheduler of the batch processor. The local scheduler may rely on more or less sophisticated approaches; in Supplement S4.1 we describe an efficient solution method (aimed at a particular case) based on dynamic programming.

Periodic scheduling

Periodic scheduling assumes that a given set of tasks must be indefinitely repeated in time. We are given a production mix, that is, a set of parts to be processed in a given proportion; this set is also known as a **Minimal Part Set**. For instance, we may have to produce repeatedly 3 parts of type A, 4 of type B and 1 of type C. We should schedule the minimal part set and repeat the schedule; this is typical of repetitive production environments. In such a situation, it does not make sense to consider objective functions based on job completion times: there are infinitely many jobs, and if C_i is the completion time of a job, we have infinite jobs of the same type completed at $C_i + kT$, where k is any integer number and T is the period of the schedule.

The objective function to consider depends on the problem at hand. We have seen an example of single-machine periodic scheduling in Example 3.1, whereby the objective function expressed the need to maintain uniform component absorption rates. Another objective function, making sense in a multi-machine environment like a job shop, is to minimize the period of the schedule, which is equivalent to maximizing the throughput. It can be shown that minimizing the period is actually a trivial problem; given a minimal part set, we can easily compute the load for each machine. The machine with the maximum load is called the **critical machine**, and its load is just the minimum period. Such a schedule assumes that a suitably large set of parts is loaded in the system, in order to keep the critical machine busy; the consequence may a large WIP level. We could consider the problem of minimizing the Work in Process, or trading off throughput and WIP. Such a problem is NP–hard. In Supplement S4.2 we give more details about periodic job shop scheduling along with a mathematical formulation of the problem.

Note that the tradeoff between throughput and WIP is basically the same tradeoff we have in nonperiodic scheduling when we compare makespan and WIP (see Section 4.3.3).

Scheduling with sequence-dependent setup structures

In simple scheduling models, the setup time is assumed either negligible or sequence-independent and hence included in the processing time. When the setup time depends on the similarity of jobs consecutively processed on the same machine, one could consider the problem of sequencing the machine in order to minimize the total setup time. Such a problem is strictly related to the Asymmetric Traveling Salesman Problem (a model for the ATSP is given in Section C.1).

A practically relevant case lies somewhere in between the no-setup and the TSP case. Consider a lathe. Parts are blocked by jaws for cutting, and a specific type of jaws is able to block parts with a given diameter range. If we process parts with similar diameters, we do not have to change the jaws, but if there is a relevant difference in the diameter, we must change them. We can think of the parts as partitioned in families. If two parts of the same family are adjacent in the sequence, we have no setup time; but if we switch from one family to another one, we have a setup time. However, this setup time is not sequence-dependent, in the sense that it depends only on the next family to be processed, and not on both families. An example will clarify the impact on modeling such a problem. We will say that we have a **family dependent** setup time.

Example 4.3
Consider a part type selection problem with family setup times. We have a single machine, available for a time C, and a set of jobs indexed by $i = 1, \ldots, N$; let p_i be the processing time of job i. To each job, we associate a weight w_i, which is related to the urgency of the corresponding order. We want to select a subset of jobs in order to maximize the "value" of the selected part types, taking into account the limited availability of the machine.

If the setup times were negligible, this would be a simple knapsack problem. We assume that jobs are partitioned in families indexed by $j = 1, \ldots, G$; let s_j be the familiy dependent setup time of family F_j. We obtain the following model:

$$\max \sum_{i=1}^{N} w_i x_i \quad (4.1)$$

$$\text{s.t.} \sum_{j=1}^{G} s_j y_j + \sum_{i=1}^{N} p_i x_i \leq C \quad (4.2)$$

$$x_i \leq y_j \quad \forall j, \forall i \in F_j \quad (4.3)$$

$$x_i, y_j \in \{0, 1\} \quad \forall i, j, \quad (4.4)$$

where x_i is set to 1 if job i is selected and y_j is set to 1 if at least one job of family j is selected.

Note that we have not modeled the job sequence; it is obviuosly optimal to sequence all the jobs within the same family consecutively, and the family sequence is irrelevant. In the case of sequence-dependent setup times, we should also model the job sequence (see Section 4.6.3); this would also be necessary if due date issues were considered. □

A further case occurs when we have a large setup time when switching families, and a small setup time when switching jobs within a given family. This case is known as **major/minor** setup structure. In Section 4.8.5 we outline a heuristic approach to cope with major/minor setup structures in a parallel machine case.

Scheduling with nonregular objective functions

The typical objective functions used in scheduling models are regular functions, i.e., nondecreasing functions of the job completion times; in this case, the earlier the a job is completed, the better. If we consider inventory issues, we see that it is not always true that an early job completion should be rewarded. A typical example of nonregular objective function is

$$\sum_{i=1}^{N}(w_i^T T_i + w_i^E E_i),$$

where both job tardiness and earliness are penalized.

Using a nonregular objective function complicates the solution of a scheduling problem. From the formal point of view, the modeling approach described in Section 4.6.2 can be easily adapted to a nonregular objective function. However, a nonregular function makes many scheduling approaches not viable. The reason is that Theorem 4.1 no longer holds, and we cannot limit ourselves to considering active schedules.

Scheduling with additional resources

Classical scheduling models consider the machine as the only resource to be managed. However, it is often the case that additional resources, such as skilled labor, tools, dies, and fixtures, constrain the schedule.

We consider different models for scheduling with multiple resources in Section 4.6.4. Another classical example is Flexible Manufacturing Systems (FMS) scheduling; in this case we have a set of identical machines that must be properly equipped by tools in order to carry out the required operations.

4.3 Polynomial complexity scheduling problems

Machine scheduling problems look deceptively simple; while almost trivial to state in logical terms, they are quite difficult to solve. Generally, scheduling problems are NP–hard; enumerative approaches are called for, such as Dynamic Programming or Branch and Bound, or heuristic methods. This is essentially due to the impossibility of finding a characterization of the optimal solution. Nevertheless, in a (very) few lucky cases a simple characterization of the optimal solution exists, resulting in polynomial optimization algorithms.

In the following we consider the most important polynomial scheduling problems, respectively solved by the SPT, EDD and Johnson algorithms (essentially requiring to sort a vector of numbers). Such algorithms are called **constructive**, since they directly build a schedule. We also outline the role of preemption in simplifying some scheduling problems.

4.3.1 The SPT sequencing rule

The $1//\sum C_i$ problem can be considered a sequencing problem, since the solution is represented by a permutation σ of the job indexes. This problem is easily solved by sequencing jobs in decreasing order of their processing time, i.e., the optimal sequence is such that

$$p_{\sigma(1)} \leq p_{\sigma(2)} \leq \cdots \leq p_{\sigma(N)}.$$

This sequencing rule is called **Shortest Processing Time (SPT)** rule. To see why, consider a permutation σ of the jobs; $C_{\sigma(k)}$ is the completion time of the kth job in the sequence. We have

$$C_{\sigma(1)} = p_{\sigma(1)}$$
$$C_{\sigma(2)} = p_{\sigma(1)} + p_{\sigma(2)}$$
$$\cdots$$
$$C_{\sigma(k)} = \sum_{l=1}^{k} p_{\sigma(l)}$$
$$\cdots$$

Summing up all such relations we can see that

$$\sum_{i=1}^{N} C_i = N p_{\sigma(1)} + (N-1) p_{\sigma(2)} + \cdots + 2 p_{\sigma(N-1)} + p_{\sigma(N)}.$$

To minimize this expression, we should choose the sequence so that $p_{\sigma(1)}$ is small and $p_{\sigma(N)}$ is large. Note that the SPT rule also minimizes the average WIP level, which is given by the formula

$$\frac{1}{C_{\max}} \int_0^{C_{\max}} N(t) dt,$$

where $N(t)$ is the number of jobs queued at the machine at time t, including the one in service; $N(t)$ is a piecewise function which is decreased by 1 at each job completion. Since the makespan $C_{\max}$ is constant, the integral is minimized by decreasing $N(t)$ as fast as possible.

Example 4.4
Consider the $1//\sum_i C_i$ problem characterized by the following data:

Job J_i	J_1	J_2	J_3	J_4	J_5
Processing time p_i	8	16	10	7	2

The optimal sequence is $(J_5, J_4, J_1, J_3, J_2)$. □

If weights are considered, i.e., we have to solve $1//\sum w_i C_i$, we can generalize the SPT rule by sorting jobs in decreasing order of the ratio w_i/p_i. This rule is known as **Weighted SPT (WSPT)** rule. To prove this result, we can use a quite common type of argument in scheduling theory, the **interchange argument**.

Consider a non-WSPT sequence. Then, there must be a position k such that

$$\frac{w_\alpha}{p_\alpha} < \frac{w_\beta}{p_\beta},$$

where $\alpha = \sigma(k)$ and $\beta = \sigma(k+1)$. If C_β is the completion time of J_β, we have, since there is no idle time in the schedule,

$$C_\alpha = C_\beta - p_\beta.$$

Now consider the sequence obtained by swapping α and β. Clearly, only the completion times of these two jobs are affected:

$$C'_\alpha = C_\beta$$
$$C'_\beta = C'_\alpha - p_\alpha = C_\beta - p_\alpha.$$

We can easily compute the difference between the objective function for the first and the second sequence:

$$\overbrace{w_\alpha(C_\beta - p_\beta) + w_\beta C_\beta}^{\text{before swap}} - \underbrace{[w_\beta(C_\beta - p_\alpha) + w_\alpha C_\beta]}_{\text{after swap}} = -w_\alpha p_\beta + w_\beta p_\alpha > 0.$$

We see that a swap of the jobs violating the WSPT rule improves the objective function; by repeating the argument we see that the WSPT rule is optimal.

Both SPT and WSPT require sorting a vector of N numbers, which can be accomplished in polynomial time (see Section 2.3). Therefore $1//\sum C_i$ and $1//\sum w_i C_i$ are polynomial problems. Unfortunately this holds under quite restrictive hypotheses: the addition of precedence constraints or nonzero release times makes the problem NP–hard.

4.3.2 The EDD sequencing rule

The $1//L_{\max}$ problem can be solved by sorting jobs according to their due dates:

$$d_{\sigma(1)} \leq d_{\sigma(2)} \leq \ldots \leq d_{\sigma(N)}.$$

This sequencing rule is called **EDD (Earliest Due Date)** rule. The optimality of the EDD rule can be established by an interchange argument as in the SPT case. Note that, due to the partial equivalence between $1//L_{\max}$ and $1//T_{\max}$, the latter problem is also solved by the EDD rule.

Example 4.5
Consider the $1//T_{\max}$ problem characterized by the following due dates:

Job J_i	J_1	J_2	J_3	J_4	J_5
Due date d_i	20	9	16	18	30

The optimal sequence is (*independently* of the processing times of the jobs) $(J_2, J_3, J_4, J_1, J_5)$. □

Note that the EDD rule does *not* solve the $1//\sum_i T_i$ problem. A counter-example is easy to build. Consider two jobs such that $p_1 > p_2$ and $d_1 < d_2$, and assume that $d_2 < p_2$. This implies that both jobs will be late. The EDD sequence is (J_1, J_2); however, since both jobs are late:

$$\sum_{i=1}^{2} T_i = \max\{0, C_1 - d_1\} + \max\{0, C_2 - d_2\} = (C_1 + C_2) - (d_1 + d_2).$$

Now, since the second term is a constant, we see that the problem boils down to a $1//\sum_i C_i$ problem, and its optimal solution is the SPT sequence (J_2, J_1).

If precedence constraints are added to $1//L_{\max}$, the problem is still polynomial and can be solved by a simple backward scheduling algorithm due to Lawler; any problem of the form $1/prec/\max_i \gamma_i(C_i)$ is solved by this algorithm, which is described, e.g., in [79, Chapter 4], provided that the γ_i are regular.

Adding release times is not harmless: we have shown in Section S2.1 that $1/r_i/L_{\max}$ is a NP–hard problem; however, unlike other NP–hard problems, it can be solved by a quite efficient Branch and Bound procedure described in Section 4.7.2.

4.3.3 The Johnson algorithm

Consider the $F2//C_{\max}$ problem. Since there are two machines, one could wonder if two different sequences on them should be considered. The following theorems show that this is not the case (see e.g. [11, pp. 139-140] for a proof).

THEOREM 4.2
Given a $F//C_{\max}$ problem, there exists an optimal solution such that the sequences on the last two machines are equal.

THEOREM 4.3
Given a $F//B$ problem, where B is any regular objective function, there exists an optimal solution such that the sequences on the first two machines are equal.

The two theorems show that the optimal solutions of both $F2//C_{\max}$ and $F3//C_{\max}$ problems are permutation schedules; however the second problem is NP-hard, and it requires a Branch and Bound procedure, such as the one described in Section 4.7.3, whereas the first one can be solved by a polynomial algorithm due to Johnson.

Note that, in this problem, the first machine is always busy, and that minimizing the makespan is equivalent to minimizing the idle time on the second machine. If there is a limited capacity buffer between the machines, the first machine can be blocked, and this property is not true. The Johnson algorithm is based on the idea that the first jobs in the sequence should have a short processing time on the first machine, in order to avoid keeping the second machine idle, and that the last jobs should have a short processing time on the second machine, in order to reduce the tail of the schedule, in which the first machine is idle and the second machine terminates the last jobs.

The Johnson algorithm can be formalized as follows:

Step 1: build the sets $\mathcal{J}_1 = \{J_i \mid p_{i1} \leq p_{i2}\}$, $\mathcal{J}_2 = \{J_i \mid p_{i1} > p_{i2}\}$.
Step 2: sequence the jobs in $\mathcal{J}_1$ in nondecreasing order of p_{i1}.
Step 3: sequence the jobs in $\mathcal{J}_2$ in nonincreasing order of p_{i2}.
Step 4: merge the two sequences, starting with the jobs in $\mathcal{J}_1$.

Example 4.6
Consider the $F2//C_{\max}$ problem:

Job J_i	J_1	J_2	J_3	J_4	J_5
p_{i1}	6	2	8	7	3
p_{i2}	5	9	4	1	10

Then $\mathcal{J}_1 = \{J_2, J_5\}$ and $\mathcal{J}_2 = \{J_1, J_3, J_4\}$; the optimal sequence is

$$(J_2, J_5, J_1, J_3, J_4). \quad \square$$

The Johnson algorithm can be generalized to the $J2//C_{\max}$ problem, provided that each job has at most two operations. Define the following sets:

$\mathcal{J}_1 = \{J_i | J_i \text{ is processed only on machine 1}\}$

$\mathcal{J}_2 = \{J_i | J_i \text{ is processed only on machine 2}\}$

$\mathcal{J}_{12} = \{J_i | J_i \text{ is processed first on machine 1, then on machine 2}\}$

$\mathcal{J}_{21} = \{J_i | J_i \text{ is processed first on machine 2, then on machine 1}\}$.

Then the problem can be solved as follows:

Step 1: sequence the jobs in $\mathcal{J}_{12}$ with the Johnson algorithm.

Step 2: sequence the jobs in $\mathcal{J}_{21}$ with the Johnson algorithm.

Step 3: sequence the jobs on machine 1 in the order $(\mathcal{J}_{12}, \mathcal{J}_1, \mathcal{J}_{21})$, where the jobs in $\mathcal{J}_1$ are sequenced arbitrarily.

Step 4: sequence the jobs on machine 2 in the order $(\mathcal{J}_{21}, \mathcal{J}_2, \mathcal{J}_{12})$, where the jobs in $\mathcal{J}_2$ are sequenced arbitrarily.

It is interesting to note that this algorithm minimizes the makespan at the expense of WIP: there is a buildup of WIP before machine 1, while machine 2 processes the jobs in $\mathcal{J}_{21}$, and vice versa. This is reasonable; minimizing the makespan is equivalent to maximizing machine utilization; a high WIP level ensures high machine utilization.

4.3.4 Complexity of preemptive scheduling

Given a generic scheduling problem P, there is a related preemptive scheduling problem $\overline{P}$, whereby the nonpreemption assumption is relaxed. The following significant cases may happen (provided that the computational complexity of P and $\overline{P}$ is known):

1. there is no advantage in allowing preemption, in the sense that there is always an optimal nonpreemptive schedule;
2. P is NP–hard but $\overline{P}$ is polynomial

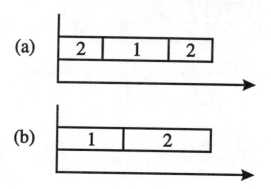

FIGURE 4.6
The preemptive schedule (a) does not improve the nonpreemptive schedule (b)

3. both P and $\overline{P}$ are NP–hard.

The first case occurs for single-machine scheduling with zero release times and a regular objective function. Figure 4.6 shows why. Two jobs, J_1 and J_2, are considered: J_2 starts processing, then it is preempted by J_1; after the completion of J_1, J_2 is completed. Now consider the nonpreemptive schedule (J_1, J_2): the completion time of J_2 does not change, and the completion time of J_1 is anticipated. Therefore, in this case, there is no gain in allowing preemption.

As an example of the second case, consider the $1/r_i/\sum_i C_i$ problem, which is NP–hard. The $1/pmtn, r_i/\sum_i C_i$ problem can be easily solved by the Shortest Remaining Processing Time (SRPT) rule: this rule is a straightforward generalization of the SPT rule, whereby, at each job completion and at each release time, the set of available jobs is considered, including the partially completed ones, and the one with the shortest remaining processing time is selected. This fact can be used in developing Branch and Bound algorithms; the preemptive problem is a relaxation of the preemptive one, and its solution yields a lower bound.

Many other preemptive problems are NP–hard, such as the preemptive versions of the flow shop and job shop scheduling problems. Another example is $1/pmtn, r_i/\sum_i w_i C_i$, which has been shown to be NP–hard; it is not possible to generalize the WSPT rule to a WSRPT rule.

4.4 Dynamic programming approaches

Dynamic Programming (DP) is most naturally applied to discrete-time models where it is natural to define a set of state variables. In the DEDS case a suitable definition of state is not so easy to obtain; we will use a specific example to illustrate the implied issues.

4.4.1 Applying Dynamic Programming to single-machine problems

We consider here single-machine scheduling problems with an additive objective function of the kind:

$$\sum_{i=1}^{N} \gamma_i(C_i),$$

where $\gamma_i(\cdot)$ are nondecreasing functions of their arguments (hence, regular functions). In particular, we apply a DP approach (due to Held and Karp) to $1//\sum T_i$, which is a NP–hard problem not solved by simple sequencing rules.

The application of DP requires adapting the Bellman Optimality Principle to the problem at hand (see Section D.2.1). Let σ be the optimal permutation of the N jobs; the first H jobs in σ are an optimal sequence for the problem obtained by considering only the first $H \leq N$ jobs. In fact, for any $H = 1, \ldots, N$ we can write the value of the optimal permutation:

$$\sum_{k=1}^{N} \gamma_{\sigma(k)}(C_{\sigma(k)}) = \sum_{k=1}^{H} \gamma_{\sigma(k)}(C_{\sigma(k)}) + \sum_{k=H+1}^{N} \gamma_{\sigma(k)}(C_{\sigma(k)}) = A + B. \quad (4.5)$$

If we consider the subsequence $(J_{\sigma(1)}, J_{\sigma(2)}, \ldots, J_{\sigma(H)})$, it is clear that it is an optimal solution for the problem involving the jobs $\{J_{\sigma(1)}, J_{\sigma(2)}, \ldots, J_{\sigma(H)}\}$. If not, the complete sequence would not be optimal, since we could decrease A without increasing B. Note that this reasoning is correct if we can associate a makespan to a subset of jobs, which is constant and independent of the sequence; in this case, the first subsequence does not influence the value of the possible completions, i.e., B is not influenced by A. For this to be true, we must have zero release times, sequence-independent setups, and a regular objective function, ruling out idle time on the machine.

Given a set Q of jobs, let $C_Q = \sum_{J_i \in Q} p_i$ be the completion time of the last job in any permutation of the jobs in Q. Let $\Gamma(Q)$ be the value of the optimal sequence of the jobs in Q. Clearly, if $Q = \{J_i\}$, then $\Gamma(Q) = \gamma_i(p_i)$. If Q consists of more than one job, considering that the completion time of

Job J_i	J_1	J_2	J_3	J_4
Processing time p_i	8	6	10	7
Due date d_i	14	9	16	16

TABLE 4.2
Data for the $1//\sum T_i$ problem

Q	$\{J_1\}$	$\{J_2\}$	$\{J_3\}$	$\{J_4\}$
$p_i - d_i$	-6	-3	-6	-9
$\Gamma(Q)$	0	0	0	0

TABLE 4.3
Values of $\Gamma(Q)$ for the single-job sets

the last job in the sequence is in any case C_Q, we can write the dynamic programming equation:

$$\Gamma(Q) = \min_{J_i \in Q}\{\Gamma(Q - \{J_i\}) + \gamma_i(C_Q)\}. \qquad (4.6)$$

To solve the recursive equation we have to compute Γ for all the possible sets consisting of one job; then for all the sets consisting of two jobs, and so on until the whole set of jobs is considered. Note that this is a *forward* DP recursion. The process is illustrated in the following Example.

Example 4.7
Consider a $1//\sum_i T_i$ problem characterized by the data shown in Table 4.2 (the example is taken from [79, p. 89]). First we compute $\Gamma(Q)$ for the four sets of one job,

$$\Gamma(\{J_i\}) = \gamma_i(C_i) = \max\{0, p_i - d_i\}.$$

We get the results summarized in Table 4.3.

Then we have to build the table of $\Gamma(Q)$ for the six possible two-job sets. For instance, if $Q = \{J_1, J_2\}$, we have $C_Q = p_1 + p_2 = 14$ and

$$\Gamma(Q) = \min\{\Gamma(\{J_1\}) + \gamma_2(14), \Gamma(\{J_2\}) + \gamma_1(14)\}$$
$$= \min\{0 + 5, 0 + 0\} = 0.$$

Q	$\{J_1, J_2\}$		$\{J_1, J_3\}$		$\{J_1, J_4\}$	
C_Q	14		18		15	
last J_i in sequence	J_1	J_2	J_1	J_3	J_1	J_4
$\gamma_i(C_Q)$	0	5	4	2	1	0
$\Gamma(Q - \{J_i\}) + \gamma_i(C_Q)$	0	5	4	2	1	0
optimal last job	*			*		*
$\Gamma(Q)$	0		1		0	
Q	$\{J_2, J_3\}$		$\{J_2, J_4\}$		$\{J_3, J_4\}$	
C_Q	16		13		17	
last J_i in sequence	J_2	J_3	J_2	J_4	J_3	J_4
$\gamma_i(C_Q)$	7	0	4	0	1	1
$\Gamma(Q - \{J_i\}) + \gamma_i(C_Q)$	7	0	4	0	1	1
optimal last job		*		*	*	
$\Gamma(Q)$	0		0		1	

TABLE 4.4
The $\Gamma(Q)$ function for the two-job sets

The results are summarized in Table 4.4. The asterisk below each column corresponds to the best job to be placed in the last position in the sequence.

Then we compute the table of the four sets of three jobs. For instance, in the case $Q = \{J_1, J_2, J_3\}$, we have $C_Q = p_1 + p_2 + p_3 = 24$, and

$$\Gamma(Q) = \min\{\Gamma(\{J_1, J_2\}) + \gamma_3(24),$$
$$\Gamma(\{J_1, J_3\}) + \gamma_2(24), \quad \Gamma(\{J_2, J_3\}) + \gamma_1(24)\} =$$
$$= \min\{0 + 8, 2 + 15, 0 + 10\} = 2.$$

We obtain the results shown in Table 4.5.

Finally, we build the table for the set of the four jobs, as shown in Table 4.6. Tracing backwards the last job in each subsequence, we obtain the optimal sequence (J_2, J_1, J_4, J_3). □

We see that the state in this example is represented by job subsets; the time at which the job subset is completed is determined since no idle time nor sequence-dependent setup times are considered. In the following Section we consider the same problem with sequence-dependent setup times.

Q	$\{J_1, J_2, J_3\}$			$\{J_1, J_2, J_4\}$		
C_Q	24			21		
last J_i in sequence	J_1	J_2	J_3	J_1	J_2	J_4
$\gamma_i(C_Q)$	10	15	8	7	12	5
$\Gamma(Q - \{J_i\}) + \gamma_i(C_Q)$	10	17	8	7	12	5
optimal last job			*			*
$\Gamma(Q)$	8			5		
Q	$\{J_1, J_3, J_4\}$			$\{J_2, J_3, J_4\}$		
C_Q	25			23		
last J_i in sequence	J_1	J_3	J_4	J_2	J_3	J_4
$\gamma_i(C_Q)$	11	9	9	14	7	7
$\Gamma(Q - \{J_i\}) + \gamma_i(C_Q)$	3	9	11	15	7	7
optimal last job	*				*	
$\Gamma(Q)$	9			7		

TABLE 4.5
The $\Gamma(Q)$ function for the three-job sets

Q	$\{J_1, J_2, J_3, J_4\}$			
C_Q	31			
last J_i in sequence	J_1	J_2	J_3	J_4
$\gamma_i(C_Q)$	17	22	15	15
$\Gamma(Q - \{J_i\}) + \gamma_i(C_Q)$	24	31	20	23
optimal last job			*	
$\Gamma(Q)$	20			

TABLE 4.6
The $\Gamma(Q)$ function for the four-job sets

4.4.2 Solving $1/s_{ij}/*$ problems by Dynamic Programming

In this case the state is not completely determined by the jobs in a subset; the way the corresponding subsequence is completed critically depends on the last job in the subsequence. Consider a partial subsequence

$$A = \left(J_{\sigma(1)}, J_{\sigma(2)}, \ldots, J_{\sigma(H)}\right),$$

and the corresponding job subset

$$Q = \left\{J_{\sigma(1)}, J_{\sigma(2)}, \ldots, J_{\sigma(H)}\right\}.$$

Assume that A is optimal for Q. Unlike the $1//\sum T_i$ case, when adding a job $J_{\sigma(H+1)}$ to this partial sequence, we must explicitly take into account the *last* job in sequence, which influences the setup time between $J_{\sigma(H)}$ and $J_{\sigma(H+1)}$. Given the subset of jobs Q, we cannot consider only the optimal subsequence A; we can only assume that the first $H-1$ jobs in Q are optimally sequenced.

The consequence is that the state definition is now more complicated, in that it consists of the pair $(Q, J_{\sigma(H)})$, where $H = |Q|$ and $J_{\sigma(H)}$ is the last job in the partial sequence of the jobs in Q. In the case of a $1/s_{ij}/C_{\max}$ problem, i.e., a single-machine problem with sequence-dependent setup times and a makespan objective function, the definition (4.6) of the Γ function should be modified as

$$\Gamma(Q,j) = \min_{i \in Q - \{j\}} \left\{\Gamma\left(Q - \{j\}, i\right) + s_{ij} + p_j\right\},$$

where s_{ij} is the setup time if i is immediately followed by j and p_j is the processing time of j, not including the setup time. Clearly, both the computational effort and the storage requirements grow considerably.

4.4.3 The curse of dimensionality

We have seen how the complexity of a DP scheduling approach grows when adding complicating features such as sequence-dependent setup times. Things get much worse when considering multiple machines: in the $1/s_{ij}/*$ case, we may assume that the time at which each subsequence is completed is implied by the pair $(Q, J_{\sigma(H)})$. This is not the case when, for instance, parallel machines are dealt with: we should keep track of what jobs are scheduled on each machine, and of the completion times of each subsequence on each machine. The resulting DP formulation is highly impractical. This is why alternative approaches are usually pursued when dealing with scheduling problems.

Nevertheless, it should be emphasized that in some peculiar situations DP can turn out to be a good approach (see Supplement S4.1). Furthermore, its performance is improved when the problem is strongly constrained, as in the case of precedence constraints or tight deadlines. Another way to adopt DP is within decomposition-based heuristic approaches. Consider, for instance, a large $1//\sum_i T_i$ problem; the idea is to sort the jobs in EDD order, and to schedule with DP the subset of the first K jobs, where K is a computationally tractable problem size; then the first part of the optimal subsequence ($H < K$ jobs) is frozen, the next H jobs are taken from the EDD list and the process is repeated.

4.5 A modeling framework based on node potential assignment

In the previous Section, we have seen that trying to adapt a state equation approach to scheduling problems is rather difficult, and it results in DP algorithms which are not easy to grasp or to implement. We turn our attention to a different modeling framework, more suitable for coping with DEDS. Basically, we have to choose the start or the completion times of certain activities (or operations): note that if the activities are characterized by a known and fixed processing time, we may just consider the problem of finding either the start or the completion times.

We will first introduce the basic idea with reference to a network describing the precedence relationships among the activities of a project: in this problem we do not consider capacity constraints, and we obtain a continuous LP model. Then we show how the modeling framework can be adapted to capacitated job shop problems; this requires introducing a new type of arc in the network; we will see that the corresponding model turns out to be a difficult MILP problem.

4.5.1 Project networks

Consider a project encompassing a set of activities. We assume no resource limitations; the only constraints we consider are the precedence constraints among the activities. We may represent the project by a directed network, whose nodes correspond to activities and the arcs to precedence constraints; it is customary to add two dummy nodes representing the start and the completion of the project. Let 0 be the start node and N the finish node. The node set is denoted by $\mathcal{N}$ and the processing time of each activity $i \in \mathcal{N}$ by p_i, with $p_0 = p_N = 0$.

An example of a project network is shown in Figure 4.7. Operations precedences are represented by directed arcs: we have an arc (i, j) if activity i must be completed before starting activity j. Let $\mathcal{A}$ denote the set of precedence arcs. Each activity with no successors is linked to the last activ-

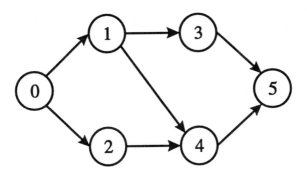

FIGURE 4.7
A project network

ity by a precedence arc; the start node is linked by a precedence arc to any node with no predecessor. This type of representation is called *activity-on-node*. In the project management literature an alternative representation, called *activity-on-arc*, is often used, whereby the nodes represent start and completion events for the activities associated to each arc. We will always use the first type of representation, which is typical of the scheduling literature.

We want to find the completion time of the overall project, assuming that each activity is started as soon as its predecessors are completed. The problem can be solved by associating with each node i of the network a *potential* C_i representing its completion time. If we have a precedence arc (i, j), we know that activity j cannot be completed earlier than the completion time of i, plus the processing time p_j:

$$C_j \geq C_i + p_j \qquad \forall (i,j) \in \mathcal{A}.$$

The completion time of the project is C_N. To compute the project makespan, we can formulate the following LP model:

$$\begin{aligned}
\min \ & C_N \\
\text{s.t.} \ & C_j \geq C_i + p_j \qquad \forall (i,j) \in \mathcal{A} \\
& C_0 = 0.
\end{aligned}$$

Note that we get a continuous LP problem. In fact, it is a peculiar LP problem: by taking its dual, we obtain a longest path problem (Section B.6.4), which can be solved by methods based on Dynamic Programming (Section D.1). Note that, since we adopted an activity-on-node representation, the processing times are weights associated with nodes rather than arcs, and the length of a path is the sum of the weights of the nodes lying on it. By computing the longest, or critical, path we collect useful information. Apart from the expected project makespan, we spot critical activities, i.e., those activities for which a duration larger than expected implies a longer makespan; such activities are associated with active inequality constraints in the above LP model. This kind of analysis can be carried out within a scheduling context in order to spot bottlenecks.

A useful electrical interpretation of the above LP model is that, in order to schedule the activities, we must assign some numbers to the nodes of the graph, the node **potentials**, subject to constraints on the minimum tension for certain pairs of nodes.

4.5.2 The disjunctive graph representation of job shop problems

In order to extend the node potential modeling framework, we must devise a way to take into account machine capacity limitations; they imply that if two operations require the same machine, they cannot overlap, i.e., one must be scheduled before the other. In formulae, we must enforce a **disjunctive** constraint,

$$\text{either} \quad C_i \geq C_j + p_i \quad \text{or} \quad C_j \geq C_i + p_j.$$

Such disjunctive constraints can be represented by **disjunctive arcs**, obtaining what is called a **disjunctive graph**. A simple example is the best way to illustrate the idea.

Example 4.8

Consider the job shop scheduling problem we considered in Example 4.1. We can associate a disjunctive graph with this problem as follows (the graph is shown in Figure 4.8):

- for each operation of each job a node is created with a weight equal to its processing time (the weight is shown within each node); the operations of the three jobs correspond to nodes (1,2,3), (4,5), and (6,7,8), respectively;
- two dummy nodes, corresponding to an 'initial' and a 'final' operation are created with null weight (nodes 0 and 9 respectively);
- each operation of each job is linked to the next operation in the process plan by an arc expressing technological precedence (for instance, for job J_1 we have arcs from node 1 to node 2, and from node 2 to node 3); such arcs are represented by continuous lines and are called **conjunctive arcs**; similar arcs are drawn linking the dummy start

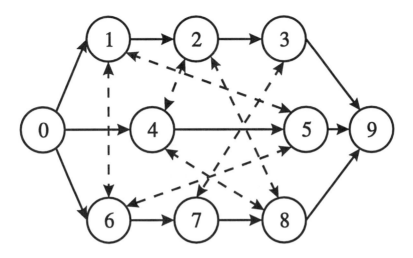

FIGURE 4.8
A disjunctive graph for a simple job shop problem

operation to the first operation of each job, and the last operation of each job to the dummy final operation;

- the nodes corresponding to operations to be executed on the same machine are linked to each other by disjunctive arcs (represented by double-arrow dotted lines); each disjunctive arc must be oriented, in order to express precedence relations due to sequencing decisions; a clique, i.e. a complete subgraph, is obtained for each machine; for example, the clique of nodes 1, 5, and 8 corresponds to machine M_1.

▫

When the direction of the disjunctive arcs is chosen, an ordinary directed graph is obtained (called an **orientation**): if the orientation is acyclic, it represents the operations precedence of a feasible schedule. Conjunctive and oriented disjunctive arcs enforce minimum tensions among nodes: the minimum difference of potential is the processing time of the successor activity in the arc. An orientation of the previous disjunctive graph is shown in Figure 4.9. A necessary condition for the graph to be acyclic is that the disjunctive arcs of each clique are set properly, in order to represent a sequence on a machine; however, it can be the case that a cyclical graph is obtained even though the single-machine sequences are feasible (see Figure 4.10). Given an acyclic orientation of the disjunctive arcs, the length of the critical path from the initial dummy node to the final dummy node is the makespan of the schedule corresponding to the orientation. Note that

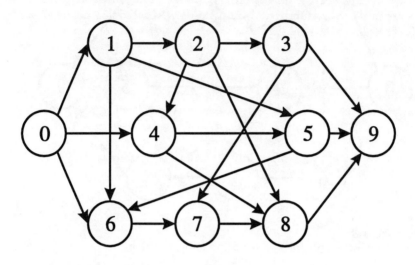

FIGURE 4.9
An orientation of a disjunctive graph

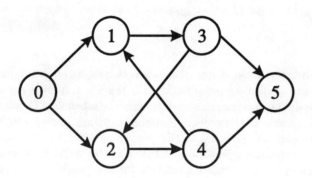

FIGURE 4.10
A cyclic orientation of a disjunctive graph

the critical path need not be unique. Furthermore, its computation can be speeded up by omitting redundant arcs. In fact, some oriented disjunctive arcs are implied by transitivity; if operation A is scheduled before B, and B is scheduled before C, then A is scheduled before C, yet, the arc (A, C) will never be on the critical path. The consequence is that each node has at most two in-going arcs and at most two out-going arcs: this can be exploited in order to streamline the computation of the critical path. The $J//C_{\max}$ problem boils down to finding an orientation whose critical path has minimal length. In the case of other (possibly nonregular) objective functions, these will depend on the potentials (completion times) associated with the last operation nodes of the jobs.

The disjunctive graph formulation can be easily translated to an MILP problem (see Section 4.6.2), and it is fundamental both for building exact and heuristic scheduling algorithms for job shop problems (see Sections 4.8.3 and 4.8.7).

4.6 MILP models for scheduling problems

We discuss here different modeling strategies for scheduling problems:

1. permutation modeling;
2. disjunctive graph based models;
3. TSP-like modeling;
4. modeling with time-indexed variables;
5. modeling with assignment and selection variables.

4.6.1 Permutation modeling

This kind of model is suitable for those cases in which job scheduling requires finding an optimal permutation of jobs, as in single-machine scheduling with regular objective function or the permutation flow shop problem. We consider here a model for the permutation $F//C_{\max}$ and $F//\sum_i C_i$ problems, proposed in [212].

We use here variables x_{ij} to denote the job position in the sequence, i.e.,

$$x_{ij} = \begin{cases} 1 & \text{if job } i \text{ is sequenced in position } j \\ 0 & \text{otherwise.} \end{cases}$$

The permutation variables must satisfy the following constraints:

$$\sum_{i=1}^{N} x_{ij} = 1 \quad \forall j$$

$$\sum_{j=1}^{N} x_{ij} = 1 \quad \forall i.$$

In this model it is convenient to consider the *start* time s_{jk} of the jth job in the sequence on machine k. By noting that there is no idle time on machine 1, nor for job 1, we can link the start times and the permutation decision variables by requiring:

$$s_{11} = 0$$

$$s_{j+1,1} = s_{j1} + \sum_{i=1}^{N} p_{i1} x_{ij} \quad j = 1, \ldots, N-1$$

$$s_{1,k+1} = s_{1k} + \sum_{i=1}^{N} p_{1k} x_{i1} \quad k = 1, \ldots, M-1.$$

For the next jobs on the next machines we must write inequality constraints, allowing idle time for machines and waiting times for jobs. By noting that the processing time of the jth job in the sequence on machine k can be expressed as $\sum_{i=1}^{N} p_{ik} x_{ij}$, we may write:

$$s_{j,k+1} \geq s_{jk} + \sum_{i=1}^{N} p_{ik} x_{ij} \quad j = 2, \ldots, N; \; k = 1, \ldots, M-1$$

$$s_{j+1,k} \geq s_{jk} + \sum_{i=1}^{N} p_{ik} x_{ij} \quad j = 1, \ldots, N-1; \; k = 2, \ldots, M.$$

If we want to minimize the makespan, we can express it as the start time of the last job on the last machine plus its processing time:

$$\min \left(s_{NM} + \sum_{i=1}^{N} p_{iM} x_{iN} \right).$$

In the $F // \sum_i C_i$ case, we may consider the equivalent objective function obtained by summing up the start times of the operations on the last machine:

$$\min \sum_{j=1}^{N} s_{jM}.$$

Note that this kind of modeling approach works here because we need not be concerned about the identity of the jobs in sequence. In an $F // \sum_i w_i C_i$ case, we would have problems in linking jobs and positions.

4.6.2 Disjunctive graph based models

We have seen in Section 4.5.2 how scheduling problems may be represented by disjunctive graphs: we can build a class of models based on binary decision variables corresponding to disjunctive arcs. Let (i,j) be a disjunctive arc relating two operations i and j; the orientation of the arc is expressed by a variable x_{ij} such that

$$x_{ij} = \begin{cases} 1 & \text{if operation } i \text{ is scheduled before operation } j \\ 0 & \text{otherwise.} \end{cases}$$

For any disjunctive arc we have the constraint

$$x_{ij} + x_{ji} = 1.$$

Obviously we might consider just *one* variable for each disjunctive arc. As we noted earlier, we must be sure to avoid cyclic orientations. We present here three modeling examples:

- a single-machine model with no idle time;
- a single-machine model with possible idle time;
- a job shop model.

In Section 4.6.4 we outline a disjunctive graph-based model for a multi-resource scheduling problem, which is compared with a time-indexed model.

Example 4.9
Consider a $1/prec/\sum w_i C_i$ problem. Clearly, no idle time needs to be considered in the optimal solution. Let $e_{ij} = 1$ for each precedence arc (i,j). Since there is no idle time in the optimal schedule, the completion time of job j can be expressed as

$$\sum_{i=1}^{N} p_i x_{ij} + p_j.$$

Based on this expression, we can build the following model:

$$\min \sum_{i=1}^{N}\sum_{j=1}^{N} p_i x_{ij} w_j$$

s.t. $x_{ij} \geq e_{ij}$

$$x_{ij} + x_{ji} = 1 \qquad \forall i,j;\ i \neq j \qquad (4.7)$$

$$x_{ij} + x_{jk} + x_{ki} \geq 1 \qquad \forall i,j,k;\ i \neq j \neq k \qquad (4.8)$$

$$x_{ij} \in \{0,1\} \qquad \forall i,j$$

$$x_{ii} = 0 \qquad \forall i.$$

Note the use of (4.8) to prevent cycles, i.e. situations in which job i precedes job k, which precedes job j, which precedes job i. Such cycles are not ruled out by the disjunctive constraints (4.7); note also that forbidding three-job cycles is sufficient to eliminate cycles with an arbitrary number of jobs.
□

Example 4.10
Consider the $1/r_i/\sum_i T_i$ problem. To build a model, we must link the completion times C_i of each job to the variables associated with disjunctive arcs. Unfortunately, we cannot express the completion times as in the previous example, since nonzero release times may induce some idle time on the machine. If J_k is scheduled before J_j, we must have

$$C_j \geq C_k + p_j;$$

if J_k is scheduled after J_j, we must have:

$$C_k \geq C_j + p_k.$$

We can enforce the disjunction of these constraints by introducing a suitably large constant M and requiring

$$C_j - p_j \geq C_k - M x_{jk},$$

$$C_k - p_k \geq C_j - M x_{kj}.$$

M should be an upper bound on the schedule makespan. If $x_{jk} = 1$ and $x_{kj} = 0$, i.e., J_j precedes J_k, the first constraint is redundant, and the second constraint is enforced; the contrary happens if $x_{jk} = 0$. We obtain the following model:

$$\min \sum_{i=1}^{N} T_i$$

$$\text{s.t.} \quad T_i \geq C_i - d_i \quad i = 1, \ldots, N$$

$$T_i \geq 0 \quad i = 1, \ldots, N$$

$$x_{jk} + x_{kj} = 1 \quad k \neq j;\ k, j = 1, \ldots, N$$

$$C_j - p_j \geq C_k - M x_{jk} \quad k \neq j;\ k, j = 1, \ldots, N \quad (4.9)$$

$$C_i \geq p_i + r_i \quad i = 1, \ldots, N$$

$$x_{jk} \in \{0, 1\} \quad k \neq j;\ k, j = 1, \ldots, N.$$

Unlike the model of example 4.9, here we do not need to add constraints to prevent cycles, since these are ruled out by (4.9). □

Example 4.11
Based on the disjunctive graph representation, it is rather easy to model the $J//C_{\max}$ problem. Let $\mathcal{N} = \{0, 1, \ldots, N-1, N\}$ be the set of operation nodes, where 0 and N are the dummy start and finish nodes. Let $\mathcal{P}$ denote the set of conjunctive arcs due to technological constraints, and $\mathcal{D}$ denote the set of disjunctive arcs.

Then we may write the following model:

$$\min C_N$$
$$\begin{aligned}
\text{s.t.} \quad & C_j \geq C_i + p_j & \forall (i,j) \in \mathcal{P} \\
& \left.\begin{aligned} C_j &\geq C_i + p_j - M(1 - x_{ij}) \\ C_i &\geq C_j + p_i - M x_{ij} \end{aligned}\right\} & \forall (i,j) \in \mathcal{D} \\
& x_{ij} \in \{0,1\} & \forall (i,j) \in \mathcal{D} \\
& C_i \geq p_i & \forall i \in \mathcal{N}.
\end{aligned}$$

Note that the model can be easily adapted to any (possibly nonregular) objective function depending on the completion times of the jobs. □

4.6.3 TSP-like modeling

We consider here the $1/s_{ij}/\sum_i w_i C_i$ problem; we want to minimize the total weighted completion time on a single-machine with significant and sequence-dependent setup times. This problem is strictly related to the TSP problem; indeed, it is sometimes referred to as the Traveling Repairman Problem. The modeling approach we follow here is quite similar to the one adopted for TSP (see Section C.1).

When job j is sequenced immediately after job i, a setup time s_{ij} must elapse before processing. If job 0 is associated with the current state of the machine, s_{0i} is the setup time spent if job i is the first one in the sequence.

The problem can be formulated as a mixed-integer programming problem by introducing a set of binary decision variables x_{ij} indicating that job j is processed *immediately after* job i. Note that the meaning of these variables is different than that of disjunctive variables, which only specify that a job is processed *after* another one. The following model is obtained:

$$\min \sum_{i=1}^{N} w_i C_i$$
$$\text{s.t.} \sum_{\substack{i=0 \\ i \neq j}}^{N} x_{ij} = 1 \qquad j = 0, \ldots, N \tag{4.10}$$

$$\sum_{\substack{j \neq i \\ j=0}}^{N} x_{ij} = 1 \qquad i = 0, \ldots, N \qquad (4.11)$$

$$C_j \geq C_i + p_j + s_{ij} + M(x_{ij} - 1)$$
$$i = 0, \ldots, N, \ j = 1, \ldots, N, \ j \neq i \qquad (4.12)$$

$$C_0 = 0$$

$$x_{ij} \in \{0, 1\} \qquad i, j = 0, 1, \ldots, N, \ i \neq j.$$

Constraints (4.10) and (4.11), stating that each job has one predecessor and one successor in the sequence, define an assignment problem substructure. The variable x_{i0} is set to 1 when i is the last job in the sequence. Constraints (4.12) enforce the correct completion times, under the usual assumption of nonpreemption and nonshareability of the machine; M is a suitably large constant, which can be set to an upper bound on the makespan:

$$M = \sum_{i=1}^{N} (p_i + \max_{\substack{j \neq i \\ j = 0, 1, \ldots, N}} \{s_{ji}\}).$$

Note that constraints (4.12) forbid the creation of job cycles, which are compatible with the assignment constraints (4.10) and (4.11). This issue is related to the problem of avoiding subtours in the solution of TSP.

4.6.4 Modeling with time-indexed variables

This chapter is focused on DEDS models for scheduling problems; this does not mean that discrete-time models cannot be adopted for such problems, as shown in Sections 3.3 and 3.4. In this section we illustrate another discrete-time modeling framework for scheduling problems. Unlike the previous discrete-time models, it cannot be interpreted as a state equation model, and this is why it has not been described in Chapter 3; furthermore, it is useful to compare the discrete-time model with DEDS models for similar problems.

We consider here a multi-resource constrained scheduling problem: a set of activities, related by precedence constraints, must be carried out in the shortest makespan, and each activity requires a given amount of a set of resources. We can think of this problem as the capacitated version of the project network problem presented in Section 4.5.1.

We formulate the problem first with a discretized time horizon; let t be the index of the time buckets, k be the resource index and i the activity index; let N be a final dummy activity with zero processing time. Note that we consider here *small* time buckets, such that we may express processing times as an integer number of time buckets. The set of activities which precede activity j is denoted by $\mathcal{P}_j$. The key feature is the introduction of a set of binary decision variables such that:

$$x_{it} = \begin{cases} 1 & \text{if activity } i \text{ completes in time period } t \\ 0 & \text{otherwise.} \end{cases}$$

The completion time of activity i is then:

$$\sum_{t=1}^{T} t x_{it}.$$

We obtain the following model:

$$\min \sum_{t=1}^{T} t x_{Nt}$$

$$\text{s.t.} \sum_{i=1}^{N} r_{ik} \sum_{\tau=1}^{t+p_i-1} x_{i\tau} \leq A_k \qquad \forall t, k \tag{4.13}$$

$$\sum_{t=1}^{T} t x_{it} + p_j \leq \sum_{t=1}^{T} t x_{jt} \qquad \forall j, \forall i \in \mathcal{P}_j \tag{4.14}$$

$$\sum_{t=1}^{T} x_{it} = 1 \qquad \forall i \tag{4.15}$$

$$x_{it} \in \{0,1\} \qquad \forall i, t,$$

where A_k is the availability of resource k, r_{ik} the amount of resource k required by activity i, and p_i the processing time of activity i.

The objective function is to minimize the completion of the last (dummy) activity, which is the makespan. Eq. (4.13) is the resource availability constraint; note that activity i is in process at time t if

$$\sum_{\tau=1}^{t+p_i-1} x_{i\tau} = 1,$$

so that (4.13) simply states that for each time bucket the resource requirements for each activity under processing cannot exceed the availability. Eq. (4.14) is written for all precedence arcs (i,j). Finally (4.15) ensures that all the activities are processed.

Now consider how a similar problem could be formulated within a DEDS modeling framework, following a model proposed in [62]. Let C_i be the completion time of activity i and x_{ij} a binary variable defined as in Section 4.6.2, i.e., a variable related to disjunctive constraints, stating the precedence relation between activities i and j. We denote by $\mathcal{T}_i$ the set of resources needed by activity i. Using the disjunctive modeling approach we get

$$\min C_N$$
$$\text{s.t.} \quad C_j \geq C_i + p_j \quad \forall (i,j) \in \mathcal{P}$$
$$C_j \geq C_i + p_j - M x_{ji} \quad \forall i,j;\ i \neq j,\ T_i \cap T_j \neq \emptyset \quad (4.16)$$
$$x_{ij} + x_{ji} = 1 \quad \forall i,j;\ i \neq j,\ T_i \cap T_j \neq \emptyset.$$

Eq. (4.16) is the key feature of the model, and it requires that two activities do not overlap if their resource sets have an intersection. Note that this model is not exactly the same as the discrete-time model; it requires that resources are a-priori assigned to activities. It is possible to model the resource assignment by introducing other binary decision variables, but this would require reformulating (4.16) by adding another big M, with a detrimental effect on model solvability.

From the point of view of model solvability and flexibility, the DEDS approach does not appear advantageous; still, it can be exploited within a Lagrangian relaxation framework as shown in Section 4.7.4.

4.6.5 Modeling with assignment and selection variables

Sometimes, the scheduling problem does not actually require part sequencing; this happens with load balancing and part type selection problems, provided that there is no sequence-dependent setup time. For instance, consider the $R//C_{\max}$ problem: clearly, the only issue is to assign parts to machines, since sequencing is irrelevant.

In this case, we need a set of binary variables modeling the decision of assigning a job to a machine, or the selection of a job.

A mathematical model for the $R//C_{\max}$ problem

Let p_{ij} denote the processing time of job i on machine j. Introducing a set of binary variables:

$$x_{ij} = \begin{cases} 1 & \text{if job } i \text{ is processed on machine } j \\ 0 & \text{otherwise,} \end{cases}$$

we obtain the following model:

$$\min C_{\max}$$

$$\text{s.t.} \sum_{i=1}^{N} p_{ij} x_{ij} \leq C_{\max} \quad \forall j \qquad (4.17)$$

$$\sum_{j=1}^{M} x_{ij} = 1 \quad \forall i \qquad (4.18)$$

$$x_{ij} \in \{0, 1\} \quad \forall i, j.$$

Note how (4.17) is used to pick up the maximum load, i.e. the makespan, over the machines; (4.18) simply states that each job must be processed on exactly one machine.

A mathematical model for part type selection in a FMS

Consider an FMS consisting of M identical machines; we assume that each job needs a set of operations that can be carried out consecutively on one machine, provided that the tool set for the job has been loaded on the limited capacity on-board tool magazine.

We assume here that it is not possible to dynamically change the tool loaded on each machine: parts are processed in batches, separated by a complete reconfiguration of the FMS. Therefore, we need to select the part types of the next batch, trading off their similarity with their urgency. We further assume that tool wear needs not to be modeled.

The urgency can be expressed by a weight w_i associated with each job $i = 1, \ldots, N$. We want to select a subset of jobs in order to maximize the total weight of the selected jobs. Another requirement is that the load is not too unbalanced. If L_j is the load on machine $j = 1, \ldots, M$, the ideal load is $\overline{L} = \sum_j L_j / M$: the balancing requirement may be stated as

$$(1 - \alpha)\overline{L} \leq L_j \leq (1 + \alpha)\overline{L} \quad \forall j,$$

where $\alpha \in (0, 1)$ is an unbalancing tolerance.

Let us introduce the following binary decision variables:

$$x_{ij} = \begin{cases} 1 & \text{if job } i \text{ is processed on machine } j \\ 0 & \text{otherwise;} \end{cases}$$

$$y_{tj} = \begin{cases} 1 & \text{if a tool of type } t \text{ is loaded on machine } j \\ 0 & \text{otherwise.} \end{cases}$$

We can model the problem as follows:

$$\max \sum_{i=1}^{N} \sum_{j=1}^{M} w_i x_{ij} \qquad (4.19)$$

$$\text{s.t.} \sum_{j=1}^{N} x_{ij} \leq 1 \quad \forall i \tag{4.20}$$

$$\sum_{j=1}^{M} y_{tj} \leq D_t \quad \forall t \tag{4.21}$$

$$\sum_{t=1}^{T} y_{tj} \leq C \quad \forall j \tag{4.22}$$

$$x_{ij} \leq y_{tj} \quad \forall i, j, \ t \in T_i \tag{4.23}$$

$$(1-\alpha)\sum_{i=1}^{N}\sum_{j=1}^{M} p_i x_{ij} \leq M \sum_{i=1}^{N} p_i x_{ij} \quad \forall j \tag{4.24}$$

$$M \sum_{i=1}^{N} p_i x_{ij} \leq (1+\alpha)\sum_{i=1}^{N}\sum_{j=1}^{M} p_i x_{ij} \quad \forall j \tag{4.25}$$

$$x_{ij}, y_{tj} \in \{0,1\} \quad \forall i, j, t,$$

where D_t is the number of available copies of tool type t, C is the capacity of the tool magazine on each machine, T_i is the set of tool types needed for processing job i, and p_i is the processing time of job i on any machine (we assume identical machines here). Eq. (4.20) states that each job can be processed on *at most* one machine, Eq. (4.21) limits the number of tools of type t that are loaded on all the machines, Eq. (4.22) limits the number of tools loaded on each machine, and Eq. (4.23) relates the decision variables (we can assign a job to a machine only if the needed tools are loaded); finally, Eqs. (4.24) and (4.25) enforce the limits on the workload unbalance ($\sum_i p_i x_{ij}$ is the load on machine j).

4.7 Branch and Bound methods

Since most scheduling problems are NP-hard, enumerative algorithms are needed to obtain the optimal solution. There is ample literature on Branch and Bound methods for machine scheduling; we can apply general integer programming methods or special purpose methods.

Unfortunately, integer programming methods based on continuous relaxation are usually limited to very small problem instances; in Section 4.7.1 we discuss why. Then we give some examples of special purpose methods. In Section 4.7.2 we describe a quite successful Branch and Bound method for the $1/r_i/L_{\max}$ problem; the main feature of this method is that the branching process is based on *feasible* solutions of the problem, and not partial ones. A very simple Branch and Bound method for the minimization of makespan in a permutation flow shop is described in Section 4.7.3; it is a good example of how bounds can be obtained by intuitive reasoning, without building mathematical models. Finally, in Section 4.7.4 we describe some applications of Lagrangian relaxation to scheduling problems: Lagrangian relaxation may yield stronger bounds than continuous relaxation, and it is most useful in devising heuristic methods.

4.7.1 Integer programming approaches to machine scheduling

It has been argued that there is no point in using integer programming techniques for solving scheduling problems (see e.g. [79, p. 135]). Computational experience with commercial Branch and Bound codes applied to DEDS machine scheduling models has indeed been quite disappointing. The reason behind such a failure is the poor quality of the bounds obtained by continuous relaxation.

To understand why, consider how disjunctive arcs have been modeled in Section 4.6.2:

$$C_j - p_j \geq C_k - M x_{jk}.$$

When the binary variable x_{jk} is relaxed, constraints of this kind define a continuous feasible region much larger than the convex hull of the integer feasible points. This is the same kind of issue we faced in Section 3.2.2, and is due to the use of the big-M. The following Example shows how this affects even simple scheduling problems.

Example 4.12
Consider the following model for the $1//\sum_i C_i$ problem:

$$\min \sum_{i=1}^{N} C_i$$
$$\text{s.t.} \quad C_i \geq p_i \quad \forall i$$
$$C_j \geq C_i + p_j - M x_{ji} \quad \forall i, \forall j \neq i$$
$$x_{ij} + x_{ji} = 1 \quad \forall i, \forall j \neq i$$
$$x_{ij} \in \{0, 1\} \quad \forall i, \forall j \neq i.$$

We know that this problem is solved by the SPT rule, but consider the continuous relaxation of this model: its optimal solution is $C_i = p_i$ and the lower bound we obtain is $\sum_i p_i$. To see why, consider a disjunctive arc(i, j) and the related couple of constraints:

$$C_j \geq C_i + p_j - M x_{ji}$$

$$C_i \geq C_j + p_i - M(1 - x_{ji}).$$

Substituting $C_i = p_i$ and $C_j = p_j$ we get

$$x_{ji} \geq \frac{p_i}{M}$$

$$x_{ji} \leq 1 - \frac{p_j}{M}.$$

These inequalities are trivially satisfied by setting $x_{ji} = p_i/M$. The resulting lower bound is very weak; for instance, if we have $p_i = i \cdot 10$ for $i = 1, \ldots, 10$, the lower bound is 550, whereas the SPT sequence has a total completion time 2200. ▯

Unfortunately, it is very difficult, if at all possible, to strengthen the formulation of scheduling problems: even in a single-machine case, strengthening the formulation requires a considerable degree of ingenuity, both for a DEDS and a discrete-time modeling approach. The interested reader is referred to the research literature listed at the end of the chapter.

4.7.2 A Branch and Bound algorithm for $1/r_i/L_{\max}$

We have shown in Supplement S2.1 that $1/r_i/L_{\max}$ is an NP-hard problem; therefore its solution calls for enumerative methods such as Branch and Bound. For this problem a quite efficient Branch and Bound algorithm has been proposed in [142]. A most interesting feature of this approach is that a *feasible solution* is associated with each node of the search tree; this feature explains the computational efficiency of this algorithm, which may be used as a subroutine of heuristic procedures (see Section 4.8.7).

The starting point is the application of a modified EDD rule to the problem: at time t we consider the set of schedulable jobs (i.e., those with $r_i \leq t$), and we select the one with the smallest due date and update t; if no job is available, we set t to the smallest release time of the jobs to be scheduled. We obtain a schedule characterized by **blocks**, i.e., sets of jobs processed during a period of machine utilization followed by idle time (see Figure 4.11). The first job in a block starts processing at its release time. The overall idea is to build a search tree whose nodes are feasible schedules obtained by this EDD-like rule. Branching is made on the basis of job swaps aimed at *improving* the feasible solution corresponding to the node; the node is fathomed if it can be shown that the solution cannot be improved.

Apart from the branching scheme, we need a way to compute a lower bound associated to a feasible schedule. Consider a job j, and let B_j be the set of jobs that precede j in its block, including j itself: the completion time C_j in the current schedule is the earliest completion time for the set

Branch and Bound methods

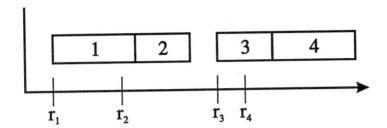

FIGURE 4.11
Blocks obtained by applying a modified EDD rule to the $1/r_i/L_{\max}$ problem

of jobs in B_j. If we take a job $k \in B_j$ and we schedule k as the last job in B_j, we have two possibilities:

1. if $d_k \leq d_j$, job k is at least as late as job j;
2. if $d_k > d_j$, then it could not complete its processing before $C_j + 1$, since a gap is left in the schedule: in fact, if $d_k > d_j$ but j has been scheduled after k, it is because of some problem with the release times. In this case the lateness of k is at least $C_j + 1 - d_k$.

We obtain the following lower bound, associated to each job j:

$$LB_j^1 = \begin{cases} C_j - d_j & \text{if } d_j = \max_{i \in B_j} d_i \\ C_j + 1 - d_k & \text{if } d_j \neq d_k = \max_{i \in B_j} d_i. \end{cases}$$

Another obvious lower bound is $LB^2 = \max_i \{r_i + p_i - d_i\}$. We therefore obtain:

$$LB = \max \left\{ \max_j LB_j^1, LB^2 \right\}.$$

Having a way to compute a lower bound, we must specify the branching scheme. Given the EDD schedule, we can try to improve it as follows. Let j be the *critical* job, i.e., the one to which the maximum lateness is associated; clearly, we can improve the solution only if we schedule the critical job earlier. If $d_j = \max_{i \in B_j} d_i$ the schedule cannot be improved and the EDD schedule is optimal. We can improve the solution only if there is a $k \in B_j$ such that $d_j < d_k$. Let G_j be the set of such jobs, called the *generating set*. The search space may be partitioned with respect to the relative order of the critical job j and $k \in G_j$; a simple way to force k to be scheduled after j is to modify its release time by setting $r'_k = r_j$.

Branching is obtained by spotting the critical job j and creating, for each job $k \in G_j$ a new node with $r'_k = r_j$ and reapplying the modified EDD rule. Nodes are fathomed when no improvement is possible.

This kind of Branch and Bound algorithm, based on some form of reasoning about *feasible* solutions, is quite powerful; indeed, the most powerful Branch and Bound methods for the $J//C_{\max}$ problem are based on this idea. We do not cover such methods here because, despite the improvements obtained with respect to early methods, Branch and Bound approaches are not yet applicable to practical size $J//C_{\max}$ problems.

4.7.3 A Branch and Bound algorithm for the permutation $F//C_{\max}$ problem

In the previous section, we have seen how lower bounds can be derived by exploiting the peculiarities of the problem at hand; we have not relied on standard bounding schemes based on mathematical modeling. Another useful bounding approach for scheduling problems is constraints elimination. In Section 4.3.4 we have seen that sometimes the preemptive version of an NP-hard problem is easy; by relaxing the nonpreemption constraint we obtain easy to compute lower bounds. Another possibility is to relax the precedence constraints among operations on the same job. Here we apply this idea to the permutation $F//C_{\max}$ problem; we describe a lower bound originally proposed in [108]; note that this is not necessarily the tightest bound for flow shop scheduling, but it is very simple and instructive.

Since we are considering a permutation problem, the solution is represented by a permutation σ of the jobs. The most natural branching scheme is to build the sequence step by step (see Section C.5.2). At each node of the search tree we have a partial sequence. Let $A = \left(J_{\sigma(1)}, J_{\sigma(2)}, \ldots, J_{\sigma(K)}\right)$ be the partial sequence, and U the set of unscheduled jobs. We can easily compute the completion time C_j^A of the partial sequence A on each machine j. The first machine, if there is no buffer limitation, can process jobs consecutively without any idle time; therefore we have

$$C_1^A = C_{\sigma(K),1} = C_{\sigma(K-1)} + p_{\sigma(K),1} = \sum_{h=1}^{K} p_{\sigma(h),1}.$$

On the remaining machines, we must consider that an operation can be scheduled if two condition are met: the machine must be ready, and the job must be available. We have the recursive relation

$$C_{\sigma(h),j} = \max\left\{C_{\sigma(h),j-1}, C_{\sigma(h-1),j}\right\} + p_{\sigma(h),j},$$

with the initial condition

$$C_{\sigma(1),j} = \sum_{h=1}^{j} p_{\sigma(1),h}.$$

Using these relations we may compute the completion time of the partial schedule A on machine j, $C_j^A = C_{\sigma(K),j}$.

We need a lower bound on the completion time of the last job on the last machine. An obvious lower bound is

$$LB_M = C_M^A + \sum_{h \in U} p_{hM}.$$

The lower bound is obtained by simply adding the processing times of the unscheduled jobs on the last machine to the makespan of the partial schedule; this assumes that the last machine is never idle waiting for a job being processed on machine $M-1$. Such an assumption is equivalent to the elimination of the precedence constraint

$$C_{ij} \geq C_{i,j-1} + p_{ij}.$$

By similar reasoning, we obtain lower bounds LB_j on each machine M_j ($j = 1, \ldots, M-1$); these lower bounds can be stronger than LB_M if an intermediate machine acts as a bottleneck. The lower bound assumes that machine j is never kept idle. Under this hypothesis the optimal makespan is obtained by choosing the last job in the sequence in order to minimize its "tail" on the machines following j:

$$LB_j = C_j^A + \sum_{h \in U} p_{hj} + \min_{h \in U} \left\{ \sum_{l=j+1}^{M} p_{hl} \right\}.$$

The last term, i.e. the tail, is based on the assumption that the last job in the sequence will not have to wait for service on any machine after visiting machine j. Then we take $LB = \max_j LB_j$ as the lower bound associated with a partial sequence. This lower bound is not particularly strong, but it illustrates very well how simple, and quite fast to compute, lower bounds can be obtained without building sophisticated mathematical models.

4.7.4 Lagrangian relaxation approaches to machine scheduling

We have seen in Section 4.7.1 that integer programming approaches based on continuous relaxation fail because the relaxation of the capacity constraints, which involve big Ms, is very weak. However, there are scheduling problems, such as $1//\sum_i w_i C_i$, which are easily solved by a direct approach, even if their MILP model is not amenable to efficient solution procedures. $1//\sum_i w_i C_i$ could be stated as

$$\min \sum_{i=1}^{N} w_i C_i$$

$$\text{s.t.} \quad \mathbf{C} = \begin{bmatrix} C_1 \\ \vdots \\ C_N \end{bmatrix} \in \mathcal{C},$$

where $\mathbf{C}$ is the vector of completion times, and $\mathcal{C}$ is the set of feasible completion times, i.e., those obtained under no preemption and finite capacity constraints; in other words, $\mathcal{C}$ is an *implicit* representation of the set obtained by enforcing capacity constraints. Lagrangian relaxation may be exploited if we are able to build models in which these constraints are left implicit; we need only to formalize the objective function and the difficult constraints, whereas the "easy" part of the model is dealt with by a specific solution method.

In the following we give four examples of Lagrangian relaxation. In the first three, the easy part of the model is solved by the WSPT rule. Note that these bounds are stronger than those obtained by continuous relaxation, since, as we have seen in Example 4.12, $1//\sum_i w_i C_i$ does not enjoy the integrality property (see Section C.4.5). In the fourth Example we apply Lagrangian relaxation to a model with time-indexed variables; the easy problem is solved by complete enumeration.

We have seen in Chapter 3 that Lagrangian relaxation applied to lot-sizing can be interpreted as a form of decomposition; dualization of the capacity constraints yields a decomposition into single-item subproblems. Here, we do not dualize capacity constraints: this would bring the big M in the dual function and would weaken the lower bound considerably. On the contrary, Lagrangian relaxation for scheduling problems can be often interpreted as a decomposition by resources: in the second and third example we schedule each resource independently by the WSPT rule. In the fourth case, we dualize the capacity constraint, but due to the different modeling framework we have no problem with the big M.

A final remark is important: although Lagrangian bounds may be sharper than those obtained by continuous relaxation, solving large scale scheduling problems by exact Branch and Bound methods is usually out of the question. Nevertheless, Lagrangian relaxation is interesting, since it is relatively easy to derive primal feasible solutions from an optimal solution of the relaxed problem. On the one hand, this may be useful within Branch and Bound algorithms; on the other hand, it paves the way for the development of dual heuristics.

Application to $1/prec/\sum_i w_i C_i$

The $1//\sum_i w_i C_i$ problem is easily solved by the WSPT rule, whereas $1/prec/\sum_i w_i C_i$ is NP-hard: this suggests the possibility of regarding $1/prec/\sum_i w_i C_i$ as an easy problem complicated by the precedence constraints. The approach presented here is fully described in [191], where it was originally proposed.

We can formalize the problem as

$$\min \sum_{i=1}^{N} w_i C_i$$
$$\text{s.t.} \quad C_j \geq C_i + p_j \quad \forall (i,j) \in \mathcal{P} \quad (4.26)$$
$$\mathbf{C} \in \mathcal{C},$$

where $\mathcal{P}$ is the set of job pairs related by precedence constraints. By associating a nonnegative multiplier μ_{ij} with each precedence constraint (4.26), we obtain

$$\min \sum_{i=1}^{N} \left(w_i + \sum_{j \in \mathcal{A}_i} \mu_{ij} - \sum_{j \in \mathcal{B}_i} \mu_{ji} \right) C_i + \sum_{i=1}^{N} \sum_{j \in \mathcal{A}_i} \mu_{ij} p_j$$
$$\text{s.t.} \quad \mathbf{C} \in \mathcal{C},$$

where $\mathcal{A}_i$ and $\mathcal{B}_i$ are the set of immediate successors and predecessors of job i. The relaxed problem is easily solved by the WSPT rule: note how the disjunctive constraints are left implicit and dealt with by a special purpose method.

In [191] special purpose methods are discussed for adjusting the Lagrangian multipliers, in such a way that an ascent direction in the space of dual variables is guaranteed. Still, it is useful to understand what a plain subgradient method would accomplish in this case. A multiplier μ_{ij} is increased if the respective constraint (4.26) is violated, i.e., if job j is scheduled before job i. However, this happens if

$$\frac{w_i + \sum_{k \in \mathcal{A}_i} \mu_{ik} - \sum_{k \in \mathcal{B}_i} \mu_{ki}}{p_i} \leq \frac{w_j + \sum_{k \in \mathcal{A}_j} \mu_{jk} - \sum_{k \in \mathcal{B}_j} \mu_{kj}}{p_j}.$$

Now, $j \in \mathcal{A}_i$ and $i \in \mathcal{B}_j$: therefore, increasing μ_{ij} tends to reverse this inequality and to make the resulting WSPT "more feasible". Hence we see that subgradient methods implicitly aim at a primal feasible solution.

Application to $F//\sum_i w_i C_i$

Using the same modeling approach used in the $1/prec/\sum_i w_i C_i$ case, we can formulate $F//\sum_i w_i C_i$ as follows:

$$\min \sum_{i=1}^{N} w_i C_{iM}$$

$$\text{s.t.} \quad C_{i,j+1} \geq C_{ij} + p_{i,j+1} \quad i=1,\ldots,N;\ j=1,\ldots,M-1 \quad (4.27)$$

with the addition of constraints ensuring that on each machine the capacity constraint is satisfied; by leaving such constraints implicit, we can dualize the interaction constraints (4.27) and obtain single-machine problems directly solvable by the WSPT rule.

We introduce a set of Lagrangian multipliers μ_{ij} for each interaction constraint. Dualization yields the objective function

$$\min \sum_{i=1}^{N} w_i C_{iM} + \sum_{i=1}^{N} \sum_{j=1}^{M-1} \mu_{ij} \left(C_{ij} + p_{i,j+1} - C_{i,j+1} \right),$$

that can be rewritten as

$$\min \sum_{i=1}^{N} (w_i - \mu_{i,M-1}) C_{iM}$$

$$+ \sum_{i=1}^{N} \mu_{i1} C_{i1}$$

$$+ \sum_{i=1}^{N} \sum_{j=2}^{M-1} (\mu_{ij} - \mu_{i,j-1}) C_{ij}$$

$$+ \sum_{i=1}^{N} \sum_{j=1}^{M-1} \mu_{ij} p_{i,j+1}. \quad (4.28)$$

We see that the objective function of the relaxed problem consists of four terms, which correspond respectively to machine M, machine 1, and intermediate machines, whereas the last one is, for a given set of multipliers, a constant term. The problem is now decomposed into a set of $1//\sum_i w_i C_i$ problems, easily solved by the WSPT rule. On machine 1 we sort the jobs in increasing order of

$$\frac{\mu_{i1}}{p_{i1}};$$

on machine M we sort the jobs in increasing order of

$$\frac{w_i - \mu_{iM}}{p_{iM}};$$

on machine $j = 2, \ldots, M - 1$ we sort the jobs in increasing order of

$$\frac{\mu_{ij} - \mu_{i,j-1}}{p_{ij}}.$$

A significant lower bound is obtained only if we enforce the following conditions on the multipliers:

$$0 \leq \mu_{i1} \leq \mu_{i2} \leq \ldots \leq \mu_{i,M-1} \leq w_i \quad \forall i;$$

if these conditions are not met, there are negative weights associated to some jobs on some machines, and the corresponding completion times are set to $+\infty$.

Note how the relaxation of the interaction constraints yields a decomposition *by resource*. From the solution of the relaxed problem, we can easily obtain a feasible solution for the original problem: in fact, putting together each single machine sequence, we cannot obtain a cyclic disjunctive graph, unlike the job shop problem. If the flow shop is of the permutation type, things are a little more complicated; we should sequence the machines in the same way. Note if the flow shop consists of just two machines, we know that the optimal solution is of the permutation type. The $F2//\sum_i C_i$ problem has been tackled in [192], where the difficulty is solved by imposing further restrictions on the multipliers and by devising a suitable multiplier adjustments scheme. In [193, Chapter 3], a generalization to M machines is outlined.

Jobs Requiring Simultaneous Use of Resources

We consider here a scheduling problem characterized by single-operation jobs which need different resources; for instance, to complete the task we might need a machine, an operator with a particular skill and a specialized tool. We present here a Lagrangian relaxation approach proposed in [62]; the objective is the minimization of the total weighted completion time. The DEDS modeling approach illustrated in Section 4.6.4 can be used. However, to pave the way for a decomposition approach, we need to reformulate the model by distinguishing the completion time C_{ij} of job i on resource j. Clearly, we have $C_i = C_{ij}$; the dualization of these constraints is the key to the Lagrangian decomposition. The model can be formulated as

$$\min \sum_{i=1}^{N} w_i C_i \qquad (4.29)$$

$$\text{s.t.} \quad C_i = C_{ij} \qquad \forall i, \forall j \in T_i \qquad (4.30)$$

$$x_{ik} = x_{ikj} \qquad \forall i \neq k, \forall j \in T_i \cap T_k \qquad (4.31)$$

$$C_{ij} - p_i \geq C_{kj} - M x_{ikj} \qquad \forall i \neq k, \forall j \in T_i \cap T_k \qquad (4.32)$$

$$x_{ikj} + x_{kij} = 1 \qquad \forall i \neq k, \forall j \in T_i \cap T_k \qquad (4.33)$$

$$C_{ij} \geq p_i \qquad \forall i, \forall j \in T_i \qquad (4.34)$$

$$x_{ikj} \in \{0, 1\} \qquad \forall i \neq k, \forall j \in T_i \cap T_k, \qquad (4.35)$$

where i, k refer to jobs, j to resources, T_i is the set of resources needed by job i, p_i is the processing time of job i, x_{ik} is the usual disjunctive variable set to 1 if job i precedes job k, and x_{ikj} is the corresponding disjunctive variable on resource j.

The suggested Lagrangian relaxation is to dualize (4.30) with multipliers λ_{ij} and to drop (4.31). This yields the following relaxed problem:

$$w(\lambda) = \min \sum_{i=1}^{N} \left(w_i - \sum_{j \in T_i} \lambda_{ij} \right) C_i + \sum_{i=1}^{N} \sum_{j \in T_i} \lambda_{ij} C_{ij}$$

$$\text{s.t.} \quad (4.32), (4.33), (4.34), (4.35).$$

Since C_i is unrestricted in sign, we must require $w_i = \sum_{j \in T_i} \lambda_{ij}$ to obtain finite values for the dual function $w(\lambda)$. The dual problem is:

$$\max_{\lambda} \sum_{i=1}^{N} \left\{ \begin{array}{l} \min_{\mathbf{x}, \mathbf{C}} \sum_{j \in T_i} \lambda_{ij} C_{ij} \\ \text{s.t.} \quad (4.32), (4.33), (4.34), (4.35). \end{array} \right\}$$

where the maximization is also subject to

$$\sum_{j \in T_i} \lambda_{ij} = w_i.$$

The relaxed problem is easily decomposed into single-resource problems. A further restriction on the multipliers is $\lambda \geq 0$: this is needed to avoid infinite completion times, and *not* to the type of dualized constraints (equality constraints do not need restricted multipliers). Multiplier adjustment schemes are described in the original reference, where an alternative method based on surrogate relaxation is also described, as well as heuristics based on both relaxations.

Lagrangian relaxation with time-indexed variables

We briefly describe the basic idea of a scheduling approach proposed in [134], which is based on a partially implicit model with time-indexed variables. Consider an identical parallel machine scheduling problem with a quadratic tardiness objective

$$\min \sum_{i=1}^{N} w_i T_i^2.$$

Note that, unlike the case of $R//C_{\max}$, building a mathematical model in the DEDS style would require using both assignment and sequencing variables: the resulting model is particularly rich in big Ms, and hence almost useless.

An alternative approach is to use time-indexed variables, as done in Section 4.6.4; the model here is a little bit different, and it is not completely explicit. The model is based on binary decision variables

$$\delta_{it} = \begin{cases} 1 & \text{if job } i \text{ is under processing during time bucket } t \\ 0 & \text{otherwise.} \end{cases}$$

If M_t is the number of available machines during time bucket t, the capacity constraint is expressed as

$$\sum_{i=1}^{N} \delta_{it} \leq M_t \qquad \forall t. \tag{4.36}$$

The completion time of job i is (under a no preemption assumption)

$$C_i = S_i + p_i - 1,$$

where S_i is the start time and p_i the processing time. The interpretation is that the job is active both during time bucket S_i and during time bucket C_i; this is why the term -1 is introduced. The problem consists of selecting the start times. Note that we should write constraints ensuring that a job is processed during *consecutive* time buckets; such constraints are left implicit, and directly dealt with by the solution procedure for the relaxed problem.

By dualizing (4.36) with multipliers $\mu_t \geq 0$, we get the relaxed problem:

$$\min_{\{S_i\}} \left\{ \sum_{i=1}^{N} w_i T_i^2 + \sum_{t=1}^{T} \mu_t \left(\sum_{i=1}^{N} \delta_{it} - M_t \right) \right\}.$$

The relaxed problem is easily decomposed *by job*:

$$\min_{1 \leq S_i \leq T - p_i + 1} \left\{ w_i T_i^2 + \sum_{t=S_i}^{S_i + p_i - 1} \mu_t \right\}.$$

This problem is solved by trying all possible values of S_i, and taking the one yielding the lowest value. Note how μ_t can be interpreted as a resource price during time bucket t.

4.8 Heuristic scheduling methods

We have seen how a MILP modeling framework may capture a wide variety of scheduling problems. Unfortunately, scheduling problems are among the hardest discrete optimization problems; for a fixed number of integer variables, there are other classes of MILP models which are by far easier to solve. Due to the computational complexity of scheduling problems, it is unlikely that an efficient method able to yield the optimal solution with a reasonable computational effort will be ever devised. In practice, we must settle for a suboptimal solution obtained by a heuristic strategy. There is a huge literature on heuristic scheduling algorithms, and it is not our aim to survey this mass of work. We just want to illustrate a set of more or less *general* principles; by "general" we mean methods which can be tailored to different situations and that are not too problem-specific. By understanding the concepts underlying these approaches, the reader, when facing a practical scheduling problem, should be able to come up with a viable solution strategy. As with discrete-time models, we want to stress the fact that mathematical modeling *may* be a valuable source of heuristics; we have seen an application of approximate discrete-time continuous-flow models in Section 3.4.2; here we will emphasize the use of dual heuristics.

The simplest class of scheduling heuristics is based on the application of simple priority rules (also known as dispatching rules); this approach is called **list scheduling**, and it is commonly adopted in commercially available scheduling software packages. List scheduling is the subject of Section 4.8.1. We may consider list scheduling as a one-shot, greedy heuristic strategy; the myopic behavior is this approach can be partially avoided by adopting beam search methods (see section C.8.2).

List scheduling is an example of a commonsense heuristic approach, which does not require any mathematical modeling; the use of mathematical modeling to devise dual heuristics is covered in Section 4.8.2.

Recently, a wide class of scheduling algorithms has been proposed, based on local search meta-heuristics (see Section C.8). We deal with scheduling by the two most common local search methods, i.e., simulated annealing (Section 4.8.3), and tabu search (Section 4.8.4).

We have seen that some problems actually involve the *assignment* or the *selection* of jobs rather than their scheduling/sequencing; a typical case is load balancing. Well-known heuristics for load balancing and similar problems are the LPT rule and the Multifit algorithm, which are described in Section 4.8.5. Load balancing may occur as a subproblem of a complex problem, such as FMS scheduling. In Section 4.8.6 we outline decomposition methods for scheduling in Flexible Manufacturing Systems.

Finally, we describe a heuristic approach based on the concept of a *bottleneck machine*. The rationale behind the approach is that we should schedule the bottleneck machine first, and then the remaining ones, which are less critical. This idea can be applied in a variety of ways; in Section 4.8.7 we describe one of them, namely the **shifting bottleneck** procedure for the $J//C_{\max}$ problem.

4.8.1 List scheduling and greedy methods

In Section 4.3 we have seen that simple rules, such as EDD, yield the optimal solution to simple scheduling problems. We may apply a rule like EDD to any scheduling problem involving due dates; we do not obtain the optimal solution in general, but we obtain a solution with a very limited computational effort. We just need to simulate a distributed real time scheduling process: each machine looks at its local queue (the *list* of waiting jobs), and, when it is idle, it picks up the most urgent job for processing. Urgency may be defined by any reasonable **priority rule**; there are tens of them proposed in the literature.

Simulating this real time scheduling process calls for a procedure to generate an active or a nondelay schedule. We will now describe a procedure able to generate an *active* schedule; it is a variation of a procedure, due to Giffler and Thompson, able to generate the set of *all* the active schedules. We assume for simplicity that all the release times are zero. The following notation is used:

- PS_t is a partial schedule with t operations already scheduled;
- S_t is the set of the schedulable operations at stage t; this set corresponds to a partial schedule PS_t;
- α_j is the (potential) earliest start time of an operation $j \in S_t$;
- ω_j is the (potential) earliest completion time of an operation $j \in S_t$.

Given a partial active schedule PS_t, α_j is the larger of the following quantities: the completion time of the direct predecessor of j (in the process plan of the corresponding job) and the latest completion time on the machine required by j. The earliest completion time is simply $\omega_j = \alpha_j + p_j$, where p_j is the processing time of j. The procedure is as follows:

> *Step 1:* t is set to 0 and PS_t is empty; S_t is initialized to the set of the first operations of all the jobs (i.e., the operations with no predecessors).

Step 2: find $\hat{\omega} = \min_{j \in S_t} \omega_j$; let $\hat{M}$ be the machine corresponding to this minimum, breaking ties arbitrarily.

Step 3: for each operation $j \in S_t$ requiring machine $\hat{M}$, such that $\alpha_j < \hat{\omega}$, calculate a priority index; add the operation with the lowest (or highest) priority index to the partial schedule, scheduling it as early as possible, to obtain PS_{t+1}.

Step 4: update S_{t+1} by removing the scheduled operation and adding its direct successor (if any); increment t.

Step 5: if S_t is empty go to *Step 2*; otherwise stop.

Note that in step 3 we make a choice based on a priority index (the lowest or the largest, depending on how priority is expressed); by generating a tree of all the possible partial schedules, we would generate all the active schedules. List scheduling may be considered a greedy strategy to explore this search tree; only one of the possible branches is explored.

This algorithm generates active schedules owing to the condition $\alpha_j < \hat{\omega}$ in step 3. Having determined $\hat{\omega}$ and $\hat{m}$, we must assign some processing to this machine before $\hat{\omega}$: if we left $\hat{m}$ idle until $\hat{\omega}$, the obtained schedule would not be active, since we could left-shift an operation into this idle period without delaying any other operation. The method can be easily modified to obtain *nondelay* schedules:

Step 1: t is set to 0 and PS_t is empty; S_t is initialized to the set of the first operations of all the jobs (i.e., the operations with no predecessors).

Step 2: find $\hat{\alpha} = \min_{j \in S_t} \alpha_j$; let $\hat{M}$ be the machine corresponding to this minimum, breaking ties arbitrarily.

Step 3: for each operation $j \in S_t$ requiring machine $\hat{M}$, such that $\alpha_j = \hat{\alpha}$, calculate a priority index; add the operation with the lowest (or highest) priority index to the partial schedule, scheduling it as early as possible, to obtain PS_{t+1}.

Step 4: update S_{t+1} by removing the scheduled operation and adding its direct successor (if any); increment t.

Step 5: if S_t is empty go to *Step 2*; otherwise stop.

The two methods yield different schedules in general; note that in practice they should be adapted to cope with nonzero release times and alternative machines per operation. The point now is to pick a suitable priority rule. We will not list all the rules proposed in the literature, but only the most significant ones. A major distinction can be drawn between *static* and *dynamic* rules. Static rules are characterized by a constant priority, whereas the priority computed according to dynamic rules changes in time. Among the most common static rules we mention the following.

- The SPT (Shortest Processing Time) rule: the operation with the smallest processing time p_i is selected.
- The WSPT (Weighted Shortest Processing Time) rule: the operation with the smallest p_i/w_i ratio is selected, where w_i is the weight associated with the job.
- The EDD (Earliest Due Date) rule: select the operation whose job has the tightest due date.
- The LWKR (Least Work Remaining) rule: select the operation associated with the job with the least work remaining (sum of the processing time of the remaining operations).
- The MWKR (Most Work Remaining) rule: select the operation associated with the job with the largest work remaining (sum of the processing time of the remaining operations).

Dynamic rules are particularly useful for problems involving due dates. A well known rule is SLACK: the highest priority is given to the operation with the minimum slack time, expressed as $d_i - rk_i - t$, where d_i is the due date of the corresponding job, rk_i is the remaining work, and t is the time at which the selection is made. Variations of the SLACK rule are the following:

- S/RMOP (Slack per remaining operation): similar to the slack rule, but we consider here the ratio
$$\frac{d_i - rk_i - t}{rp_i},$$
where rp_i is the remaining number of operations.
- S/RMWK (Slack per remaining work): similar to the slack rule, but we consider here the ratio
$$\frac{d_i - rk_i - t}{rk_i}.$$

There are other, more complex, dynamic rules for scheduling with due dates. Here we just mention the ATC (Apparent Tardiness Cost) rule, which was proposed in [201] for solving the $J//\sum_i w_i T_i$ problem. The expression of the priority of job i on machine j at time t is:

$$\frac{w_i}{p_{ij}} \exp\left(-\left[\frac{d_i - t - p_{ij} - \sum_{q=j+1}^{m_i}(W_{iq} + p_{iq})}{k\bar{p}}\right]^+\right),$$

where: w_i is the weight of job i and p_{ij} is its processing time on machine j, the notation $[x]^+$ is equivalent to $\max\{0, x\}$, W_{iq} is an estimate of the waiting time of job i on machine q, m_i is the number of machines to be visited by job i (note that by summing from $q = j+1$ to $q = m_i$ we consider

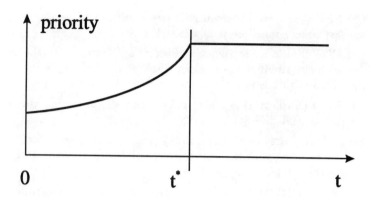

FIGURE 4.12
The priority index of the ATC rule as a function of time

the machines to be visited after j), k is a parameter to be selected, and $\bar{p}$ is the average processing time of the waiting jobs. It is interesting to interpret the rule. The expression

$$t^* = d_i - p_{ij} - \sum_{q=j+1}^{m_i} (W_{iq} + p_{iq})$$

can be interpreted as the maximum allowed start time of job i on machine j, obtained by subtracting the expected lead time of i from the order due date d_i. If $t \leq t^*$ we are on schedule, as far as job i is concerned; otherwise we are late. If we are on schedule, the term between $[\cdots]^+$ is positive, and the priority grows as an exponential function. If we are late, the priority is w_i/p_{ij}, which is simply the priority index of the WSPT rule. A plot of the priority index as a function of t is shown in Figure 4.12. Why should we switch to the WSPT rule if we are late? A justification is the following: if we are late on many jobs, we have

$$\sum_i w_i T_i = \sum_i w_i \max\{C_i - d_i, 0\} \approx \sum_i w_i C_i - \sum_i w_i d_i.$$

We see that the objective function is in this case the total weighted completion time, which, for a single machine, is solved by the WSPT rule. Now, we should estimate the waiting times W_{iq}; a suggested rule is to set $W_{iq} = bp_{iq}$, where b is a parameter to be selected. We see that the ATC rule is *parametrized*; its application requires the tuning of parameters k and b.

For any scheduling problem we may come up with a priority rule. If sequence-dependent setup times or costs are involved, a greedy rule is to select the next job as the one yielding the lowest time/cost. If we have a problem involving blocking due to buffer saturation, we may apply a Longest Processing Time rule (the opposite of SPT) if the downstream buffer if almost full (in order to allow the downstream machine to reduce the buffer level) and an SPT rule if the downstream buffer is almost empty (in order to avoid starving the downstream machine). Priority rules are a flexible approach that can be applied to problems with nonzero release times and with additional resources. They may also be used for real time production scheduling in uncertain environments or when controlling the Material Handling System is important. Their flexibility is the reason they are used in most commercial scheduling packages.

The disadvantage of priority rules is that their behavior is rather erratic. They may yield very bad solutions, whereas a random priority rule is not as bad as one could imagine compared with other rules. The insufficiency of priority rules for complex scheduling problems has been pointed out in [158, pp. 51-52]. This is due to their one-shot and greedy nature; this problem may be partially overcome by applying beam search principles. In general, better results are obtained by the application of a more specific and less greedy approach. This, however, requires developing special purpose software and it requires a larger computational effort.

4.8.2 Dual heuristics for scheduling problems

We have considered the development of dual heuristics for lot-sizing problems in Sections 3.2.3 and 3.3.2. The idea is to solve a sequence of relaxed problems with multipliers adjusted according to a subgradient procedure or to a special purpose scheme, and to derive a primal feasible solution from the (generally infeasible) solution of the relaxed problem.

The same approach can be applied to all the Lagrangian relaxation methods that we have considered in Section 4.7.4 for scheduling problems: in fact, some of them were actually developed just for this purpose, and not to obtain optimal solutions by a Branch and Bound algorithm. The key points are multipliers adjustment and the procedure to patch up the dual solution to get a primal feasible one. This last issue may be more or less complicated; it may even happen that the solution of the relaxed problem is always feasible, as in the following example.

Example 4.13
We outline here a dual heuristic for the $R//C_{max}$ problem, originally proposed in [194]. Consider again the model we have formulated in Section 4.6.5:

$$\min \ C_{max}$$

$$\text{s.t.} \sum_{i=1}^{N} p_{ij} x_{ij} \leq C_{\max} \quad \forall j \qquad (4.37)$$

$$\sum_{j=1}^{M} x_{ij} = 1 \quad \forall i \qquad (4.38)$$

$$x_{ij} \in \{0,1\} \quad \forall i,j, \qquad (4.39)$$

and dualize (4.37) with nonnegative multipliers μ_j. The relaxed problem is

$$w(\boldsymbol{\mu}) = \min \ \left(1 - \sum_{j=1}^{M} \mu_j\right) C_{\max} + \sum_{i=1}^{N} \sum_{j=1}^{M} \mu_j p_{ij} x_{ij}$$
$$\text{s.t.} \ (4.38), (4.39).$$

Note that we get a finite value of $w(\boldsymbol{\mu})$ if and only if

$$\sum_{j=1}^{M} \mu_j = 1.$$

In this case, by defining 'processing costs' $\tilde{c}_{ij} = \mu_j p_{ij}$, we obtain the relaxed problem

$$\min \ \sum_{i=1}^{N} \sum_{j=1}^{M} \tilde{c}_{ij} x_{ij}$$
$$\text{s.t.} \ (4.38), (4.39),$$

which can be decomposed by job, and is trivially solved by selecting, for each job i, the machine j with the minimum processing cost.

On the one hand, we obtain a lower bound; since the relaxed problem clearly enjoys the integrality property, this lower bound cannot be stronger than the lower bound obtained by continuous relaxation, but it is cheaper to compute (see Section C.4.5). Apart from this consideration, the most interesting feature of this relaxation is that any solution of the relaxed problem is feasible for the original one: we have dualized a constraint whose purpose is to link the objective function to the decision variables, but the 'physical' requirements are still satisfied (each job must be processed on one machine).

As to the multiplier adjustment, we should modify the subgradient method in order to cope with the restriction $\sum_j \mu_j = 1$. However, it is possible to exploit the peculiar structure of the problem to overcome the difficulty, and to obtain a dual ascent method. In [194] the following surrogate relaxation is suggested for this purpose:

$$\min\ C_{\max}$$
$$\text{s.t.}\ C_{\max} \geq \frac{\sum_i \sum_j \mu_j p_{ij} x_{ij}}{\sum_j \mu_j}$$

(4.38), (4.39).

We see that this surrogate relaxation is, for fixed multipliers, equivalent to the Lagrangian relaxation. We will not go into details, but the key consideration of the approach is that the search for the optimal value of μ can be limited to the nondifferentiability points of the dual function $w(\mu)$; such points correspond to multiplier values such that two machines are equivalent for a certain job, i.e., there are multiple solutions for the relaxed problem. At these points, it is possible to find an expression of the directional derivatives of the dual function: we know that directional derivatives are subgradients, and we may spot directional derivatives corresponding to ascent directions. The step length is selected in such a way as to end up at another nondifferentiability point of the dual function. ☐

The $F//\sum_i w_i C_i$ problem is another case in which it is easy to obtain a feasible solution. From the relaxed problem we obtain a set of sequences, one for each machine; putting them together we get a feasible overall schedule. This is not the case with the $J//\sum_i C_i$ problem, since, putting single-machine sequences together, we may end up with a cyclic precedence graph as shown in Figure 4.10.

Going into deeper detail about dual heuristics for scheduling problems is beyond the scope of an introductory textbook; nevertheless, we want to point out that the literature on this subject is growing and that it illustrates how good heuristics may be developed through mathematical modeling. For more information on specific applications, the reader is referred to the literature listed at the end of the chapter.

4.8.3 Scheduling by simulated annealing

Simulated annealing is a local search meta-heuristic (see Section C.8.4). The term "meta-heuristic" is due to the relative generality of the approach. In practice, the application of local search methods to a specific problem requires some customization: a particularly critical issue is the selection of the neighborhood structure, i.e., the set of rules by which candidate solutions are generated from the current one. Here we describe only the application of simulated annealing to the permutation $F//C_{\max}$ problem. In the next section we will apply tabu search both to $F//C_{\max}$ and $J//C_{\max}$: the remarks we make about the application of tabu search to $J//C_{\max}$ apply to the simulated annealing case too.

The first step in applying a local search method is the definition of a suitable neighborhood structure. In this case it is rather easy; the solution is simply represented by a permutation σ of the jobs, and all we need is to perturb this permutation. We may consider a neighborhood $\mathcal{N}$ generated by the following perturbation operators:

1. pairwise swaps of (not necessarily adjacent) jobs:

$$(J_{\sigma(1)}, \ldots, J_{\sigma(k)}, \ldots, J_{\sigma(l)}, \ldots, J_{\sigma(n)})$$
$$\Downarrow$$
$$(J_{\sigma(1)}, \ldots, J_{\sigma(l)}, \ldots, J_{\sigma(k)}, \ldots, J_{\sigma(n)})$$

2. shift and insert moves like the following:

$$(\ldots, J_{\sigma(l)}, \ldots, J_{\sigma(k-1)}, J_{\sigma(k)}, J_{\sigma(k+1)}, \ldots)$$
$$\Downarrow$$
$$(\ldots, J_{\sigma(k)}, J_{\sigma(l)}, \ldots, J_{\sigma(k-1)}, J_{\sigma(k+1)}, \ldots).$$

The implementation of the method requires specifying

- how the starting solution is obtained,
- a cooling schedule,
- and a termination rule.

A trivial, but quite easy to implement, algorithm is obtained by randomly selecting the initial permutation, by stopping the execution after a given number of iterations, and by setting the temperature T_k at iteration k according to the rule $T_k = \alpha T_{k-1}$, where α is a decrease parameter ranging in the interval $(0, 1)$. The effect of the decrease parameter α is shown in Figures 4.13 and 4.14 for a 20-job, 3-machine problem. The figures show a plot of the current and the currently optimal makespan for each iteration. In Figure 4.13 the initial temperature is very high ($T_1 = 300000$), but cooling is very fast ($\alpha = 0.99$). The solution obtained after 8000 iterations has a makespan of 600. Note that, after a few iterations, the noise contribution disappears and the plot descends monotonically. In Figure 4.14 the initial temperature is much lower ($T_1 = 15$), but cooling is slower ($\alpha = 0.9995$). We obtain a better makespan (587). Note that good cooling schedules do not decrease the temperature at each step; the temperature is kept constant for a certain number of steps to let the system reach a thermodynamic equilibrium. The application of simulated annealing took about one minute on a 386-based PC; we applied to the same problem instance a depth-first version of the Branch and Bound algorithm illustrated in Section 4.7.3. We were not able to find a better schedule in 30 hours (this was a particularly difficult case; larger problems can be solved in a matter of minutes). A fundamental feature of simulated annealing for flow shop problems is that changing the objective function requires a

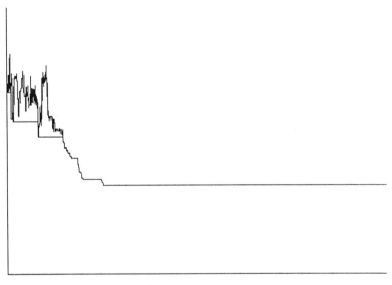

current optimal makespan = 600
stopped

FIGURE 4.13
Trace of a $F//C_{max}$ problem with fast cooling

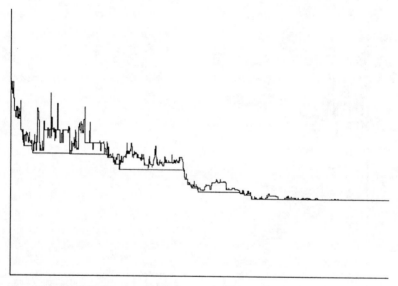

current optimal makespan = 587
stopped

FIGURE 4.14
Trace of a $F//C_{max}$ problem with slow cooling

minimal modification of the program. It must be observed that the application of simulated annealing to permutation problems is straightforward; for more complex problems the customization of the meta-heuristic may be more difficult, and the running times may grow considerably. For difficult combinatorial optimization problems running times of the order of some hours are reported in the literature; this may be acceptable for design problems, but probably not for scheduling problems. Recent research has shown that, in general, better results are obtained by the application of alternative methods such as tabu search.

4.8.4 Scheduling by tabu search

We outline the application of tabu search to two classical problems: the permutation $F//C_{\max}$ and $J//C_{\max}$. We will see that the first case is rather easy to deal with, whereas the second one requires some care.

Application to the Permutation $F//C_{\max}$ problem

To solve $F//C_{\max}$ by tabu search, we could use the same neighborhood structure used in the annealing case. Note, however, that tabu search requires fully exploring the neighborhood of the current solution; a rich neighborhood may be too expensive from the computational point of view. In general, one must try to restrict the neighborhood by some problem-dependent consideration. For illustrative purposes, we consider a (rather poor) neighborhood $\mathcal{N}$ generated by the following perturbation operators:

1. **pairwise exchange of** *adjacent* **jobs**:

$$(J_{\sigma(1)}, \ldots, J_{\sigma(k)}, J_{\sigma(k+1)}, \ldots, J_{\sigma(n)})$$
$$\Downarrow$$
$$(J_{\sigma(1)}, \ldots, J_{\sigma(k+1)}, J_{\sigma(k)}, \ldots, J_{\sigma(n)})$$

2. **left rotation of the current sequence**:

$$(J_{\sigma(1)}, J_{\sigma(2)}, \ldots, J_{\sigma(n-1)}, J_{\sigma(n)})$$
$$\Downarrow$$
$$(J_{\sigma(2)}, \ldots, J_{\sigma(n-1)}, J_{\sigma(n)}, J_{\sigma(1)})$$

3. **right rotation of the current sequence**:

$$(J_{\sigma(1)}, J_{\sigma(2)}, \ldots, J_{\sigma(n-1)}, J_{\sigma(n)})$$
$$\Downarrow$$
$$(J_{\sigma(n)}, J_{\sigma(1)}, J_{\sigma(2)}, \ldots, J_{\sigma(n-1)}).$$

The rotation operators are justified by the following observation: if the current first job in the sequence and the last one must be exchanged to obtain the optimal sequence, they must travel a long way to reach the correct position if only pairwise exchanges of adjacent jobs are allowed. Note that we are using such a small neighborhood (it consists of $n+1$ sequences) only for didactic purposes; a small neighborhood allows us to fully illustrate the execution of the algorithm step by step. More effective neighborhood structures are reported in the literature.

At each iteration, the best nontabu sequence in the neighborhood of the current sequence is chosen as the new current sequence. The reverse of the accepted perturbation is marked as tabu. In our case the tabu is characterized by the following attributes:

- a pair of jobs which cannot be exchanged;
- an initial (final) job which cannot be moved to the end (beginning) of the sequence by left (right) rotation.

Let us consider a small example involving four jobs and three machines. The processing times on each machine are the following:

Job	p_{i1}	p_{i2}	p_{i3}
J_1	1	8	4
J_2	2	4	5
J_3	6	2	8
J_4	3	9	2

This example is proposed in [79], where it is shown that the optimal makespan is 28.

Step 1. Let us start from the initial sequence

$$(J_4, J_3, J_2, J_1)$$

whose makespan is 31. Let us consider the whole set of neighboring sequences with their makespans:

$$(J_3, J_4, J_2, J_1) \to 34$$
$$(J_4, J_2, J_3, J_1) \to 33$$
$$(J_4, J_3, J_1, J_2) \to 31$$
$$(J_3, J_2, J_1, J_4) \to 31$$
$$(J_1, J_4, J_3, J_2) \to 33$$

The selected sequence is

$$(J_4, J_3, J_1, J_2)$$

whose makespan is 31. The tabu list is now

$$\{(\texttt{exch}, J_2, J_1)\}$$

Step 2. The neighborhood of the new sequence is:

$$(J_3, J_4, J_1, J_2) \to 35$$
$$(J_4, J_1, J_3, J_2) \to 37$$
$$(J_4, J_3, J_2, J_1) \to 31 \quad \text{(tabu)}$$
$$(J_3, J_1, J_2, J_4) \to 31$$
$$(J_2, J_4, J_3, J_1) \to 29$$

The selected sequence is

$$(J_2, J_4, J_3, J_1)$$

whose makespan is 29. The updated tabu list is

$$\{(\texttt{exch}, J_2, J_1), (\texttt{rotl}, J_2)\}$$

Step 3. Now we consider the sequences

$$(J_4, J_2, J_3, J_1) \to 33$$
$$(J_2, J_3, J_4, J_1) \to 32$$
$$(J_2, J_4, J_1, J_3) \to 35$$
$$(J_4, J_3, J_1, J_2) \to 31 \quad \text{(tabu)}$$
$$(J_1, J_2, J_4, J_3) \to 32$$

With a local improvement scheme, we would stop with the locally optimal makespan 29. The best sequence in this neighborhood is

$$(J_4, J_3, J_1, J_2)$$

which is excluded since it results from the application of a tabu perturbation. The best nontabu sequence is

$$(J_2, J_3, J_4, J_1)$$

whose makespan is 32 (we increase the makespan). The new tabu list is

$$\{(\texttt{exch}, J_2, J_1), (\texttt{rotl}, J_2), (\texttt{exch}, J_3, J_4)\}.$$

Step 4. Finally we have the set

$$(J_3, J_2, J_4, J_1) \to 33$$

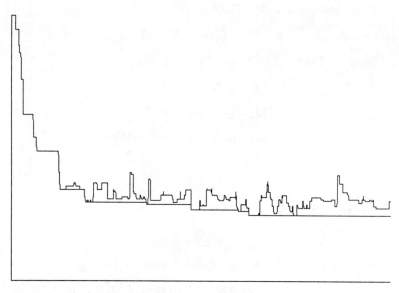

current optimal makespan = 587
stopped

FIGURE 4.15
Trace of a $F//C_{max}$ problem solved by tabu search

$$(J_2, J_4, J_3, J_1) \to 29 \quad \text{(tabu)}$$
$$(J_2, J_3, J_1, J_4) \to 29$$
$$(J_3, J_4, J_1, J_2) \to 35$$
$$(J_1, J_2, J_3, J_4) \to 28$$

and we end up with an optimal sequence

$$(J_1, J_2, J_3, J_4)$$

In Figure 4.15 a plot is shown of the current solution and the best solution found when using the tabu search algorithm for the 20-job, 3-machine problem previously solved with the annealing method. Note how the current makespan does not descend monotonically. The figure shows 2000 iterations, and the same solution is found as in the annealing case. We use fewer iterations than with simulated annealing, but each iteration is more costly; the running time for this example is practically identical to that required by simulated annealing (about one minute on a 386-based PC).

Heuristic scheduling methods

Application to the $J//C_{\max}$ problem

The application of a local search approach to a job shop problem requires more care than the flow shop case. There are two basic issues to deal with:

1. the size of the neighborhood: we may perturb the sequences on each machine using the flow shop neighborhood structure, but this results in a huge neighborhood to explore; furthermore a large number of moves are ineffective, in the sense that they do not change the makespan of the schedule;

2. unlike the flow shop case, we have no guarantee that perturbed solutions are feasible; we have seen that swapping disjunctive arcs may lead to cyclic orientations (see Figure 4.10).

When solving $J//C_{\max}$, both issues are solved at once by defining a neighborhood structure based on the critical path on the disjunctive graph. The critical path enjoys the following properties:

- given a schedule, it is necessary to reverse a disjunctive arc lying on the critical path in order to reduce the makespan; this observation restricts the neighborhood of the current schedule that must be explored;

- given an acyclic orientation, it is impossible to obtain a cyclical orientation by reversing disjunctive arcs lying on the critical path; only feasible solutions are obtained. To see this, consider the portion of a disjunctive graph shown in Figure 4.16, where a disjunctive arc (i, j) is reversed, obtaining an arc (j, i). Assume that (i, j) is on the critical path, which can be decomposed into a path $0 \to i$, the arc (i, j), and a path $j \to N$. If reversing (i, j) results in a cycle, there must be a path from $i \to j$, which becomes a cycle when (j, i) is added. However, if we consider the path formed by $0 \to i$, $i \to j$, and $j \to N$, this is clearly longer than the path formed by $0 \to i$, (i, j), and $j \to N$, which cannot be the critical path.

It is therefore natural to devise a neighborhood structure based on the reversal of disjunctive arcs lying on the critical path. A simple tabu attribute is obtained by forbidding the reversal of that arc for the next few iterations. In the literature, alternative neighborhood structures are described, but they are all based on the critical path.

4.8.5 LPT and Multifit algorithms for load balancing

In this section we describe two heuristic principles originally proposed for the $P//C_{\max}$ problem. Note that this parallel machine scheduling problem is equivalent to a load balancing problem if the objective is expressed

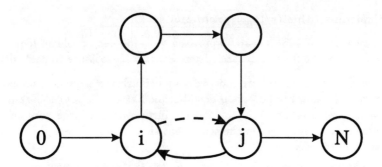

FIGURE 4.16
Reversing a critical disjunctive arc cannot result in a cyclic orientation

as the minimization of the maximum load. Both problems are basically assignment problems; no sequencing need be considered.

The Longest Processing Time (LPT) heuristic is simply a list scheduling method: first the data are sorted in such a way that $p_1 \geq p_2 \geq \ldots \geq p_n$, then jobs are assigned one at a time to the minimally loaded machine.

Example 4.14
Consider a $I3//C_{\max}$ problem involving eight jobs with the following processing times:

Job J_i	J_1	J_2	J_3	J_4	J_5	J_6	J_7	J_8
Processing time p_i	2	1	4	3	7	8	5	6

The application of the LPT heuristic would proceed as summarized in Table 4.7. For each step, we list the selected assignment and the load L_i on each machine. We can see the rationale behind the approach; by assigning the shorter jobs later, we avoid creating large unbalances during the last assignments. □

The LPT algorithm is an example of a heuristic method for which a worst-case bound is known. The bound, originally presented in [94], is

$$\frac{C_{\max}^{\text{LPT}}}{C_{\max}^*} = \frac{4}{3} - \frac{1}{3M},$$

step	assignment	L_1	L_2	L_3
1	J_6 on M_1	8	0	0
2	J_5 on M_2	8	7	0
3	J_8 on M_3	8	7	6
4	J_7 on M_3	8	7	11
5	J_3 on M_2	8	11	11
6	J_4 on M_1	11	11	11
7	J_1 on M_1	13	11	11
8	J_2 on M_2	13	12	11

TABLE 4.7
An application of the LPT rule

where M is the number of machines.

Another simple heuristic for load balancing is the Multifit method. The idea is to find a good solution by a binary search on the value of the objective function. Given a tentative makespan, we look for a feasible solution not exceeding that makespan, and, depending on the outcome, we increase or decrease the tentative value. The tentative value is the midpoint of an interval bracketed by a lower and an upper bound on the makespan. Finding a feasible solution for a given makespan value is actually a bin packing problem (see Section C.1). If the solution of the bin packing problem uses no more than M bins, it is a feasible loading. The bin packing problem is solved by the First Fit Decreasing (FFD) heuristic. First the jobs are sorted in such a way that $p_1 \geq p_2 \geq \ldots \geq p_n$; then they are assigned one at a time to the lowest indexed machine that can process the job without exceeding the makespan value. The loading problem is infeasible if we load a job on a bin with an index larger than M.

The Multifit algorithm can be outlined as follows:

Step 1. Compute initial lower and upper bounds C_{low} and C_{up}. To obtain such bounds, compute the ideal load $C^* = \sum_i p_i/M$ and set $C_{up} = \max\{p_1, 2C^*\}$, $C_{low} = \max\{p_1, C^*\}$.

Step 2. Set a tentative makespan value $C = \lceil \frac{1}{2}(C_{low} + C_{up}) \rceil$

Step 3. Apply FFD with capacity C.

Step 4. If a feasible solution has ben found, set $C_{up} = C$. Otherwise set $C_{low} = C + 1$.

Step 5. If the two bounds coincide, stop. Otherwise go to step 2.

The Multifit approach is more sophisticated than LPT, but there is no guarantee that it obtains better results. A combination of the two approaches has been proposed in [130]. Worst case bounds are known for Multifit; refinements of worst-case bounds for parallel machine scheduling are described, e.g., in [27, Chapter 5].

Both LPT and Multifit have interesting applications for load balancing and minimum makespan problems not involving sequencing. Here we illustrate the application of a Multifit-based method to parallel machine scheduling with a major/minor setup structure. Load balancing plays an important role in scheduling Flexible Manufacturing Systems as discussed in Section 4.8.6.

A Multifit approach to parallel machine scheduling with major/minor setup times

The Multifit logic can be applied to scheduling problems not involving sequencing. We describe here a heuristic for makespan minimization on identical parallel machines with a major/minor setup structure; the method has been proposed in [214].

We have a set of part types, which are partitioned in families; there is a *sequence-independent* major setup time to switch to a certain family, and a *sequence-independent* minor setup time to switch to a certain part type within the same family. It is possible to split a batch of parts on different machines, at the cost of additional setups. Since setup times are not sequence-dependent, the problem involves just the assignment of families and part types to machines and batch splitting, but no sequencing must be considered.

The heuristic is structured on three levels:

1. the first level consists of a Multifit loop setting a target makespan;
2. the second level deals with the assignment of families in order to obtain a feasible solution;
3. the third level deals with the assignment of part types in order to obtain a feasible solution.

The following notation is used: index i refers to part types, j to families, k to machines, and C is the makespan that should not be exceeded in finding a feasible solution. At each step of the algorithm, the remaining (i.e., not yet allocated) processing time of families and part types is considered; a family is unallocated if its remaining processing time is positive.

The second level is a loop that at each step selects a family j and a machine k and calls the third level in order to allocate to k as much as possible of the remaining processing time of family j. The unallocated family with the largest major setup time is selected; the rationale is to avoid repeating this setup on more than one machine if possible. The selection of the machine may follow two logics: let $\mathcal{K}$ be the set of machines to which family j can be *completely* allocated (without exceeding C). If $\mathcal{K}$ is not empty, it is advisable to select the machine with the largest load. In this way, more room is left on other machines for other families. If $\mathcal{K}$ is empty, it is advisable to select the machine with the smallest load; this machine is able to absorb the largest remaining processing time of family j.

The third level allocates family j to machine k one part type at a time. First, the algorithm finds the set $\mathcal{I}$ of the part types of family j that can be completely allocated to machine k; these part types are allocated, starting from the part type with the largest minor setup time (in this way we avoid carrying out this setup more than once). This first loop stops when $\mathcal{I}$ is empty; at this point, if family j has not been completely allocated yet, we try to partially allocate a part type to machine k, in order to saturate its capacity. In this case, the part type with the smallest minor setup is selected, since this setup time will be incurred at least once more on some other machine.

The procedure stops either when a feasible allocation has been found or when the second level is not able to allocate any processing time on any machine: note that this does not mean that a feasible allocation with makespan not larger than C cannot be computed.

4.8.6 Decomposition approaches to FJS and FMS scheduling

Classical job shop problems assume that there is a only one machine able to carry out a certain operation; only scheduling issues need to be addressed. On the contrary, the parallel machine case is characterized by a routing problem, but scheduling is relatively simple. In practice, joint routing and scheduling problems must be solved; this the case with flexible job shops (FJS) and Flexible Manufacturing Systems (FMS). The FJS problem is a job shop problem in which different machines may be selected for certain operations. The FMS case is also characterized by the need for loading tools on the machines; more often than not, FMS scheduling is more similar to parallel machine scheduling problems; the flexibility is such that one or two machine visits are needed to complete the job.

Solving a joint routing and scheduling problem is rather difficult. The most common solution approach is based on the decomposition of the two subproblems. First the routing problem is solved, then the scheduling one. For instance, in the FJS case, we first assign the operations to the machines, then we solve a classical job shop problem.

An obvious issue is the determination of the objective function to be used in the solution of the routing subproblem: we cannot use the same objective as in the scheduling problem. Usually, a load balancing objective is adopted; the idea is that a balanced workload allows obtaining a good quality schedule, by avoiding the occurrence of bottlenecks. The relationship between load balance and schedule quality should not be taken for granted; load balancing is a static objective, whereas bottlenecks may shift in time. Furthermore, workload balance, if expressed by the minimization of maximum load, may be thought as a surrogate of makespan, obtained by relaxing operations precedence constraints; this suggests a certain correlation between makespan and the minimization of the load on the critical machine. However, if the scheduling objective is related to due date satisfaction, the relationship is not that obvious. In such cases, apart from adopting priority rules, one may devise a more complex hierarchical decomposition approach. The simplest hierarchical approach, whereby the routing problem is solved once and for all before solving the scheduling problem, can be called *one-way*; if the routing problem is solved again after solving the scheduling problem, we get a *two-way* method. A two-way method is clearly more complex and time-consuming, but it may be much more effective to cope with dynamic bottlenecks.

Typical examples of one-way methods are those for FMS scheduling under the hypothesis that tools cannot be dynamically moved. By merging the models for the $R//C_{\max}$ problem and the part type selection problem in an FMS (see Section 4.6.5), we obtain a model to allocate tools and parts to the machines ensuring a workload-balanced solution. Workload balancing may be itself a difficult problem. Heuristic methods based on LPT and Multifit algorithms are described in [115]. Related approaches for FMS scheduling are described in [51, 69]. Scheduling an FMS with dynamic tool movements may be more difficult; one possible approach is to invert the hierarchy and to solve the scheduling problem first, and then cope with the tool dispatching (possibly in real time). An interesting mathematical model for the minimization of tool traffic was proposed in [99], but computational experience showed that, due to the dynamic nature of tool movement, there was no significant gain with respect to simple rules. FMS scheduling with dynamic tool movements is also described in [207].

An example of a two-way approach to FJS scheduling (with $C_{\max}$ objective) is described in [31]. The idea is to exploit two nested tabu search loops. For a given routing, the lower level solves a job shop scheduling problem by the tabu search method based on the critical path, as outlined in Section 4.8.4; the higher level exploits the information provided by the critical path in order to reallocate critical operations to alternative machines. The same principle has been adopted within a beam search algorithm in [48]. The approach is based on the repeated application of an SPT priority rule for the joint routing and scheduling problem; after obtaining the first SPT solution, the critical path is computed and the critical operations are considered. Nodes of a beam search tree are created by reallocating the operations on the critical path, one at a time and in chronological order. With each node, a partially frozen schedule is associated; the first part of the schedule is fixed according to the rerouting decisions, and the second part is left free. The nodes are evaluated by completing the partial schedules with the SPT rule.

4.8.7 The Shifting Bottleneck procedure for $J//C_{\max}$

Another heuristic principle that can be exploited to solve scheduling problems is that one should take particular care in scheduling bottleneck resources; their capacity is the tightest constraint in determining the quality of the resulting schedule. Bottleneck machines should be scheduled first, enforcing constraints on the scheduling of less critical machines. This principle is at the heart of software packages such as OPT, and it can be exploited in a variety of ways. Here we present a bottleneck-based heuristic originally proposed in [1] for the $J//C_{\max}$ problem. The approach, which is called the **shifting bottleneck** procedure, is based on the decomposition of $J//C_{\max}$ into a set of $1/r_i/L_{\max}$ subproblems, where release times and due dates have been suitably selected.

Consider a job shop scheduling problem represented as a disjunctive graph. We assume here, for simplicity, that each job visits a machine at most once. Let a certain machine M_b be the bottleneck; for now we do not have a precise way of spotting a bottleneck, but this will be clarified in the following. If M_b is the bottleneck, we may (in a rough-cut approximation of the problem) relax the disjunctive arcs related to the other machines, which are treated as infinite capacity resources; the resulting disjunctive graph is shown in Figure 4.17.

Scheduling the bottleneck is a single-machine scheduling problem. To see that this problem can be reduced to a $1/r_i/L_{\max}$ problem, consider the node O_{ib} representing the operation of job i on machine b, and let C_{ib} be its completion time. Consider also the critical path from the dummy start node to O_{ib} and let h_{ib} be its length, which is in this case simply the sum of the processing times of the operations of job i preceding O_{ib}. This length can be considered as the "head" of job i on machine b, and it can be regarded as the release time of i in the scheduling problem on the bottleneck machine. Similarly, let t_{ib} be the length of the critical path from node O_{ib} to the dummy final node, which is in this case simply the sum of the processing times of the operations of job i following O_{ib}; t_{ib} may be interpreted as the "tail" of job i, and it helps us in defining the due date of job i on machine b. In fact, we can reformulate the objective function $C_{\max}$ as

$$C_{\max} = \max_i \{C_i\} \approx \max_i \{C_{ib} + t_{ib}\}.$$

This expression depends on the fact that the operations following O_{ib} are performed on infinite-capacity machines. If we subtract any suitably large

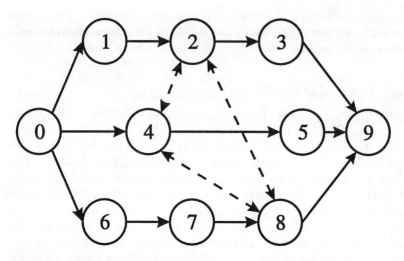

FIGURE 4.17
A relaxed disjunctive graph for a job shop problem with a single bottleneck machine

constant K from this objective function, we obtain an equivalent problem:

$$\max_i\{C_{ib} + t_{ib}\} - K = \max_i\{C_{ib} + (K - t_{ib})\} = \max_i\{C_{ib} - d_{ib}\},$$

where we have introduced a due date $d_{ib} = K - t_{ib}$ for job i on machine b. Note that the longer the tail, the tighter the due date. The objective function is clearly a maximum lateness. Summarizing, we have reduced the minimum makespan problem to a minimum max lateness problem on the bottleneck machine, with release times and due dates given by the heads and the tails of each job. This problem can be efficiently solved by the Branch and Bound procedure described in Section 4.7.2.

Until now, we assumed that a specific machine b was the bottleneck. Defining what we mean by "bottleneck" is not so easy. In flow shop problems with a repetitive mix, the bottleneck machine is clearly the machine with the largest processing time; in a job shop problem, the bottleneck is more of a *dynamic* concept, since it may shift in time. A suitable way to spot a bottleneck is to solve the $1/r_i/L_{\max}$ problem for all the machines, where each machine has its set of heads and tails (computed assuming that the other machines have infinite capacity). The bottleneck machine will be the one with the largest value of $L^*_{\max}$ (obviously, the constant K must be the same for all the machines). After solving the single-machine problem on all machines, we know which machine is the bottleneck, and we have a schedule for it. We orient the disjunctive arcs for the bottleneck according to this schedule, and we solve again the single-machine problems for the remaining $M-1$ machines. Now, the heads and the tails are not simply the sum of processing times, since in computing the critical paths we must consider the already oriented disjunctive arcs. The schedule on the bottleneck enforces some constraints on the schedule of the other machines. Among the remaining $M-1$ machines we select the bottleneck, we freeze its schedule, and the process is repeated until all the machines are eventually scheduled. Note that, due to the additional information provided by the first schedule, the new bottleneck need not be the second machine in the first ranking.

A possible difficulty may occur if, by putting all the single machine schedules together, we end up with a cyclic precedence graph. In [1] a way to overcome this difficulty is described, along with possible improvements of the algorithm.

4.9 Discussion

The size of this chapter is an indicator of the amount of work that has been devoted over the years to scheduling problems. Yet, the practical relevance of this mass of work is by no means accepted; [64, 65] are only a few example of papers showing some criticism on this point.

Artificial Intelligence and knowledge-based methods have been proposed as a practical tool to cope with scheduling problems (see [49] for a survey). Indeed, there is room for the application of such techniques within production management. Consider for instance a parallel machines problem with sequence-dependent setup times; to solve it by whatever method, we need a matrix of setup times; but if there are some hundreds of items, creating and maintaining the *data* for the scheduling algorithm is difficult (thousands of numbers must be calculated and entered into the system). In such a situation, a rule-based system to compute the setup times is useful. However, despite much effort and enthusiasm, Artificial Intelligence approaches do not seem to be a practical solution method [141]; they may be inefficient, and often they boil down to simple heuristics which are called knowledge-based just for the fun of it.

In fact, the difficulty with scheduling problems does not actually lie in the *solution* algorithm; as we have shown, there is an array of heuristic principles that can be exploited to devise better solution methods than list scheduling (the solution approach of most scheduling packages). The real issue is whether an *off line* and *detailed* scheduling approach is suitable. There are cases in which the complexity and the level of uncertainty make a detailed schedule virtually useless; a nice schedule is rather vulnerable to machine failures or uncertainty in the processing time. Consider, for instance, the problem of electronic chips manufacturing. Two basic stages are involved: first silicon slices (containing several integrated circuits) are processed, then they are cut, and the integrated circuits are assembled on a chip. However, the processing of silicon slices is very complex, and it requires tens of operations carried out on a few machines. The problem is characterized by reentrant flows and by a high level of uncertainty. There seems to be little point in specifying a detailed schedule for both silicon processing and chip assembly. A better approach is to associate a standard lead time with silicon processing and to coordinate production on the two stages at an aggregate level [20, Chapter 8]. Then, it is the task of local production control of silicon processing not to exceed the standard lead time, by using real time priority rules. It may even be the case that, by proper reorganization of the layout of the manufacturing system (e.g. by Group Technology) and by setup time reduction, the importance of scheduling is considerably reduced. By this techniques, WIP may be decreased, and therefore there is less room (and need) for sequencing decisions.

However, this does not imply that scheduling problems cannot be effectively addressed in other cases. Furthermore, it does not imply that mathematical modeling cannot be useful to this purpose; successful experience with real life problems tackled by dual heuristics is reported in [134]. Even when detailed scheduling approaches are unsuitable, mathematical models based on continuous flows may be adopted (see Sections 3.4.2 and 6.2).

As a final remark, it must be pointed out that any scheduling algorithm should *support* the decision making of the user; a good scheduling software must allow the human scheduler to modify the solution to accomplish objectives which may be difficult to state formally (both in terms of a mathematical model and of a knowledge base). The so-called **leitstand** approach is aimed just at this purpose; scheduling packages with *interactive* Gantt charts are now commercially available [2, 167]; there is no reason why they cannot be integrated with sophisticated heuristic methods.

S4.1 Dynamic Programming for scheduling a batch processor

In Section 4.4 we have considered the application of Dynamic Programming (DP) to machine scheduling problem, and we have seen that in general it turns out to be rather impractical. However, this does not imply that DP should be dismissed. In this supplement we tackle a nonstandard scheduling problem: we consider a batch processor, i.e., a machine able to process a batch of parts together, with the objective of minimizing maximum tardiness. Such a problem is typical of burn-in operations in the electronic industry. The approach presented here is due to [131].

The aims of this supplement are:

- to show how DP can yield polynomial time algorithms in particular, but interesting, cases;
- to show how the applicability of DP is not restricted to minsum objective functions: in fact, DP is a *decomposition* principle that can be applied to minmax objective functions;
- to show how an optimization problem can be transformed into a set of feasibility problems: this same idea has been adopted in Section 3.4.2, to obtain a polynomial algorithm for the minimum $L_{\max}$ problem on a flexible flow Line modeled by continuous flows, and in Section 4.8.5 within the Multifit algorithm.

The problem we consider here can be coded as $1/r_i, p_i = p, B/T_{\max}$: we want to minimize $T_{\max}$ on a single batch processor (hence the B in the code), with nonzero release times and a common processing time p for all the jobs. The last assumption may be justified or not, depending on type of operation carried out by the batch processor. A further assumption we make is that the problem is characterized by *agreeable* due dates and release times: this means that $r_i \leq r_j$ implies $d_i \leq d_j$, and it is a reasonable assumption if the input data of the scheduling problem are derived by an upper level production management module. Actually, it is often the case that a standard lead time is associated with batch processor from the point of view of production management. In such a case the difference $d_i - r_i$ is equal for all the jobs. Let B denote the capacity of the machine, i.e., the maximum number of jobs that can be processed together.

Under the above hypotheses, an important property can be shown. If a solution with $T_{\max} = 0$ exists, then there is a solution with $T_{\max} = 0$ which is, in some sense, in EDD order. More precisely, if we consider a pair of batches P and Q such that P is scheduled before Q, then there are no two pair of jobs i and j such that $i \in P, j \in Q$ and $d_i > d_j$. Given this property, we can index the jobs in ascending order of their due dates. The problem actually involves to determining a *consecutive partition* of the jobs, i.e., determining where each batch starts and finishes in the EDD-ordered sequence of jobs. Note that batching jobs in such a way to obtain full capacity batches may result in suboptimal schedules; this requires waiting until the release time of the last job in each batch, with a corresponding delay for the first job in the batch.

The problem can be tackled by binary search on the possible values of $T_{\max}$. Given an upper bound $\hat{T}_{\max}$, we try to find a feasible schedule with a maximum tardiness given by $\lceil \hat{T}_{\max}/2 \rceil$, and so on. The idea is that due dates are relaxed by a variable quantity and are transformed into deadlines;

the feasibility problem consists in finding a solution meeting the deadlines. If we are able to solve the feasibility problem in polynomial time, then a polynomial algorithm for the overall problem is obtained.

The following DP recursion finds a feasible schedule with minimum makespan if one exists. The value function is $f(j)$, which denotes the minimum makespan for scheduling jobs $1, \ldots, j$ if a feasible schedule exists for them, $+\infty$ otherwise. The initial conditions are

$$f(0) = 0, \qquad f(j) = \infty \text{ for } j < 0,$$

and the functional equation is

$$f(j) = \min_{\max\{1, j-B+1\} \leq i \leq j} f_i(j), \qquad (4.40)$$

where

$$f_i(j) = \begin{cases} \max\{f(i-1), r_j\} + p & \text{if } \max\{f(i-1), r_j\} + p \leq d_i \\ \infty & \text{otherwise} \end{cases} \qquad (4.41)$$

denotes the completion time of jobs $1, \ldots, j$, where the last batch consists of jobs i through j. The minimization in (4.40) is carried out over values of i ranging in the interval $[\max\{1, j-B+1\}, j]$ because at most B jobs are processed in a batch. To understand (4.41), note that the batch consisting of jobs i through j can be started either at the end of the previous batch, which is completed at time $f(i-1)$, or at the release time r_j of job j (the last job to arrive for processing). The schedule is feasible if the start time of the batch plus the processing time p is compatible with the due date of the most urgent job of the batch, i.e., d_i. This DP recursion has a complexity $O(nB)$ where n is the number of jobs; in fact, there are n states and each one is evaluated in $O(B)$ operations. Then, DP yields a polynomial algorithm for $1/r_i, p_i = p, B/T_{\max}$ with agreeable due dates and release times.

S4.2 Periodic scheduling problems

In this supplement we present a mathematical model for the periodic job shop problem originally proposed in [174]; the reader is referred to [168] for a thorough treatment of periodic scheduling problems. The model is based on the assignment of potentials to nodes of a disjunctive graph, and it shows the power of the modeling framework introduced in Section 4.6.2. The reader should compare this DEDS model with the discrete-time periodic model presented in Section 3.5.1. Writing a discrete-time model for periodic lot-sizing is almost trivial; a DEDS periodic model is not so easy to grasp, but it does not suffer from the modeling errors associated with time-discretization.

In a periodic scheduling problem each machine repeats the same operation after a constant time interval; let T denote this period. The problem makes sense within a repetitive production environment characterized by a production mix. The production mix can be translated into a **Minimal Part Set** (MPS), which is the number of parts fed into the system during each period. Due to the periodic character of the problem, we must distinguish the following two concepts: each element of the MPS is called a job; the *physical* piece entering the system is called a part. Each job has a set of operations, related by technological precedence constraints. The problem can be represented by an extension of the disjunctive graph: conjunctive and disjunctive arcs have the usual interpretation, but there is a conjunctive arc from the last operation of each job to the first one; here there is no need to introduce dummy start and finish nodes. We have therefore $|M|$ disjoint cliques (of nodes connected by disjunctive arcs) and $|J|$ disjoint circuits (of conjunctive arcs), where M is the set of available machines and J is the set of jobs. Let j be the job index; the circuit corresponding to job j is denoted by C_j. We will refer to operations by the subscripts $a, b \in N$, where N is the set of operation nodes in the graph.

It can be shown that the WIP of any job j is constant over time and it is equal to the number of periods n_j each part of job j spends within the system. This holds under the assumption that a part leaves the system exactly when a new part of the same job enters the system; note that the flow time of a part is $n_j T$.

When dealing with a periodic problem, the objective functions based on job completion times make no sense; the two typical performance measures of interest are the period and the WIP level. Minimizing the period means maximizing the throughput; note that throughput and WIP are negatively correlated. It can be shown that minimizing the period is a trivial problem; the minimum period is the load on the critical machine

$$T^* = \max_{m \in M} \sum_{a | m_a = m} p_a,$$

where m_a is the machine on which operation a must be executed and p_a is its processing time. This solution is characterized by a high WIP level, which is needed to keep the critical machine busy. The minimization of WIP is also trivial; it can be shown that the minimum WIP is given by $|J|$. The minimum WIP level is associated with a large period and low throughput. Applying the constraints method for multiobjective problems, a suitable compromise between productivity and WIP level can be found by minimizing the period under a constraint on the WIP. Unfortunately this problem is NP-hard [174]; we describe here a mathematical programming model for its solution. Let s_a denote the start time of operation a; due to periodicity, in the model we must require

$$0 \leq s_a < T. \tag{4.42}$$

The operations are started on the parts at instants $s_a + zT$, $z \in \mathbf{Z}$.

We must express the technological constraints (corresponding to conjunctive arcs), the capacity constraints (corresponding to disjunctive arcs), and the limit on the WIP level in terms of the operation start times within the periodic framework. This is not trivial to accomplish. Consider, for instance, a precedence constraint expressed by a conjunctive arc $(a, b) \in \mathcal{C}_j$; in a nonperiodic framework we would simply write $s_a + p_a - s_b \leq 0$. Here this condition would be wrong, since the start times are to be intended modulo-T; we might well have $s_b < s_a$. To overcome this difficulty, let us introduce the quantity v_{ab}, which is the number of time instants zT, $z \in \mathbf{Z}$, occurring within the time interval from the beginning of a and the beginning of b (on the *same* part); the length of this time interval is then expressed by $s_b - s_a + v_{ab}T$. The quantity v_{ab} is linked to the start times by the following relation:

$$v_{ab} = \left\lceil \frac{s_a + p_a - s_b}{T} \right\rceil. \tag{4.43}$$

Note that (4.42) and $p_a \leq T$ imply $v_{ab} \in \{0, 1, 2\}$. Eq. (4.43) can be rewritten as

$$s_b - s_a - p_a + v_{ab}T \geq 0; \tag{4.44}$$

this constraint can be interpreted as a nonnegativity condition on the waiting time of a part after completing a and before starting b.

The quantity v_{ab} is also useful to express the WIP level n_j for each job; we have seen that the flow time of a part of job j is $n_j T$, but it may also be expressed by

$$\sum_{(a,b) \in \mathcal{C}_j} v_{ab} T;$$

hence, we have

$$n_j = \sum_{(a,b) \in \mathcal{C}_j} v_{ab}. \tag{4.45}$$

Now we have to model capacity constraints: operations a and b must not overlap if $m_a = m_b$. Note that, in the periodic framework, we must require that the execution of a must not overlap with the both the previous and the following occurrence of b. It turns out that this can be accomplished by the following constraint:

$$s_a \leq s_b - p_a + u_{ab}T, \tag{4.46}$$

where $u_{ab} \in \{0,1\}$ for each disjunctive arc (a,b), and $u_{ab}+u_{ba} = 1$. Suppose $u_{ab} = 0$, $u_{ba} = 1$; writing the above constraint for operations a and b, we get

$$s_a + p_a \leq s_b$$
$$(s_b - T) + p_b \leq s_a,$$

which is exactly what is required. There is a strong similarity between (4.44) and (4.46); they can be written in a uniform way as

$$p_a \leq s_b - s_a + z_{ab}T,$$

where $z_{ab} \in \{0,1\}$ for disjunctive arcs, and $z_{ab} \in \{0,1,2\}$ for conjunctive arcs.

Putting it all together, we may write the model for the minimization of the period under WIP constraints:

$$\min\ T$$
$$\begin{aligned}
\text{s.t.}\ & p_a \leq s_b - s_a + z_{ab}T & \forall (a,b) \in \mathcal{A} \\
& \sum_{(a,b) \in \mathcal{C}_j} z(a,b) \leq \hat{n}_j & \forall j \in J \\
& 0 \leq s_a < T & \forall a \in N \\
& z_{ab} + z_{ba} = 1 & \text{for all disjunctive arcs} \\
& z_{ab} \in \{0,1\} & \text{for all disjunctive arcs} \\
& z_{ab} \in \{0,1,2\} & \text{for all conjunctive arcs},
\end{aligned}$$
(4.47)

where $\mathcal{A}$ is the set of arcs of the disjunctive graph and $\hat{n}_j$ is the upper bound on the WIP of job j.

S4.3 A multiobjective approach to machine loading

In this chapter we present a variety of scheduling models, all aimed at solving a *single* objective problems. Often, scheduling problems require finding a suitable compromise between conflicting objectives, and should be tackled as multiobjective problems. We have seen in Supplement S2.2 how a vector optimization problem can be solved by transforming it into a set of scalar problems. Here we show how these ideas can be applied to a bicriterion machine loading problem, involving load balancing and cost objectives.

We have seen in Section 4.8.6 that load balancing is a widely used objective in decomposition approaches to FJS/FMS scheduling. First, a routing subproblem is solved by assigning operations to the machines with the objective of load balancing; then, a classical scheduling problem is solved. Here we consider machine loading problems where load balancing is traded off against cost. The machine loading problem we consider is actually a process plan selection problem; given a set of alternative process plans for a job, we must select one taking into account their different costs while insuring a workload-balanced solution. In the following, the requirement of balancing the load is expressed as the minimization of the load on the critical machine (i.e., the machine with the maximum load).

We consider here two cases: the case of single machine visits and the case of multiple machine visits. The first problem is tackled by the constraints method and the second one by the scalarization method.

Single machine visits

We have a set of jobs (indexed by $i \in \mathcal{I}$) and a set of machines (indexed by $m \in \mathcal{M}$). Since each job requires one machine visit, the problem is essentially a bicriterion parallel machine scheduling problem. For each (job, machine) pair we have a processing time p_{im} and a cost c_{im}. If we introduce binary decision variables x_{im} set to 1 if job i is loaded on machine m, 0 otherwise, we obtain the following vector problem:

$$\text{``min''} \begin{bmatrix} L_{\max} \\ C \end{bmatrix}$$

$$\text{s.t.} \quad C = \sum_{i \in \mathcal{I}} \sum_{m \in \mathcal{M}} c_{im} x_{im}$$

$$\sum_{i \in \mathcal{I}} p_{im} x_{im} \leq L_{\max} \quad \forall m \in \mathcal{M} \quad (4.48)$$

$$\sum_{m \in \mathcal{M}} x_{im} = 1 \quad \forall i \in \mathcal{I}$$

$$x_{im} \in \{0, 1\} \quad \forall i \in \mathcal{I}, \forall m \in \mathcal{M},$$

where $L_{\max}$ is the maximum load and C is the cost; eq. (4.48) selects the maximum load over the machines. The problem can be tackled by the constraints method: we may optimize the cost with varying constraints on the maximum load. If we set a limit value $\bar{L}$ on the maximum load, we obtain the following scalar problem:

$$\min \sum_{i \in \mathcal{I}} \sum_{m \in \mathcal{M}} c_{im} x_{im}$$

$$\text{s.t.} \sum_{i \in \mathcal{I}} p_{im} x_{im} \leq \bar{L} \quad \forall m \in \mathcal{M}$$

$$\sum_{m \in \mathcal{M}} x_{im} = 1 \quad \forall i \in \mathcal{I}$$

$$x_{im} \in \{0, 1\} \quad \forall i \in \mathcal{I}, \forall m \in \mathcal{M}.$$

This problem is easily seen to be an instance of the Generalized Assignment Problem (GAP; see Section C.1). The GAP is an NP-hard problem for which good heuristics are known. A difficulty with the constraints approach is that when the upper bound on the machine load gets tight, there is no guarantee that a feasible solution is found by the heuristic algorithm for solving GAP, even if one exists. This is not the case when using a scalarization approach, as in the next section.

Multiple machine visits

In this case, the process plan requires visiting different machines. The process plan flexibility may be expressed by creating completely separate routings; the whole process plan is considered as one "macro-operation" with alternative options. We can extend this notion to a more modular representation, by thinking of the process plan as a sequence of macro-operations; for each macro-operation, a set of options is given, consisting in turn of a sequence of "micro-operations". This is not always the best representation of flexible process plans; see [32] for a discussion about this kind of representation and its limitations.

Here we relax the precedence constraints, and we are concerned only with the problem of selecting one option for each macro-operation (in the following, since we are not concerned with detailed scheduling, we refer to macro-operations simply as operations). For each job i a set $\mathcal{J}_i$ of operations (indexed by $j \in \mathcal{J} = \bigcup_{i \in \mathcal{I}} \mathcal{J}_i$) is given, and for each operation a set of options (indexed by $k \in \mathcal{O}_j$) is given. For each option k of operation j, we know the load p_{jkm} on machine m and the cost c_{jk}. By introducing a set of binary decision variables x_{jk} set to 1, if option k is selected for operation j, we obtain the following model:

$$\text{``min''} \begin{bmatrix} L_{\max} \\ C \end{bmatrix}$$

s.t. $C = \sum_{j \in \mathcal{J}} \sum_{k \in \mathcal{O}_j} c_{jk} x_{jk}$

$$\sum_{j \in \mathcal{J}} \sum_{k \in \mathcal{O}_j} p_{jkm} x_{jk} \leq L_{\max} \qquad \forall m \in \mathcal{M}$$

$$\sum_{k \in \mathcal{O}_j} x_{jk} = 1 \qquad \forall j \in \mathcal{J} \tag{4.49}$$

$$x_{jk} \in \{0, 1\} \qquad \forall j \in \mathcal{J}, \forall k \in \mathcal{O}_j. \tag{4.50}$$

The interpretation of the constraints is practically identical to the single machine visit case.

We tackle the vector problem by scalarization, based on a modified Tchebycheff norm. We have to solve the scalarized problem (P_1):

$$(P_1) \quad \min\{\max[\lambda(L_{\max} - L_{\max}^*), (1-\lambda)(C - C^*)]\} \qquad 0 < \lambda < 1, \tag{4.51}$$

subject to (4.49) and (4.50), where $L_{\max}^*$ is the optimal load and C^* is the optimal cost. Problem (P_1) can be reformulated by noting that

$$L_{\max} = \max_{m \in \mathcal{M}} L_m = \max_{m \in \mathcal{M}} \sum_{j \in \mathcal{J}} \sum_{k \in \mathcal{O}_j} p_{jkm} x_{jk},$$

where L_m is the load on machine m; then objective function (4.51) can be rewritten as follows:

$$\min\{\max[\lambda(\max_{m \in \mathcal{M}} L_m - L_{\max}^*), (1 - \lambda)(C - C^*)]\}$$
$$= \min\{\max[\lambda(L_1 - L_{\max}^*), \ldots, \lambda(L_M - L_{\max}^*), (1 - \lambda)(C - C^*)]\}.$$

Therefore, we may reformulate (P_1) as:

min z

s.t. $z \geq \lambda \left(\sum_{j \in \mathcal{J}} \sum_{k \in \mathcal{O}_j} p_{jkm} x_{jk} - L_{\max}^* \right) \qquad \forall m \in \mathcal{M} \tag{4.52}$

$$z \geq (1 - \lambda) \left(\sum_j \sum_k c_{jk} x_{jk} - C^* \right) \tag{4.53}$$

with the addition of constraints (4.49) and (4.50). Recall that a convex combination approach of the two objectives would not yield all the efficient solutions for our problem. These theoretical considerations may be of limited value if the scalarized problem is tackled by a heuristic approach (as is the case here, since it is an NP-hard problem). However, a Tchebycheff norm approach lends itself to the development of interesting heuristics. The most interesting feature of the last formulation is that the cost can be considered as just another 'machine'. If we introduce machine loads $\hat{p}_{jk0} = (1-\lambda)c_{jk}$, where 'machine' 0 corresponds to the cost, and $\hat{p}_{jkm} = \lambda p_{jkm}$, we can reformulate problem (P_1) as problem (P_2):

(P_2) min z

s.t. $z \geq \left(\sum_{j \in \mathcal{J}} \sum_{k \in \mathcal{O}_j} \hat{p}_{jkm} x_{jk} - B_m \right)$ $\forall m \in \mathcal{M}'$ (4.54)

$\sum_{k \in \mathcal{O}_j} x_{jk} = 1$ $\forall j \in \mathcal{J}$ (4.55)

$x_{jk} \in \{0, 1\}$ $\forall j \in \mathcal{J}, \forall k \in \mathcal{O}_j,$ (4.56)

where $\mathcal{M}' = \{0\} \bigcup \mathcal{M}$, $B_m = \lambda L^*_{\max}$ for $m \neq 0$ and $B_0 = (1-\lambda)C^*$. This formulation is particularly interesting since it is very similar to a formulation of a minimum makespan scheduling problem on parallel machines with 'bonus' capacity B_m for each machine. Indeed, the dual heuristics described in Example 4.13 for $R//C_{\max}$ can be adapted to this problem [32].

For further reading

There is a huge amount of literature on scheduling problems, that cannot be catalogued here. In the body of this chapter we have already given some important references; here we list a few more that the reader should consider just as starting points for further bibliographical research.

Notable textbooks on classical machine scheduling theory are [11] and [79], where the reader may find some theorem proofs that we have omitted. The theoretically inclined reader may find a recent overview in [128]. Another recent book where recent developments in machine scheduling theory are presented is [27]. Plenty of heuristics for solving practical scheduling problems are described in [147], which is a book completely devoted to this topic. In these general references, the reader may find further references to original works of "historical" interest; here we just mention [102], where the DP approach to single-machine scheduling was proposed, and [13], where the disjunctive graph was introduced as a useful representation of job shop scheduling problems. Further tutorial references are [63] and [195], which are particularly oriented to the application of Lagrangian relaxation to machine scheduling.

A survey of mathematical programming models for scheduling problems can be found in [28]. Some work on strong formulations has been carried out: for instance, see [70] for a strong DEDS model, and [173] for a strong model relying on time-indexed variables.

In this chapter we have adopted an activity-on-node representation of scheduling problems; the reader interested in learning more on project management and activity-on-arc representation is referred to [71, Chapter 11] or [67, Chapter 6]).

As to Branch and Bound methods for scheduling, we have completely neglected methods for the $J//C_{\max}$ problem; an important reference is [42]. We have presented a relatively old algorithm for $1/r_i/L_{\max}$; further developments can be found in [41]. Bounding methods for the flow shop problem are described in [12, 123].

List scheduling, despite its lack of academic appeal, is an important topic. As an original reference, the reader may consult [11, Chapter 7]. Recent and useful references are [114], where dispatching rules for batches of identical jobs with alternative routings are discussed, and [146], where dispatching rules are applied to FMS scheduling.

As to more sophisticated heuristics, several examples of the application of beam search are described in [147]; see also [152]. A tutorial survey on local search methods for machine scheduling is [30]; a more specific survey on scheduling by tabu search can be found in [14]. Simulated annealing is applied to flow shop scheduling in [153], and to job shop scheduling in [198]. Tabu search is applied to flow shop scheduling in [209], and to job shop scheduling in [57]. An example of how the idea of the basic idea of the shifting bottleneck procedure can be applied to complex problems can be found in [190].

In this chapter, we have basically dealt with relatively standard scheduling problems; here we list a few interesting references for nonstandard problems. Scheduling with lot streaming (i.e., with transfer batches) is discussed in [84, 188]. In [215] a tabu search approach is described for a single-machine problem with sequence-dependent setups; an interesting feature of this approach is that some (low-priority) jobs are scheduled only if possible, which is a situation encountered in practice. Scheduling with nonregular measures is dealt with in [124, 160, 175]. Scheduling with batch processors is discussed in [85]; heuristics for this problem are discussed in [147, Chapter 11].

Finally, a survey on multiobjective approaches to machine scheduling can be found in [149]. Most literature on this topic deals with single-machine scheduling; multiobjective flow shop scheduling is described in [55].

5
Evaluative models

In the preceding chapters we have considered methods for solving *generative* models of the type $\min_{\mathbf{x} \in S} f(\mathbf{x})$. Unfortunately, it is not always easy or even possible to formulate an explicit link between the decision variables $\mathbf{x}$ and the objective function f. Among the complicating factors that may make a generative modeling approach unsuitable we mention:

- the occurrence of unpredictable events, such as machine failures or issues related to quality inspection and rework;
- uncertainty in the processing times;
- dynamic arrival of jobs;
- need to schedule considering the interactions between multiple resources (operators, fixtures, tools, and the Material Handling System);
- blocking due to buffer saturation;
- routing flexibility.

In such cases, we need some way to *evaluate* a performance measure $f(\mathbf{x})$, *given* a decision policy $\mathbf{x}$. In this chapter we deal with such **evaluative** models; we can exploit an evaluative model within an optimization framework, but we defer this topic to Chapter 6. Evaluative models are usually associated with manufacturing system *design* rather than management; nevertheless, we should note that, in a dynamically evolving system, there is a strict interplay between these two tasks.

Evaluative models may be classified as **experimental** or **analytical**. Analytical models are based on more or less approximate mathematical representations of systems. The analytical models we consider in this chapter are based on **queueing theory**; they require some familiarity with stochastic modeling, to which Appendix E is devoted. Experimental models are basically simulation models aimed at capturing the working of a system within a computer program. A treatment of simulation models is largely beyond the scope of this book, because it would involve a working knowledge of programming languages and statistics. Here we limit the treatment to a brief overview of **Petri nets**. Our choice is justified, on the one hand, by the fact that understanding the issues behind the representation of a DEDS by a Petri net is a good starting point to understand the issues of building a simulation model relying on a special purpose simulation language; on the other hand, Petri nets enjoy some interesting mathematical properties, which can be exploited to analyze the qualitative properties of a manufacturing system (e.g., guaranteeing that an automated system is deadlock-free). All the models dealt with here are DEDS models. However, it should be noted that sometimes discrete-time systems are simulated, for instance, to evaluate inventory control policies in an uncertain environment.

We start in Section 5.1 with an introduction to queueing models; after introducing some basic terminology, we present the important result known as Little's law (Section 5.1.1). Then, we show how, under certain hypotheses, queueing models result in Markov chain models (Section 5.1.2); in general, the analysis of queueing systems is quite complicated, but useful approximations can be derived (Section 5.1.3). Then we deal with open and closed queueing networks in Section 5.2. Their analysis is rather simple under hypotheses insuring a **product form** solution; in Supplement S5.1 we give some insights of why and when product form solutions arise. The analysis of open networks in product form is trivial, whereas the analysis of closed networks raises some computational issues which are covered in Section 5.3. We present the convolution algorithm in Section 5.3.1 and the mean value analysis method in Section 5.3.2. Then, in Section 5.4.2, we outline approximate methods for more complicated (and realistic) models. Experimental models are discussed in Section 5.5, which is devoted to Petri nets; we give some examples of how Petri nets capture the dynamics of DEDS, and how they can be generalized. Finally, in Section 5.6, we compare analytical and experimental models.

5.1 Introduction to queueing models

Queueing systems are stochastic systems characterized by an arrival process of customers needing some processing at service centers or stations. A single queue is depicted in Figure 5.1. There are different stochastic processes involved in a queueing system. If we denote by C_k the kth customer, arriving at time t_k, we may consider the following quantities:

- S_k: its service time;
- D_k: its delay time (i.e., the time spent in the queue *before* service);
- $W_k = D_k + S_k$: its waiting time.

Introduction to queueing models

FIGURE 5.1
A single queue system

There are two counting processes associated with a queue: the arrival process $n_A(t)$ and the departure process $n_D(t)$, which count the customers arrived and departed up to time t. We consider also the number of customers in the queue $N_q(t)$, the number of customers in service $N_s(t)$, and the number of customers in the system $N(t) = N_q(t) + N_s(t)$. A possible evolution of the process $N(t)$ was depicted in Figure E.2; note how the system changes its state at discrete time instants corresponding to an arrival or departure event. Based on these processes, we may introduce the following quantities:

arrival rate:
$$\lambda = \lim_{t \to \infty} \frac{N_A(t)}{t};$$

service rate: μ, where
$$\frac{1}{\mu} = \lim_{n \to \infty} \sum_{k=1}^{n} S_k$$

is the average time in service;

average number of customers in system:
$$L = \lim_{t \to \infty} \frac{1}{t} \int_0^t N(\tau) d\tau;$$

mean queue length:
$$Q = \lim_{t \to \infty} \frac{1}{t} \int_0^t N_q(\tau) d\tau;$$

average server utilization:
$$U = \lim_{t \to \infty} \frac{1}{t} \int_0^t N_s(\tau) d\tau;$$

note that for a single server queue $N_s(t) \in \{0,1\}$;
average waiting time in the system:

$$W = \lim_{n\to\infty} \frac{1}{n} \sum_{k=1}^{n} W_k;$$

average waiting time in the queue:

$$W_q = \lim_{n\to\infty} \frac{1}{n} \sum_{k=1}^{n} D_k.$$

Based on these quantities, we can estimate the performance measures of interest for a manufacturing system. The lead time corresponds to the waiting time W, and the WIP level to the queue length L. Note the odd correspondence between L and WIP, W and lead time. This is due to the fact that queueing models were born for telephone traffic applications, rather than for manufacturing. The throughput, in an equilibrium condition, must equal the arrival rate λ. Finally, machine utilization is obviously given by U. To characterize a queueing system we need to specify:

- the arrival process;
- the queueing discipline;
- the service process.

The simplest queueing systems are those characterized by Poisson arrivals, exponentially distributed service times and a first in, first out (FIFO) queueing discipline. In this case, we have a memoryless system which can be modeled by a continuous-time Markov Chain (CTMC). For a CTMC it is fairly easy to carry out a steady-state analysis in order to evaluate the performance measures of interest; evaluating performance measures in a transient situation calls for experimental models.

By connecting single queues together and by defining a routing mechanism, we obtain **queueing networks**. When all the customers have the same processing requirements at all stations and the same routing mechanism, we have a **single-class** network; otherwise, we must cope with a **multiclass** queueing network. Under certain hypotheses, queueing networks may be analyzed as CTMCs with a simple solution.

Notation

Due to the variety of queueing systems, a suitable notation has been devised in order to clearly identify them. This notation is due to Kendall and it is similar in spirit to Graham's $\alpha/\beta/\gamma$ notation for scheduling problems.

Queueing systems are characterized by the a descriptor of the form $A/S/m/K/N/Z$ where:

- A specifies the type of arrival process: M (i.e. memoryless) denotes a Poisson arrival (with independent and exponentially distributed interarrival times); D stands for a deterministic process; G corresponds to an arbitrary distribution (sometimes the notation GI is used to stress the fact that interarrival times are independent);
- S denotes the type of service process, using the same convention as the arrival process; M is used to specify exponentially distributed service times, and G a general service time distribution;
- m denotes the number of processors serving the queue;
- K is the maximum queue length (when the queue length reaches this limit, customers are rejected);
- N is the size of the population of customers;
- Z is the service discipline.

The simplest single queueing systems are described by the first three fields only, and it is understood that $K = \infty$, $N = \infty$, and $Z =$FIFO. Typical queueing systems are $M/M/1$ (which is the easiest queueing system to analyze) and $GI/G/1$ (which is much harder to deal with).

5.1.1 Little's law

Queueing systems are very difficult to analyze; nice results hold for a limited class of systems. It is therefore interesting to note the following result, known as Little's law, which holds for a queueing system under fairly general conditions:

$$L = \lambda W. \qquad (5.1)$$

It is interesting to interpret this result from a manufacturing point of view: basically it states, for a given throughput λ, a direct proportionality relation between WIP and lead time. It is important to note that this result holds both for a single queue and a queueing network; we will exploit this fact when dealing with queueing networks.

5.1.2 Elementary Markovian queueing models

In this section we consider some simple examples of queueing systems. Some queueing systems are quite easy to analyze, since they result in simple Markov chain models. This is the case when the system is memoryless; conditions ensuring this are exponential interarrival and service times and FIFO service discipline. We consider here $M/M/1$, $M/M/1/N$, $M/M/m$ systems to show the relationship with Markov models.

The general case is practically intractable. Still, it is worthy to consider approximations to systems such as the $GI/G/1$ queue; the reason is that they are the building block of decomposition methods to analyze complex queueing networks.

The $M/M/1$ queue

The arrival process is Poisson with rate λ, therefore interarrival times are exponentially distributed with mean $1/\lambda$. Service times are exponentially distributed with mean $1/\mu$. It is natural to describe the evolution of this system by the probability π_k of having k customers in the system (including both the queued ones and the one under service). Due to the lack of memory of exponentially distributed random variables, the system can be modeled by a simple CTMC, i.e., an infinite birth-death process (see Section E.7), with parameters

$$\lambda_k = \lambda \qquad k \geq 0$$
$$\mu_k = \mu \qquad k \geq 1.$$

Using the equations of a birth-death process, we obtain

$$\pi_0 = \frac{1}{1+\sum_{k=1}^{\infty}\frac{\lambda^k}{\mu^k}} = \frac{1}{\sum_{k=0}^{\infty}\left(\frac{\lambda}{\mu}\right)^k} = 1-\rho,$$

where we have defined $\rho = \lambda/\mu$, the **traffic intensity**; hence

$$\pi_k = \rho^k(1-\rho) \qquad k \geq 0.$$

For the above equations to make sense, we must have $\rho < 1$; this condition is known as a **stability condition**, and it insures that the queue remains limited.

Now we can compute the performance measures of interest. The utilization can be computed as

$$U = P\{k > 0\} = 1 - P\{k = 0\} = 1 - (1-\rho) = \rho.$$

The mean number of customers in the system is

$$L = \sum_{k=0}^{\infty} k\pi_k = \frac{\rho}{1-\rho}.$$

The mean number of customers in the queue is

$$Q = \sum_{k=1}^{\infty}(k-1)\pi_k = L - (1-\pi_0) = \frac{\rho^2}{1-\rho}.$$

We can also obtain the mean waiting time in the system W and the mean waiting time in the queue W_q. Since

$$W = W_q + \frac{1}{\mu},$$

it suffices to obtain W from Little's law:

$$W = \frac{L}{\lambda} = \frac{1}{\mu - \lambda}.$$

The $M/M/1/N$ queue

The $M/M/1/N$ queue is characterized by a constraint on the WIP; the buffer can contain at most $N-1$ customers. The birth-death process is therefore finite, and includes only states 0 through N. The steady-state probabilities are obtained by analyzing a *finite* state birth-death process

$$\pi_k = \begin{cases} \frac{(1-\rho)\rho^k}{1-\rho^{N+1}} & k = 0, 1, \ldots, N \\ 0 & k > N. \end{cases}$$

Unlike the $M/M/1$ case, there is no need to require $\rho < 1$, since stability is ensured by the limited capacity buffer.

To obtain the performance measures of interest, one proceeds as in the $M/M/1$ case. The utilization is easily obtained:

$$U = 1 - \pi_0 = \frac{\rho(1-\rho^N)}{1-\rho^{N+1}}.$$

The queue length requires some calculations:

$$L = \sum_{k=1}^{N} k\pi_k = \frac{\rho\left[1 - \rho^N - N\rho^N(1-\rho)\right]}{(1-\rho)(1-\rho^{N+1})},$$

and the waiting time is readily obtained by applying Little's law:

$$W = \frac{L}{\lambda} = \frac{1 - \rho^N - N\rho^N(1-\rho)}{\mu(1-\rho)(1-\rho^{N+1})}.$$

The $M/M/m$ queue

In the $M/M/m$ we have m parallel and identical servers as illustrated in Figure 5.2; each server has a mean service time $1/\mu$. In this case we have

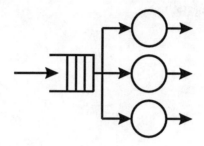

FIGURE 5.2
Graphical illustration of the $M/M/m$ queue

again a birth-death model. We must just take into account that the 'death' rates depend on the number of customers in the system:

$$\lambda_k = \lambda \qquad k = 0, 1, 2, \ldots$$
$$\mu_k = \begin{cases} k\mu & k = 0, 1, \ldots, m \\ m\mu & k > m. \end{cases}$$

We have state-dependent service rates. The application of the formulae for the birth-death process yields

$$\pi_k = \begin{cases} \frac{\pi_0 (m\rho)^k}{k!} & k = 0, 1, \ldots, m \\ \frac{\pi_0 (m\rho)^k}{m! m^{k-m}} & k > m, \end{cases}$$

where in this case $\rho = \lambda/m\mu$, and

$$\pi_0 = \left[\frac{(m\rho)^m}{m!(1-\rho)} + \sum_{k=0}^{m-1} \frac{(m\rho)^k}{k!} \right]^{-1}.$$

We get an expression of the queue length:

$$L = \frac{(m\rho)^m \rho \pi_0}{m!(1-\rho)^2} + m\rho.$$

5.1.3 Approximations for the $GI/G/1$ queue

The $GI/G/1$ queue cannot be modeled within the CTMC framework. As a result, its analysis is much more complicated; apart from particular cases, it cannot be carried out exactly. Nevertheless, many approximate formulas have been proposed in the literature. Typically, such approximations are nonlinear expressions involving only expected values and variances of random characterizing the arrival and service process; the approximations may be more or less accurate, depending on the traffic level of the system. An example is the formula

$$W_q \approx \frac{\rho^2(1+C_s^2)}{1+\rho^2 C_s^2} \left\{ \frac{C_a^2 + \rho^2 C_s^2}{2\lambda(1-\rho)} \right\} \tag{5.2}$$

which gives the delay in the queue as a function of the squared coefficients of variation of the interarrival time C_a^2 and of the service time C_s^2, the mean service time $1/\mu$, the throughput λ and the traffic intensity $\rho = \lambda/\mu$.

Approximate expressions of this type are most useful for devising approximate analysis methods for difficult queueing networks, based on node decomposition (see Section 5.4.2).

5.2 Queueing networks

Queueing networks are obtained by connecting a set of single queues and by establishing a routing pattern. The routing pattern is formalized by a **routing matrix**, whose entries r_{ij} are the probability that a part will visit machine j after a service at machine i. We consider here only a **Markovian** routing mechanism, in the sense that the routing probabilities are not affected by the past story of a part.

Queueing networks may be classified as **open** or **closed**. In the case of open networks, clients arrive from and depart to the external environment (see Figures 5.3.a and b); therefore, there is no limit to the population of clients in the system, and instability may arise. If the arrival rates from the outside are compatible with the service capacities, the network will be stable and we may compute performance measures such as lead time, WIP and machine utilization; note that in this case the system throughput must be equal to the total arrival rate. In a closed network the number of clients is kept constant; parts never leave the system (see Figure 5.3.c). This may sound odd, since in a manufacturing system parts (hopefully) get to depart from the system after processing; still, the population is kept constant if the number of parts is limited by the availability of fixtures and pallets. The case of a closed network is, in a sense, dual to that of an open one; here the total WIP level is fixed, but the throughput must be computed.

We characterize the state of a queueing network by the vector of customers at each queue:

$$\mathbf{n} = \begin{bmatrix} n_1 \\ n_2 \\ \vdots \\ n_m \end{bmatrix}$$

Under suitable hypotheses, the steady-state analysis of a queueing network can be carried out by analyzing the underlying CTMC; essentially, such hypotheses are Poisson arrivals from the outside (for an open network), Markovian routing, and exponentially distributed service times. By writing the balance equations for the steady-state analysis, we obtain the stationary probability distribution of the states, $P(n_1, n_2, \ldots, n_m)$, from which the performance measures can be computed. The link between queueing networks and CTMCs is illustrated in Section 5.2.1 for some simple cases.

Under further restrictions, we obtain a particularly simple expression of the stationary distribution: this is the case of **product form networks**, i.e., networks where the probability of a state $P(\mathbf{n})$ is expressed, possibly within a normalization constant, as the product of the marginal probabilities of each queue length $P_i(n_i)$:

$$P(\mathbf{n}) = \prod_{i=1}^{m} P_i(n_i).$$

We deal with open and closed networks in product form in Sections 5.2.2 and 5.2.3.

Conditions insuring a product form solution are Poisson arrivals, exponential service times, FIFO service discipline, infinite capacity buffers and Markov routing, failure-free machines and no assembly. Actually, we consider here simpler cases than necessary, since we consider single-class networks of load-independent servers, and we assume that each queue is served by a single machine. The results we show can be somewhat generalized: for instance, some queueing networks with blocking do admit a product form solution.

5.2.1 Two-machine queueing networks

We consider here some simple queueing networks consisting of two machines. Our aim is to establish a link between queueing networks and CTMCs.

- In Figure 5.3.a we show a series arrangement of two machines; similar arrangements of machines are known as **tandem queueing networks**. Parts arrive according to a Poisson process with rate λ, and the processing times on the two machines are exponentially distributed with rates μ_1 and μ_2 respectively.
- In Figure 5.3.b we have again a two machine open network, but there is *feedback*; after the service on machine 2, parts are routed back to machine 1 with a probability p_f, and leave the system with probability $p_e = 1 - p_f$.

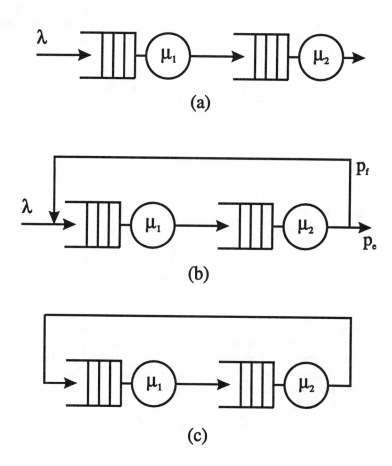

FIGURE 5.3
A two-machine tandem queueing network (a), a two-machine queueing network with feedback (b), and a two-machine closed queueing network (c)

- In Figure 5.3.c we have a closed network consisting of two machines; there are n parts in the system, and after service at machine 2, a part is immediately fed back to machine 1.

Despite the apparent differences among these three systems, they all can be analyzed as CTMCs. In these cases, the state of the system is represented by the pair (n_1, n_2) consisting of the number of parts queued at machines 1 and 2 (including those in service).

The CTMC corresponding to the tandem system is shown in Figure 5.4, whereas the CTMC for the open network with feedback is shown in Figure 5.5. These networks can be analyzed by computing the steady-state equilibrium probabilities $P(n_1, n_2)$. While writing the equilibrium conditions is trivial, it may seem that actually solving them is difficult, since they are an infinite system of equations with an infinite number of unknowns. For instance, in the case with feedback, the balance equation for a generic state (n_1, n_2) is

$$(\lambda + \mu_1 + \mu_2)P(n_1, n_2) = \lambda P(n_1 - 1, n_2) + \mu_1 P(n_1 + 1, n_2 - 1) + \\ \mu_2 p_e P(n_1, n_2 + 1) + \mu_2 p_f P(n_1 - 1, n_2 + 1).$$

This equation is easy to interpret; it equals the probability flow out of state (n_1, n_2) and the probability flow into that state. We depart from (n_1, n_2) if a customer arrives, or if a service is completed at one station; these events happen with a rate $\lambda + \mu_1 + \mu_2$. We may reach (n_1, n_2) from $(n_1 - 1, n_2)$ if a customer arrives (with rate λ), from $(n_1 + 1, n_2 - 2)$ if a customer leaves machine 1 and queues at machine 2 (with rate μ_1), from $(n_1, n_2 + 1)$ if a customer finishes its service at machine 2 and it is not fed back (with rate $p_e \mu_2$), and from $(n_1 - 1, n_2 + 1)$ if a customer finishes its service at machine 2 and it is fed back (with rate $p_f \mu_2$). Different equations should be written for states of the form $(n_1, 0)$ and $(0, n_2)$. In the next section we will see that a closed form solution for such a system of equations can be easily found.

The CTMC for the closed system is shown in Figure 5.6. At first sight, this case is easier to deal with; after all, we have a finite system of equations. In fact, dealing with closed systems is more difficult from a computational point of view. The fundamental reason is that in this case enforcing the normalization condition on the stationary probabilities is not trivial, and it requires computing a normalization constant.

5.2.2 Product form open queueing networks

We show here how the performance measures of interest may be computed in the case of an open network of m machines satisfying the product form conditions. We use the following notation:

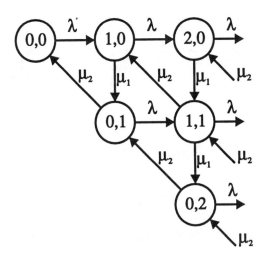

FIGURE 5.4
CTMC corresponding to a two-machine tandem network

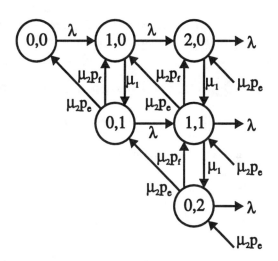

FIGURE 5.5
CTMC corresponding to a two-machine open network with feedback

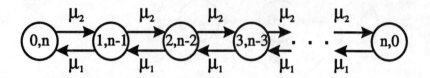

FIGURE 5.6
CTMC corresponding to a two-machine closed network

- λ_i: the arrival rate to machine i; under equilibrium, this is the throughput of machine i;
- α_i: the arrival rate to machine i *from the outside*; this is a Poisson process;
- $\lambda = \sum_{i=1}^{m} \alpha_i$: the total arrival rate into the network;
- r_{ij}: the routing probabilities; note that we admit $r_{ii} \neq 0$ since immediate feedback is allowed.

Each queue i is served by a single server with exponentially distributed service times, with mean $1/\mu_i$.

The solution approach is based on two phases: the analysis of the interactions among the queues, and then the analysis of each queue. Analyzing the interactions entails the solution of the following system of **traffic equations**:

$$\lambda_i = \alpha_i + \sum_{j=1}^{m} r_{ji}\lambda_j \qquad i = 1,\ldots,m. \tag{5.3}$$

It can be shown that this system has a unique solution yielding the machine throughputs λ_i.

The stationary probabilities are readily obtained by defining the utilization $\rho_i = \lambda_i/\mu_i$:

$$P(\mathbf{n}) = \prod_{i=1}^{m} P_i(n_i) \tag{5.4}$$

$$P_i(n_i) = (1 - \rho_i)\rho_i^{n_i}. \tag{5.5}$$

We will not prove these equations; a justification of why a product form solution is found is given in Supplement S5.1.

Now, the performance measures can be obtained by analyzing each queue independently. Formally, the system behaves like an interconnection of independent $M/M/1$ queues.

Example 5.1

Consider the tandem network depicted in Figure 5.3.a. Here the traffic equations are trivial:

$$\lambda_1 = \lambda$$
$$\lambda_2 = \lambda_1.$$

Then we have

$$\rho_1 = \frac{\lambda}{\mu_1}; \qquad \rho_2 = \frac{\lambda}{\mu_2}.$$

ρ_i is the utilization of machine 1; the network is stable if and only if $\rho_i < 1$ for all the machines. The machine with the largest utilization is the bottleneck. The system throughput is obviously λ. The queue lengths at each station are found by applying the formulae of an $M/M/1$ queue:

$$L_i = \frac{\rho_i}{1 - \rho_i}.$$

The system WIP is then

$$L = L_1 + L_2.$$

The waiting time for each station is obtained by applying Little's law:

$$W_i = \frac{L_i}{\lambda_i}.$$

The system lead time is, in this case, simply given by

$$W = W_1 + W_2. \quad \square$$

In the previous example, system performance measures have been obtained simply by summing up the measures for each machines; in general, this is valid for the WIP, but not for the lead time.

Example 5.2

Consider the open network depicted in Figure 5.3.b. The traffic equations are

$$\lambda_1 = \lambda + p_f \lambda_2$$
$$\lambda_2 = \lambda_1.$$

Therefore we have

$$\lambda_2 = \lambda_1 = \frac{\lambda}{1 - p_f} = \frac{\lambda}{p_e}$$

and
$$\rho_1 = \frac{\lambda}{p_e \mu_1} \qquad \rho_2 = \frac{\lambda}{p_e \mu_2}.$$

If stability holds, we may compute L_i, L and W_i as in the previous example. However, due to the feedbacks, it is not true that the system lead time is simply the sum of the two waiting times. We can solve the problem by applying Little's law to the overall network:

$$W = \frac{L}{\lambda} = \frac{L_1 + L_2}{\lambda}. \quad \Box$$

Another way to obtain the system lead time is the following. Let us define the **visit ratio** v_i, i.e., the average number of times a part visits machine i; we have

$$v_i = \frac{\lambda_i}{\lambda}.$$

The visit ratios can also be obtained by solving a slightly modified version of the traffic equations:

$$v_i = q_i + \sum_{j=1}^{m} v_j r_{ji} \qquad i = 1, \ldots, m, \tag{5.6}$$

where $q_i = \alpha_i/\lambda$. Then we obtain the overall lead time as

$$W = \sum_{i=1}^{m} v_i W_i.$$

Note that this expression is identical to what we obtain by applying Little's law:

$$W = \sum_{i=1}^{m} v_i W_i = \sum_{i=1}^{m} \frac{\lambda_i}{\lambda} \frac{L_i}{\lambda_i} = \frac{L}{\lambda}.$$

From Equation (5.5) we see that the marginal probabilities are given *as if* each queue behaves as a $M/M/1$ queue. It is important to note that this is not actually the case; the arrival processes to the queues are *not* Poisson in general. We show a simple and intuitive counterexample in the following; this statement can be formally proved.

Example 5.3
Consider the single queue with feedback depicted in Figure 5.7. We know that a Poisson process is a counting process with independent increments (see Section E.2.1); an arrival event does not influence future arrivals. Assume that the arrival process from the outside enjoys such a property. Now

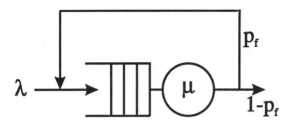

FIGURE 5.7
A single queue with feedback

consider the arrival process *to the queue*; assume also that the arrival rate λ is low, the service rate μ is high and the feedback probability p_f is high. Then, an arrival to queue will be immediately followed with high probability by another arrival of the same customer to the same queue; the counting process does not have independent increments and cannot be Poisson. □

5.2.3 Closed queueing networks

Closed networks are characterized by a constrained state space, since we have

$$\sum_{i=0}^{m} n_i = n, \qquad (5.7)$$

where n is the *fixed* total WIP level. Note that we index the machines starting from 0; machine 0 is a sort of *reference* station. This is useful to define the *system* throughput, which is problematic since, from the model's point of view, no part enters or leaves the system. In practice, there is a point in the system where parts enter and depart, and this is usually a load/unload station; therefore, it is reasonable to assign the number 0 to this station and to use its throughput as the system throughput.

Example 5.4
A classical type of closed network is shown in Figure 5.8. This type of network is called the **central server** model, and it is a common model for an FMS with an Automated Guided Vehicle (AGV). Stations 1 through m are machines, whereas station 0 is the AGV. The AGV routes a part to machine i with probability p_i. With probability p_0 the part is immediately fed back to the AGV; this corresponds to a part leaving the system. In this

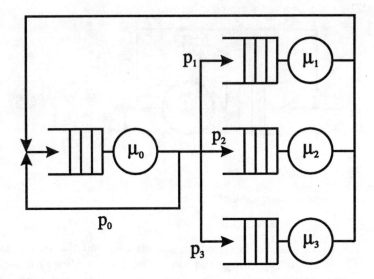

FIGURE 5.8
A central server queueing network

model, the Load/Unload station is not explicitly modeled; we could add a station on the feedback branch. Now, what is the *system* throughput in this case? The system throughput is the throughput through the Load/Unload station, which can be obtained as $U_{AGV}\mu_0 p_0$. □

Under the usual hypotheses, closed networks have a product form solution. The approach is similar to the case of open networks, but there is some additional difficulty. The visit ratio equations are in this case

$$v_i = \sum_{i=0}^{m} r_{ji} v_j \qquad i = 0, 1, 2, \ldots, m, \qquad (5.8)$$

since here $\alpha_i = 0$. Unlike the case of open networks, these equations do not have a unique solution: given a solution **v**, we can build infinite solutions of the form $\alpha \mathbf{v}$, where α is an arbitrary real number. However, we may interpret the v_i as *relative* visit ratios; v_i/v_j is the average number of visits at machine i per visit at machine j; this quantity is also the ratio of throughputs of machines i and j.

Given an arbitrary solution of the traffic equations, it can be verified that we still have a product form solution

$$P(\mathbf{n}) = \frac{1}{C(n)} \prod_{i=0}^{m} \rho_i^{n_i}, \qquad (5.9)$$

where $\rho_i = v_i/\mu_i$ is the *relative* utilization of machine i, and $\sum_{i=0}^{m} n_i = n$. A justification of (5.9) is given in Supplement S5.1. We just note here that this product form result is remarkable; the queues are clearly statistically dependent, due to the finite number of customers. $C(n)$ is a **normalization constant** such that all the probabilities sum up to one:

$$C(n) = \sum_{\mathbf{n} \in S} \left(\prod_{i=0}^{m} \rho_i^{n_i} \right), \qquad (5.10)$$

where S is the feasible states set, i.e., the set of states such that $\sum_{i=0}^{m} n_i = n$. Note that the apparent degree of freedom in solving the traffic equations is lost when computing the normalization constant. To compute $C(n)$, we should simply sum the probabilities. Unfortunately, the size of the state space may be very large. If we have a closed network of m stations with n customers the number of possible states is

$$\binom{m+n-1}{m-1}. \qquad (5.11)$$

To see this, consider the following binary encoding of a state $\mathbf{n}$:

$$\underbrace{1\ldots 1}_{n_1 \text{ times}} 0 \underbrace{1\ldots 1}_{n_2 \text{ times}} 0 \ldots 0 \underbrace{1\ldots 1}_{n_m \text{ times}}.$$

The state is encoded as a string of $m + n - 1$ symbols; for each station n_i we write a substring of n_i 1s, and each substring is separated by a 0. Since there are $m - 1$ zeros to place in the string the result follows. It is easy to see that this number grows very quickly. Computing the normalization constant by summing over states is therefore impractical; a further issue is that such a process is prone to numerical difficulties.

Conceptually, analyzing closed networks requires computing the normalization constant, deriving the stationary probabilities, and evaluating the performance measures. In the next section we show how these three issues can be handled in one step.

5.3 Computational methods for closed networks

In the following we consider two computational methods for solving closed queueing network models: the **convolution algorithm** in Section 5.3.1

and **Mean Value Analysis** in Section 5.3.2. Knowing the probability distribution of a system with some million states is of little use from a managerial point of view. The efficient numerical methods developed to compute the normalization constant can be used to directly obtain the performance measures we are interested in.

5.3.1 The convolution algorithm

The value of the normalization constant $C(n)$ defined in (5.10) can be obtained by an efficient recursive computation. Let us consider the following function:

$$G(z) = \prod_{i=0}^{m} \frac{1}{1-\rho_i z} = (1 + \rho_0 z + \rho_0^2 z^2 + \ldots) \cdots (1 + \rho_m z + \rho_m^2 z^2 + \ldots).$$

If we think of $G(z)$ as an infinite-term polynomial, we see that $C(n)$ is actually the coefficient of the z^n term; this coefficient is the sum of product terms of the form $\rho_0^{n_0} \cdots \rho_m^{n_m}$, such that the n_i sum up to n. We may also see the polynomial as

$$G(z) = \sum_{n=0}^{\infty} C(n) z^n,$$

i.e., a polynomial whose coefficients are the normalization constants for the same network with a varying number of parts.

If we consider one machine at a time, we can write $G(z)$ in recursive form:

$$G_i(z) = \prod_{k=0}^{i} \frac{1}{1-\rho_k z} = \sum_{j=0}^{\infty} C_i(j) z^j \qquad i = 0, \ldots, m.$$

Note that index i refers to machines and j to parts. From the above formula we see that

$$G_m(z) = G(z), \qquad C_m(n) = C(n).$$

If we unfold the recursion, we get

$$G_i(z) = \frac{1}{1-\rho_i z} G_{i-1}(z) \qquad i = 1, \ldots, m, \qquad (5.12)$$

with the initial condition

$$G_0(z) = \frac{1}{1-\rho_0 z}.$$

From (5.12) we get an equation involving polynomials in z

$$G_i(z) = \rho_i z G_i(z) + G_{i-1}(z),$$

and, equating the coefficients of z^j,

$$C_i(j) = \rho_i C_i(j-1) + C_{i-1}(j) \qquad j = 1, \ldots, m. \tag{5.13}$$

This recursive equation, together with the initial conditions

$$C_i(0) = 1 \qquad i = 0, 1, \ldots, n,$$
$$C_0(j) = \rho_0^j \qquad j = 0, 1, \ldots, m,$$

yields an efficient way of computing the normalization constant. The computational complexity of the convolution algorithm is roughly of the order nm; this should be compared with the complexity of directly computing the sum over all the possible states, whose number is given by (5.11).

Being able to compute the normalization constant is of little value without a way to compute performance measures. This was easy in the open case, but here the finiteness of the sum involved makes it more difficult. However, the following property can be shown:

$$P\{N_i(n) \geq k\} = \rho_i^k \frac{C(n-k)}{C(n)},$$

where $N_i(n)$ is the number of parts queued at station i. All the normalization constants $C(n)$ are easily obtained by the convolution (see Example 5.5). Then, computing the utilization of machine i is easy:

$$U_i(n) = P\{N_i(n) \geq 1\} = \rho_i \frac{C(n-1)}{C(n)}.$$

The throughput of machine i is obtained by multiplying the utilization by the service rate:

$$\Theta_i(n) = \mu_i U_i(n) = v_i \frac{C(n-1)}{C(n)}.$$

In the above relation, there is a degree of freedom in selecting the visit ratio, but this is eliminated by the ratio of the normalization constants.

Getting the individual queue length requires a little more calculation:

$$L_i(n) = \sum_{k=1}^{n} k P\{N_i(n) = k\}$$
$$= \sum_{k=1}^{n} k P\{N_i(n) \geq k\} - \sum_{k=1}^{n} k P\{N_i(n) \geq k+1\}$$
$$= \sum_{j=1}^{n} P\{N_i(n) \geq j\} = \frac{1}{C(n)} \sum_{j=1}^{n} \rho_i^j C(n-j).$$

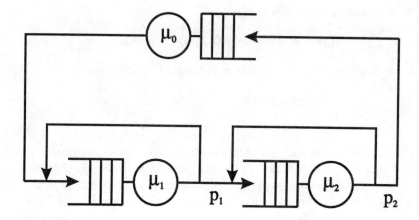

FIGURE 5.9
Closed queueing network for the example

Finally, by applying Little's law, we obtain the waiting time at machine i:

$$W_i(n) = \frac{L_i(n)}{\Theta_i(n)}.$$

Note that here we do not use the *relative* throughputs, that can be obtained from the traffic equations, but the true ones.

Example 5.5
We apply here the convolution algorithm for the closed network of Figure 5.9; it is natural to assume that station 0 is the "reference" station, i.e., its throughput is the system throughput. We have $\mu_0 = 10$, $\mu_1 = 6$, $\mu_2 = 7$, $p_1 = 0.8$, $p_2 = 0.9$ and $n = 3$.

The first step is to write the traffic equations to obtain the relative visit ratios:

$$v_0 = 0.9v_2$$
$$v_1 = v_0 + 0.2v_1$$
$$v_2 = 0.8v_1 + 0.1v_2.$$

This system has infinite solutions of the form

$$v_0 = a, \quad v_1 = 1.25a, \quad v_2 = 1.11a,$$

where a is an arbitrary constant. The relative utilizations are $\rho_i = v_i/\mu_i$. The calculations of the convolution algorithm are easily carried out exploiting a table, whose rows refer to the number of customers and the columns to the stations. To make the calculations easier, it is convenient to assume $\rho_0 = 1$; in this way, we may initialize the table as follows:

	0	1	2
0	1	1	1
1	1		
2	1		
3	1		

This is obtained by choosing $a = \mu_0 = 10$; then we have

$$\rho_0 = 1, \qquad \rho_1 = 2.08, \qquad \rho_2 = 1.58.$$

Then, we fill the table, column by column, applying (5.13):

	0	1	2
0	1	1	1
1	1	3.08	4.66
2	1	7.4064	14.7692
3	1	16.4053	39.7406

Then, the normalization constant is, for our choice of a, $C(3) = 39.74$; note that in the right column we read the normalization constants for the same network with a varying number of customers. Then we may compute the station utilizations:

$$U_0(3) = \rho_0 \frac{C(2)}{C(3)} = 0.3716$$

$$U_1(3) = \rho_1 \frac{C(2)}{C(3)} = 0.773$$

$$U_2(3) = \rho_2 \frac{C(2)}{C(3)} = 0.587.$$

We see that station 1 is the bottleneck. The system throughput is

$$\Theta(3) = \mu_0 U_0(3) = 3.716.$$

The calculation of the queue lengths is left as an exercise for the reader. The reader should check that they add up to 3. □

5.3.2 The Mean Value Analysis algorithm

The convolution algorithm allows us to obtain the performance measures of interest by a straightforward recursion; the numbers we obtain during the computation depend on how the relative visit ratios are scaled. Theoretically, the performance measures do not depend on this choice; unfortunately, the choice of the numbers may adversely affect the numerical stability of the algorithm. The Mean Value Analysis method (MVA for short) was proposed as an alternative way of obtaining average performance measures easing such numerical difficulties. One of its advantages is that it is based on "physically intuitive" formulae; this allows to introduce correction terms in order to get formulae for the approximate analysis of nonproduct form networks (see Section 5.4.1).

The foundation of the method is the **arrival theorem**: the theorem states that (under suitable hypotheses) a customer arriving at a station behaves as a random observer, in the sense that it "sees" a state whose distribution is that of the same network at equilibrium with one customer less. The consequence is that the waiting time of a part is the sum of its service time plus the service time for the $L_i(n-1)$ parts that it finds already in the queue:

$$W_i(n) = \frac{1}{\mu_i}[1 + L_i(n-1)]. \qquad (5.14)$$

The reader is warned against the intuitive appeal of this equation; it holds only under rather restrictive hypotheses, and it is not trivial to prove.

Eq. (5.14) is interesting since it suggests the possibility of obtaining the performance measures by a recursive computation. Since, obviously, $L_i(0) = 0$, we immediately obtain the waiting times $W_i(1)$. To go on, we should find a way of computing $L_i(1)$. This can be obtained by applying Little's law:

$$L_i(n) = \Theta_i(n) W_i(n). \qquad (5.15)$$

Now, we must find a way of computing the station throughputs. Since we might also be interested in the system throughput $\Theta(n)$, it is advisable to spot a reference machine, say M_0. Assume that we have solved the traffic equations in such a way that $v_0 = 1$. Now, for each visit to M_0, we have v_i visits to M_i; hence we have

$$\Theta_i(n) = v_i \Theta_0(n) = v_i \Theta(n).$$

By summing up Eq. (5.15) for all machines, we have

$$\sum_{i=0}^{m} L_i(n) = n = \sum_{i=0}^{m} v_i \Theta(n) W_i(n),$$

and
$$\Theta(n) = \frac{n}{\sum_{i=0}^{m} v_i W_i(n)}. \qquad (5.16)$$

Note that the condition $v_0 = 1$ is not necessary to compute queue lengths and waiting times; it is necessary if we want to interpret $\Theta(n)$ as the system throughput.

The iterative application of Eqs. (5.14), (5.16), and (5.15), with the initial conditions $L_i(0) = 0$, allows to compute $W_i(1)$, $T(1)$, $L_i(1)$ at the first iteration, $W_i(2)$, $T(2)$, $L_i(2)$ at the second iteration, and so on.

Example 5.6

Here we apply MVA to the closed network of Example 5.5. In this case, we should solve the traffic equations choosing $a = 1$. Then

$$v_0 = 1, \quad v_1 = 1.25, \quad v_2 = 1.11.$$

MVA is initialized by computing

$$W_0(1) = \frac{1}{\mu_0} = 0.1, \quad W_1(1) = \frac{1}{\mu_1} = 0.17, \quad W_2(1) = \frac{1}{\mu_2} = 0.14.$$

Then, the throughput with one customer is

$$\Theta(1) = \frac{1}{0.1 + 1.25 \cdot 0.17 + 1.11 \cdot 0.14} = 2.14.$$

We complete the first iteration obtaining

$$L_0(1) = 0.21, \quad L_1(1) = 0.45, \quad L_2(1) = 0.34;$$

note that the sum of the queue lengths is 1. Now we compute the performance measures for the network with two customers:

$$W_0(2) = 0.1 \cdot (1 + 0.21) = 0.12$$
$$W_1(2) = 0.17 \cdot (1 + 0.45) = 0.25$$
$$W_2(2) = 0.14 \cdot (1 + 0.34) = 0.19$$
$$\Theta(2) = \frac{2}{0.12 + 1.25 \cdot 0.25 + 0.19 \cdot 1.11} = 3.11$$
$$L_0(2) = 0.37, \quad L_1(2) = 0.97, \quad L_2(2) = 0.66.$$

For the network with three customers we have

$$W_0(3) = 0.14, \quad W_1(3) = 0.33, \quad W_2(3) = 0.23,$$

and

$$\Theta(3) = \frac{3}{0.14 + 0.33 \cdot 1.25 + 1.11 \cdot 0.23} = 3.7,$$

which is the value we obtained in Example 5.5. □

5.4 Approximate analysis of nonproduct form queueing networks

There are two basic ways to cope with difficult queueing networks. One is to modify the formulae we obtain in the product form case by introducing correction terms. The Mean Value Method lends itself to this kind of approach; in Section 5.4.1 we apply an approximate MVA method to a closed multiclass network. Alternatively, we may take a step backwards forgetting all we have accomplished with Markov models; the network is decomposed into a set of $GI/G/1$ queues, and approximations on their behavior are applied; the **node decomposition** approach is outlined in Section 5.4.2.

5.4.1 Approximate Mean Value Analysis

We describe here an approximate MVA method originally proposed in [178]; the algorithm was named MVAQ by its authors. The starting point of MVA is (5.14), which is a direct consequence of the arrival theorem and may be approximated as

$$W_i(n) \approx \frac{1}{\mu_i}\left[1 + \frac{n-1}{n}L_i(n)\right].$$

This equation does not lend itself to a recursive solution approach; we can tackle it by a successive approximation method. We start with a *guess* $\hat{L}_i(n)$ for the value of the queue lengths, and we carry out an iteration of the MVA algorithm, which yields an updated estimate of the queue lengths. We repeat the process iteratively, starting with a new guess given by the updated estimate, and we stop when the updated estimates are close to the starting guesses.

Step 0. Set the inital queue lengths

$$\hat{L}_i = \frac{n}{m+1}.$$

Step 1. Carry out an MVA iteration:

$$W_i = \frac{1}{\mu_i}\left[1 + \frac{n-1}{n}\hat{L}_i\right]$$
$$\Theta = \frac{n}{\sum_{i=0}^m v_i W_i}$$
$$L_i = v_i \Theta W_i.$$

Step 2. If
$$\sum_{i=0}^{m} |L_i - \hat{L}_i| < \epsilon$$
stop; otherwise, update $\hat{L}_i = L_i$ and go to step 1.

This approximate MVA method can be easily adapted to a *multiclass* closed network. Let the subscript p refer to the different part types; each part type is characterized by its service characteristics, its routing matrix, and its WIP level. To apply MVA to this case, we must first write the waiting time of part type p at station i; assuming a FIFO service discipline, we have

$$W_{ip} = \frac{1}{\mu_{ip}} \left(1 + \frac{n_p - 1}{n_p} L_{ip}\right) + \sum_{r \neq p} \frac{L_{ir}}{\mu_{ir}},$$

where the last term takes into account the waiting time for the service of the parts of other types r which were in the queue when the part of type p joined the queue. The other MVA equations are easily adapted.

$$\Theta_p = \frac{n_p}{\sum_{i=0}^{m} v_{ip} W_{ip}}$$
$$L_{ip} = v_{ip} \Theta_p W_{ip}.$$

The visit ratios v_{ip} are obtained by solving the traffic equations for each part type.

5.4.2 A node decomposition approach for general queueing networks

We consider here an open network with Poisson arrivals but nonexponential servers: such a network cannot be solved in product form. Still worse, it cannot be modeled by a CTMC. We must settle for an approximate solution approach; in Section 5.1.3 we have considered approximate expressions for the waiting time in the queue in a $GI/G/1$ network (Eq. 5.2). The approximation relies on the squared coefficients of variation of the interarrival and service times. In our open network, we may assume that the service time characteristics are known and expressed by the expected service time $1/\mu_i$ ($i = 1, \ldots, M$) and by the squared coefficient of variation (SCV) C_{si}^2. What we need to compute is the SCV C_{ai}^2 linked to the arrival process at station i; this may be obtained by analyzing the interactions among the queues. Note that this is not different in spirit from the product form case, which is based on the solution of the traffic equations, accounting for interactions among the queues, and on the application of the formulae of the $M/M/1$ queue. Here the traffic equations are still applied to obtain the *average* values of the quantities of interest, i.e., the throughputs λ_i and the utilizations ρ_i. We must find a way to compute C_{ai}^2, i.e., the information contained in the second order moments.

Let α_i be the arrival rate from the outside to station i and r_{ik} the entries of the routing matrix. Let C_{aik}^2 be the SCV of the interarrival times at station i *from* station k; if we denote the external environment as "station" 0, we have $C_{ai0}^2 = 1$, since the interarrival times from the outside are exponentially distributed. The arrivals at station i from station k depend on the departure process from station k; let C_{dk}^2 be the SCV of the interdeparture times from station k. What we need is a set of equations linking C_{dk}^2, C_{aik}^2 and C_{ai}^2.

C_{aik}^2 is the SCV of the random variable T_{ik}, which is the time elapsing between two consecutive arrivals at station i from station k. This variable can be expressed as

$$T_{ik} = \sum_{l=1}^{n} S_l^k$$

where S^k is the interdeparture time at station k (S_l^k is the value of the lth interdeparture time); n is also a random variable, and it is the number of departures from station k that occur between two departures routed to station i. It is easy to see that n is geometrically distributed with mean $1/r_{ki}$ (see Section E.1.2). A strong hypothesis underlying this expression is that the departure process can be approximated by a *renewal* process.

Applying the formula for the sum of a random number of random variables (see Example E.3), we get:

$$E(T_{ik}) = E(n)E(S^k) = \frac{1}{r_{ki}\lambda_k}$$

$$\mathrm{Var}(T_{ik}) = \mathrm{Var}(S^k)E(n) + \mathrm{Var}(n)E^2(S^k) =$$

$$\frac{\mathrm{Var}(S^k)}{r_{ki}} + \left(\frac{1-r_{ki}}{r_{ki}^2}\right)E^2(S^k)$$

$$C_{aik}^2 = \frac{\mathrm{Var}(T_{ik})}{E(T_{ik})} = r_{ki}C_{dk}^2 + 1 - r_{ki}. \tag{5.17}$$

It is also possible to link the departure and the arrival processes by the following approximation:

$$C_{di}^2 = \left(1 - \rho_i^2\right)C_{ai}^2 + \rho_i^2 C_{si}^2. \tag{5.18}$$

To justify this approximation, note that, if the utilization ρ_i is close to 1 the station is almost always busy, and the departure process depends on the service characteristics; if the utilization is low, then the station may be idle, and the departures also depend on the arrivals.

The last piece we need is a link between C_{ai}^2 and C_{aik}^2. This may be obtained by expressing the former SCV as a weighted average of the latter:

$$C_{ai}^2 = \frac{\alpha_i}{\lambda_i} + \sum_{k=1}^{m} \frac{\lambda_k r_{ki}}{\lambda_i} C_{aik}^2 \qquad (5.19)$$

The first term takes the external arrivals into account. Putting (5.17), (5.18), and (5.19) together, we get a system of linear equations yielding the SCV C_{ai}^2. Now, applying an approximate formula such as (5.2), we get an estimate of the waiting times in each queue W_{qi}. From these quantities, we obtain the other performance measures of interest. The lead time is computed as

$$\sum_{i=1}^{m} v_i W_i,$$

where $W_i = W_{qi} + 1/\mu_i$ and the visit ratios are obtained from the traffic equations. The traffic equations also yield the station throughputs and utilizations. Finally the WIP level is computed by applying Little's law.

5.5 Petri net models

Petri nets are a mathematical formalism developed in 1962 by C.A. Petri to represent DEDS characterized by concurrency and synchronization issues. A nice feature of Petri nets is that they couple a rigorous mathematical foundation with a simple graphical representation. A drawback is that, due to their graphical nature, they may be unreadable when the modeled system is complex; in such a case, textual modeling languages for simulation can be adopted. Here we use Petri nets as a way to illustrate the typical DEDS features we need to account for in an experimental performance evaluation model.

A Petri net is a directed bipartite graph; the nodes are divided into two sets, **places** and **transitions**. A simple Petri net is depicted in Figure 5.10(a). Places are represented by circles and transitions by bars or rectangular boxes.

More formally, we can define a Petri net as a quadruple (P, T, IN, OUT), where:

- P is the set of places (denoted by p_i);
- T is the set of transitions (denoted by t_j);
- $IN: (P \times T) \to \mathbf{N}$ is the input function, which specifies the number of arcs directed from each place to each transition; we have an arc from a place p_i to a transition t_j if $IN(p_i, t_j) > 0$;

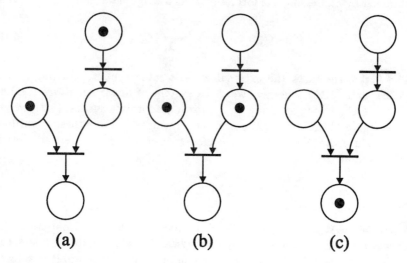

FIGURE 5.10
Token game for a simple Petri net

- $OUT: (P \times T) \to \mathbf{N}$ is the output function, which specifies the number of arcs directed from each transition to each place.

The graph is bipartite since there is no arc from a transition to a transition, nor from a place to a place, and $P \bigcap T = \emptyset$. In the following, we assume $IN(p_i, t_j), OUT(p_i, t_j) \in \{0, 1\}$, i.e., we rule out multiple arcs.

The number of tokens in each place is called a **marking**; formally, the marking of a Petri net is a function $M: P \to \mathbf{N}$. A marked Petri net is a Petri net with an initial marking M_0. The marking of a net represents the state of the system, which changes due to activities being carried out or to logical conditions; this is modeled by the firing of transitions, which remove tokens from the input places of the transition and put tokens into the output places. A transition t_j is enabled in a marking if there is at least one token for each place in the set $IP(t_j)$ of its input places:

$$M(p_i) \geq IN(p_i, t_j) \quad \forall p_i \in IP(t_j).$$

After firing we obtain a new marking

$$M'(p_i) = M(p_i) + OUT(p_i, t_j) - IN(p_i, t_j) \quad \forall p_i \in P.$$

What we obtain is a *token game* that can be used to model the behavior of DEDS. A token game is shown in Figures 5.10(a), (b), and (c).

Figure 5.11 shows how Petri nets can be used to express some typical features of DEDS, such as:

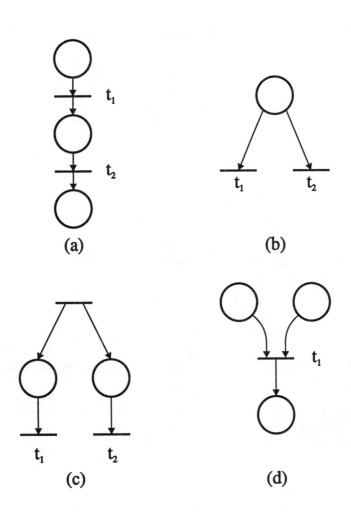

FIGURE 5.11
Typical DEDS features modeled by Petri nets

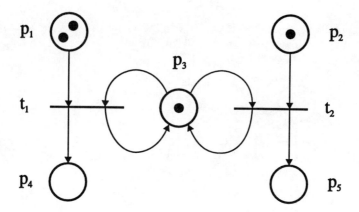

FIGURE 5.12
Modeling a simple manufacturing system by Petri nets

sequential execution: t_2 can fire only after t_1 (Fig. 5.11.a); this can be used to model causal relationships between activities and precedence constraints;

conflict: either t_1 or t_2 can be fired (Fig. 5.11.b); this can be used to model the assignment of a resource to activities;

concurrency: t_1 and t_2 may fire independently (Fig. 5.11.c);

synchronization: t_1 can fire only if a token is present in each input place (Fig. 5.11.d); this can be used to model the simultaneous use of different resources to carry out an operation.

In Figure 5.12 we illustrate a typical use of a Petri net to model manufacturing systems. We have two part types and one machine; a token in p_1 or p_2 represents a part of type 1 or 2 waiting for processing, the token in p_3 represents the availability of the machine, and the transitions t_1 and t_2 represent the processing of a part, which is then moved to places p_4 or p_5. Note that the machine processes one part at a time.

Until now, we have only considered *logical* conditions modeled by the conditions enabling transitions; this is useful to analyze the *qualitative* properties of a system, such as the possibility of getting stuck in a deadlock. However, to evaluate the *quantitative* performance of a manufacturing system we need to represent time; this is accomplished by **timed** Petri nets. A time is associated with transitions, in the sense that firing takes a finite time: at the beginning of firing tokens in input places are removed; only after the time associated with the transition are they released to the output places. If we assume that the transition times are stochastic, we have a **stochastic** Petri net. If transition times are exponentially distributed, the

Petri net is equivalent to a CTMC. The states of the CTMC are the possible markings of the net. The steady-state analysis of a stochastic net with exponential transition times can be carried out with the usual techniques; we may see stochastic Petri nets as a (relatively) high-level formalism to specify Markov models.

A further generalization of Petri nets is represented by **Generalized Stochastic Petri Nets** (GSPN); in such a net we have both immediate and exponential transitions. Immediate transitions (represented by bars) are used to model logical conditions, and exponential transitions (represented by rectangular boxes) are used to model activities being carried out. In the following example we illustrate a nontrivial application of such nets.

Example 5.7

Here we present a GSPN model of a kanban control system; see [59] or [202, pp. 514-522] (from which the model has been adapted) for further information. Consider a two-stage manufacturing system, processing a single part type: stage 1 consists of m_1 machines having exponential processing time with rate μ_2; similarly, we have m_2 machines in stage 2, with rate μ_2. Customers arrive with an exponential rate λ, and there is limit M on the customers waiting in the queue. An infinite supply of raw materials is assumed. After each stage there is a buffer for WIP and finished products; the aim of kanban control is to keep inventory levels low, while allowing for a smooth material flow with no interruptions.

Kanban control is a *pull* production control, in the sense that production is triggered by material removal, and not by an off-line schedule. WIP is limited by the number of cards (i.e. kanbans); we cannot produce a part without a kanban. If we assign K_1 kanbans to stage 1, then the WIP level between stage 1 and 2 will never exceed K_1 parts; similarly, we enforce an upper bound K_2 on the end product inventory. The GSPN shown in Figure 5.13 helps in understanding how this is accomplished.

Places p_6 and p_{12} represent the output buffer of stages 1 and 2 respectively; in the figure, they contain a number of tokens corresponding to the maximum inventory levels K_1 and K_2. Place p_{13} represents the arriving customers; the exponential transition t_8 fires with rate λ (accounting for exponential interarrival times); when this happens, we have a waiting customer in place p_{11} (note that up to M customers can be queued). Now transition t_7 may fire, with two effects: the customer is served, and a kanban is moved to place p_{10}. This represents a production request signalled to stage 1.

Now transition t_4 fires, representing the removal of a semifinished unit from the output buffer of stage 1; place p_7 represents a semifinished unit waiting for processing on stage 2. If there is an available machine in place

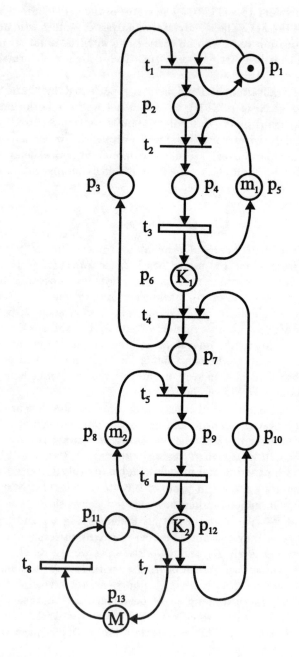

FIGURE 5.13
A GSPN model of a kanban control system

p_8, transition t_5 fires and processing is started. The exponential processing time is accounted for by transition t_6; this transition has an exponential rate given by $M(p_9)\mu_2$, where $M(p_9)$ is the marking of p_9, i.e., the number of active machines (see the $M/M/m$ queue).

Another effect of the firing of t_4 is that a production request is signalled to stage 1. Stage 1 is modeled just like stage 2. The only thing to notice is that place p_1 and the immediate transition t_1 are used to explicitly model the infinite supply of raw materials; actually, they do not influence the timed behavior of the exponential transitions.

A model like this one can be used to evaluate the tradeoff between the waiting time of customers and the WIP levels; the right number of kanbans can be determined in this way. □

The representational power of Petri nets may be further enhanced by *inhibitor arcs* linking a place to a transition; the transition is disabled if there is a token in the input place. Finally, we mention that attributes may be associated with tokens, which become complex data structures; transition firings are conditioned by predicates on such attributes. In this way we obtain **high-level** Petri nets, by which it is possible to build complex simulation models.

5.6 Discussion

In this chapter we have dealt with analytical and experimental models for performance evaluation. Though the discussion of experimental models has been very limited, a basic tradeoff between the two classes of models is clear. On the one hand, experimental models are able to capture every peculiarity of the manufacturing system, whereas analytical models only give estimates of the steady-state performance measures under some limiting assumptions; on the other hand, analytical models are more efficient, require a limited amount of data entry, and do not need statistical analyses of the results.

In recent years, the limitations of analytical models have been overcome to a large extent by considerable progress in approximate methods. By node decomposition techniques it is possible to obtain reliable performance estimates for networks of nonexponential servers. Using multiclass networks, it is possible to overcome the memoryless character of Markovian routing. A new term has been coined, Rapid Modeling Technology (RMT) [179], to point out the departure from the rather academic character of older queueing models.

In fact, both types of models are useful, and the selection of one or the other depends on the purpose of the analysis. Analytical models allow carrying out a large number of "what if" experiments which are fundamental in the early phases of the design of a manufacturing system. This type of analysis is also important for management purposes; for instance, it is possible to evaluate the impact on lead time of setup time reductions or of the introduction of transfer batches. Commercial packages such as MPX[1] are available for this purpose. The use of costly simulation models can be limited to the last phases of parameters fine-tuning, or when a rough-cut steady-state analysis is not suitable. This occurs with significant sequence-dependent setup times or with systems characterized by complex interactions between the resources (e.g., complex material handling systems, blocking, and tool management in an FMS). In the next chapter we will deal with the use of evaluative models within an optimization framework; due to their efficiency, analytical models should be used whenever possible.

S5.1 Product forms and local balance equations

In this supplement we provide some hints on why product form (PF) results arise in some cases. Such an understanding is fundamental when facing a new queueing network model and asking if it admits a PF solution; in fact, the class of models for which PF solutions are known is much larger than the one we have considered in this chapter. It is also possible to establish bounds on system performance by modifying the model so that it can be solved in PF. Here we just point out the role of *local* rather than *global* balance equations in finding the stationary probability distribution of a CTMC. A deeper understanding of this matter would require a treatment of *time-reversible* Markov chains and quasi-reversible queueing systems; the interested reader is referred to [196] or [110, Chapter 6].

Consider the $M/M/1$ queueing system. We have seen that it can be easily analyzed as a birth-death process in steady state (Section E.3.3). To carry out the analysis, we first write the balance equations

$$\lambda \pi_0 = \mu \pi_1 \tag{5.20}$$

$$(\lambda + \mu)\pi_k = \lambda \pi_{k-1} + \mu \pi_{k+1} \qquad k \geq 1, \tag{5.21}$$

where π_k is the probability of having k customers in the system. From (5.21) no closed form solution can be easily obtained. Eq. (5.21) expresses a balance condition on the global probability flow going into and out of a node of the underlying CTMC; in this sense is a global balance equation. By writing the global balance equation for $k = 1$, and using (5.20) we get

---

[1] MPX is a registered trademark of Network Dynamics, Inc., Burlington, MA.

Product forms and local balance equations

$\lambda \pi_1 = \mu \pi_2$. Going on, we may write a simpler balance equation:

$$\lambda \pi_k = \mu \pi_{k+1} \qquad k \geq 1. \tag{5.22}$$

This equation can be seen as a disaggregation of the global balance equation (5.21); its simpler form readily suggests the recursion

$$\pi_k = \frac{\lambda}{\mu} \pi_{k-1} = \rho \pi_{k-1} = \rho^k \pi_0.$$

Then π_0 is obtained by enforcing the normalization condition and the problem is solved.

Note that the local balance equations imply the global balance equations, but not vice versa, in general; in fact, we get a product form result when we may disaggregate global balance equations into simpler local balance equations. Let us see how this happens for the case of closed queueing networks. We first write the global balance equations for a generic state n:

$$\sum_{i=0}^{m} \mu_i P(\mathbf{n}) = \sum_{i=0}^{m} \sum_{j=0}^{m} \mu_j r_{ji} P(\mathbf{n} + \mathbf{1}_j - \mathbf{1}_i), \tag{5.23}$$

where we have used the notation

$$\mathbf{1}_i = [0, 0, 0, \ldots, 0, 1, 0, \ldots 0]^T$$

to denote the unit vector with a 1 in position i. These equations simply state that we leave state n when a service is completed at any queue, and we reach state n from state $(\mathbf{n} + \mathbf{1}_j - \mathbf{1}_i)$ when a customer leaves queue j and joins queue i. Now consider the visit ratio equations (5.8), rewritten as

$$1 = \sum_{j=0}^{m} r_{ji} \frac{v_j}{v_i}.$$

Using this identity, we can rewrite μ_i in the left side of (5.23) as $\mu_1 \cdot 1$:

$$\sum_{i=0}^{m} \sum_{j=0}^{m} \mu_i r_{ji} \frac{v_j}{v_i} P(\mathbf{n}) = \sum_{i=0}^{m} \sum_{j=0}^{m} \mu_j r_{ji} P(\mathbf{n} + \mathbf{1}_j - \mathbf{1}_i),$$

which can be rewritten as

$$\sum_{i=0}^{m} \sum_{j=0}^{m} r_{ji} \left\{ \frac{v_j}{v_i} \mu_i P(\mathbf{n}) - \mu_j P(\mathbf{n} + \mathbf{1}_j - \mathbf{1}_i) \right\} = 0. \tag{5.24}$$

It is easy to see that (5.24) holds if the following local balance equation holds:

$$\frac{v_j}{v_i} \mu_i P(\mathbf{n}) = \mu_j P(\mathbf{n} + \mathbf{1}_j - \mathbf{1}_i),$$

which can be rewritten as

$$P(\mathbf{n}) = \left(\frac{v_i}{\mu_i}\right)\left(\frac{v_j}{\mu_j}\right)^{-1} P(\mathbf{n} + \mathbf{1}_j - \mathbf{1}_i).$$

This suggests trying a solution of the form

$$P(\mathbf{n}) = \left(\frac{v_i}{\mu_i}\right) \hat{P}(\mathbf{n} - \mathbf{1}_i),$$

where the notation $\hat{P}$ is used to stress the fact that it is not a true probability (clearly, $P(\mathbf{n} - \mathbf{1}_i) = 0$ in a closed network). Recursively unfolding the above equation, we get

$$P(\mathbf{n}) = \left(\frac{v_i}{\mu_i}\right)^{n_i} \hat{P}(n_0, n_1, n_2, \ldots, 0, \ldots, n_m).$$

Repeating it for each station, we get

$$P(\mathbf{n}) = \prod_{i=0}^{m} \left(\frac{v_i}{\mu_i}\right)^{n_i} \hat{P}(\mathbf{0}).$$

The product form result (5.9) is finally obtained by enforcing a normalization condition.

The product form result for open queueing networks can be proved in a similar way. We see the close relationship between the simple product form result and the disaggregation of global balance equations into simpler local balance equations. Obviously, such a disaggregation is not always possible; a topological interpretation of why and when this is legitimate can be found in [159, Chapter 3].

For further reading

Performance evaluation of manufacturing systems is a well developed area, particularly as far as simulation models are concerned. We have not considered such models, but there are plenty of books on simulation: [127] is a good source for all the statistical issues involved in simulation; [43] is more specifically aimed at building simulation models for manufacturing systems. As to analytical models, a good introductory text-book is [202]. An overview on performance evaluation with an extensive bibliography can be found in [180]. The reader interested in advanced stochastic models will find [36] a most comprehensive (and challenging) reading. Queueing models of manufacturing systems are also dealt with in [6, Chapter 11], to which we are indebted for the exposition of the node decomposition algorithm in Section 5.4.2. This node decomposition was originally proposed in [170]; see also [120, 208]. Approximations of the $GI/G/1$ queue are discussed in [169]; a survey of approximate open queueing network models for manufacturing systems can be found in [24]. The original references for the convolution and MVA algorithms are [37] and [157] respectively. Readers wishing a deeper understanding of the stochastic issues involved in queueing network models are referred to [205].

As to Petri net models of manufacturing systems, we refer the reader to [61, 216]. A particularly studied class of timed Petri nets are the so-called timed event graphs: a tutorial introduction can be found [156]. The theoretically inclined reader may learn more on DEDS analysis by such methods in [8]. An extensive bibliography on kanban control systems can be found in [19].

6
Putting things together

In the previous chapters we have separately dealt with generative and evaluative models. In the first part of this chapter we "put together" different kind of models. An evaluative model can be used to evaluate the objective function given a set of policy parameters for a complex manufacturing system; by using numerical optimization methods, it is possible to look for the optimal setting of parameters. We may integrate optimization methods both with simulation and queueing network models (Section 6.1). It is also possible to consider optimization models for systems modeled partly by deterministic state equations, and partly by a stochastic model. In Section 6.2 we deal with the optimal control of a failure-prone manufacturing system whose failure and repair dynamics are modeled by a Markov chain.

Another aim of this chapter is to stress the implementation aspects of optimization models. Apart from formulating the model in a computationally convenient form and possibly devising an approximate solution method, we must translate the model into a format suitable for the software; loading the data structures in a computer-oriented format is a time-consuming and error-prone task. This task is considerably eased by model generators, i.e., systems accepting a model described in a high-level algebraic language; we give an example of a model generation language in Section 6.3.

6.1 Integrating optimization methods and evaluative models

Consider a generic optimization problem $\min_{\mathbf{x} \in S} f(\mathbf{x})$. We must resort to evaluative models whenever it is impossible to express the objective function f as an explicit function of the decision variables $\mathbf{x}$; this happens with complex manufacturing systems, particularly when affected by random phenomena. In section A.3 we have described gradient-based numerical optimization algorithms. They are natural candidates for integration with evaluative models. The gradient method is an iterative method of the form

$$\mathbf{x}^{(k+1)} = \Pi_S \left(\mathbf{x}^{(k)} - \alpha^{(k)} \nabla f(\mathbf{x}^{(k)}) \right),$$

where Π_S is the operator projecting the new point onto the feasible set S. Apart from selecting the step $\alpha^{(k)}$ in order to insure proper convergence properties, we have to estimate the gradient of the objective function. Since f is not known, a possible approach is to approximate its gradient by finite differences

$$\frac{\partial f(\mathbf{x})}{\partial x_i} \approx \frac{f(\mathbf{x} + h_i \mathbf{1}_i) - f(\mathbf{x})}{h_i},$$

where $\mathbf{1}_i$ is the ith unit vector. Central differences can also be adopted:

$$\frac{\partial f(\mathbf{x})}{\partial x_i} \approx \frac{f(\mathbf{x} + h_i \mathbf{1}_i) - f(\mathbf{x} - h_i \mathbf{1}_i)}{2h_i}.$$

There are different issues to cope with:

1. The **efficiency** of the method: finite differences require many simulation runs. To improve the efficiency of the procedure, we can use analytical rather than experimental models. Another possible strategy is to derive, from a *single* simulation run, an estimate of the gradient.

2. The **convergence** properties: on the one hand, convergence depends on the choice of the step length $\alpha^{(k)}$; on the other hand, on the quality of the estimate $\hat{\nabla} f^{(k)}$ of the gradient. When h_i gets small, several difficulties arise. The most thorny issue is of a statistical nature; more often than not, we evaluate the *expected value* of a performance measure, i.e., $f(\mathbf{x}) = E[L(\mathbf{x}; \omega)]$, where ω takes into account the stochastic nature of the system. Clearly, we can only find more or less reliable estimates of the performances, which are affected by some "noise"; when comparing two values obtained by slightly different parameter settings, this noise may severely corrupt the estimates of the gradient. In fact, central differences yield better estimates than simple finite differences, at the cost of increased computational burden; in general, longer simulation runs improve the quality of the estimates.

In the following we give some examples of how such difficulties may be dealt with. A thorough treatment would require a robust mathematical background; therefore we just hint at the fundamental issues, so that the reader may have a useful starting point for further study.

6.1.1 The Response Surface Methodology

A simulation model can be seen as a way to build an *empirical* model of the manufacturing system: a simple form of $f(\mathbf{x})$ is postulated and its parameters are identified by regression. For instance, we may consider a first order polynomial form

$$f(\mathbf{x}) = \alpha + \sum_{i=1}^{n} \beta_i x_i = \alpha + \boldsymbol{\beta}^T \mathbf{x}.$$

We may estimate α and β by evaluating f for a set of test values $\mathbf{x}^j$, and by minimizing the errors. Let $\hat{f}_j$ be the estimate of f corresponding to $\mathbf{x}^j$ ($j = 1, \ldots, m$). We have

$$\hat{f}_j = \alpha + \boldsymbol{\beta}^T \mathbf{x}^j + \epsilon_j,$$

where ϵ_j is an error term. Now we may find α and β in such a way that $\sum_j \epsilon_j^2$ is minimized; this is called the Least Squares method, and it is the simpler form of regression. Let us define the matrix

$$\mathbf{X} = \begin{bmatrix} 1 & x_1^1 & x_2^1 & \cdots & x_n^1 \\ 1 & x_1^2 & x_2^2 & \cdots & x_n^2 \\ \vdots & \vdots & \vdots & \ddots & \vdots \\ 1 & x_1^m & x_2^m & \cdots & x_n^m \end{bmatrix},$$

where x_i^j is the jth setting of x_i. It can be shown that the sum of the squared errors is minimized by

$$\begin{bmatrix} \hat{\alpha} \\ \hat{\beta} \end{bmatrix} = (\mathbf{X}^T \mathbf{X})^{-1} \mathbf{X}^T \hat{\mathbf{f}},$$

where $\hat{\mathbf{f}}$ is the vector of the m estimates of f. Then we may set $\hat{\nabla} f^{(k)} = \beta$, and use it within a gradient optimization method. A first-order fit is suitable when we are not close to the optimum; usually, nonlinear programming methods, such as quasi-Newton methods, take advantage of second-order information to improve convergence. When we are approaching the minimizer, a quadratic polynomial can be fitted:

$$f(\mathbf{x}) = \alpha + \boldsymbol{\beta}^T \mathbf{x} + \mathbf{x}^T \boldsymbol{\Gamma} \mathbf{x},$$

where Γ is a square matrix. Then it is easy to find the optimum of the fitted model.

The Response Surface Methodology, with respect to finite difference schemes, has the advantage of relying on a robust statistical background. Note that much care must be taken in the selection of the points $\mathbf{x}^j$. An obvious disadvantage is that it requires costly simulation runs; in the next section we outline a method for obtaining an estimate of the performance gradient with a *single* simulation run.

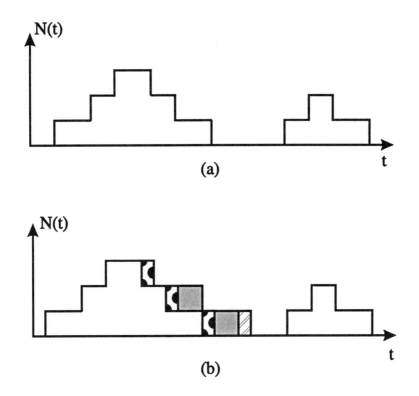

FIGURE 6.1
Busy periods in a single-queue system (a) and perturbed trajectory (b)

6.1.2 Perturbation Analysis

Different approaches have been proposed in the literature to obtain an estimate of the performance measure and of its gradient with a single simulation run. Here we just hint at **Perturbation Analysis**; other approaches are the **likelihood ratio** method (also known as the score function method) and **frequency-based experimentation** (see [81]).

Applying Perturbation Analysis involves implementation and theoretical issues. The simplest way to get a feeling for such issues is to consider a $GI/G/1$ queue; a typical state trajectory is shown in Figure 6.1(a). We have two busy cycles, i.e., two time intervals in which the machine is busy, which are separated by an idle time. Now consider what happens if the mean service time is perturbed. We might want to examine how the mean service time affects the mean waiting time of the customers (including service), in order to find a suitable tradeoff between the cost of increasing the service speed and the need to reduce the WIP level. If the perturbation is small, the busy cycles remain separated, as shown in Figure 6.1(b). We speak of Infinitesimal Perturbation Analysis (IPA) when the perturbations do not change the order of the events.

Let A_k be the interarrival time between the $(k-1)$th and the kth customer; S_k and W_k are the service and waiting time of customer k. The decision variable is the mean service time x of the server. The following recursive formula holds:

$$W_{k+1} = S_{k+1} + \begin{cases} W_n - A_{n+1} & \text{if } S_n \geq A_{n+1} \\ 0 & \text{if } S_n < A_{n+1}. \end{cases} \quad (6.1)$$

Differentiating (6.1) with respect to the decision variable, we get

$$\frac{dW_{k+1}}{dx} = \frac{dS_{k+1}}{dx} + \begin{cases} dW_n/dx & \text{if } S_n \geq A_{n+1} \\ 0 & \text{if } S_n < A_{n+1}. \end{cases} \quad (6.2)$$

It can be shown that

$$\frac{dS}{dx} = -\frac{dF(S;x)/dx}{dF(S;x)/dS},$$

where $F(S;x)$ is the cumulative distribution function of the service times S, depending on x. We may estimate the derivative of the mean waiting time as

$$\left(\frac{d\overline{W}}{dx}\right)_{IPA} = \frac{1}{N} \sum_{k=1}^{N} \frac{dW_k}{dx}, \quad (6.3)$$

where N is the number of customers that have been served. From (6.2) we see that

$$\left(\frac{d\overline{W}}{dx}\right)_{IPA} = \frac{1}{N} \sum_{m=1}^{M} \sum_{i=1}^{n_m} \sum_{j=1}^{i} \frac{dS_{jm}}{dx},$$

where M is the number of busy periods (indexed by m), n_m is the number of customers served in the mth busy period ($N = \sum_m n_m$), and S_{jm} is the service time of the jth customer of the mth busy period. This expression can also be interpreted by looking at Figure 6.1(b). Note that to obtain this estimate in practice, we just need to add some bookkeeping to the simulator; therefore implementing IPA is rather easy. Different estimators have been proposed in the literature for more complex queueing systems; they require relatively simple adjustments of the simulation program, too.

From the theoretical point of view, the matter is much more complicated. The fundamental issue is whether the following equivalence holds:

$$E\left[\nabla_{\mathbf{x}} L(\mathbf{x}; \omega)\right] = \nabla_{\mathbf{x}} E[L(\mathbf{x}; \omega)].$$

Indeed, the optimization procedure needs the first quantity, but we actually compute the second. When commuting expectation and differentiation is not legitimate, we get incorrect estimates (technically speaking, *biased* estimates). Analyzing such issues is by no way trivial, and we refer the reader to the literature. Variants of Perturbation Analysis have been proposed to overcome technical difficulties when Infinitesimal Perturbation Analysis is not applicable.

6.1.3 Optimization with queueing models

In the previous sections we have considered the integration of optimization and simulation models. Unfortunately, the Response Surface Methodology is rather inefficient from the computational point of view, whereas gradient estimation techniques such as Perturbation Analysis are perhaps still more of an academic research field than a practical tool. Queueing models, though approximate in nature, enjoy efficiency characteristics that make them suitable for integration within an optimization procedure. We outline here the integration of a Linear Programming model and a Closed Queueing Network model adapted from [7].

Consider an FMS consisting of M machines, on which we produce N part types; let K_i be the number of operations for part type i. For each part type we have a profit w_i and an upper bound d_i on demand. The problem is to determine the production mix in order to maximize the profits. The problem has also a routing component; we must decide, for each operation of each part type, how many parts visit each machine. The following LP model can be used:

$$\max \sum_{i=1}^{N} w_i x_i \quad (6.4)$$

$$\text{s.t.} \sum_{i=1}^{N} \sum_{k=1}^{K_i} p_{ijk} x_{ijk} \leq C_j \quad \forall j \quad (6.5)$$

$$\sum_{j=1}^{M} x_{ijk} = x_i \quad \forall i, \ k = 1, \ldots, K_i \quad (6.6)$$

$$x_i \leq d_i \quad \forall i \quad (6.7)$$

$$(1+\alpha) \sum_{j=1}^{M} \sum_{i=1}^{N} \sum_{k=1}^{K_i} p_{ijk} x_{ijk} \geq M \sum_{i=1}^{N} \sum_{k=1}^{K_i} p_{ijk} x_{ijk} \quad \forall j \quad (6.8)$$

$$x_{ijk}, x_i \geq 0.$$

The production mix is specified by the production quantities x_i for part type i during a given time period, p_{ijk} is the processing time of operation k of part type i on machine j, and x_{ijk} is the number of parts of type i that visit machine j for operation k. Eq. (6.5) is a capacity constraint; C_j is the capacity of machine j. Eq. (6.6) relates the production mix to the routing variables, and Eq. (6.7) relates the production mix to the upper bounds d_i on the demand for part type i. Finally, Eq. (6.8) is a load balancing constraint, allowing for a maximum percentage deviation α from the ideally balanced routing.

Note that the decision variables are assumed continuous; this is a suitable approximation if the involved numbers are large enough and allows for a quick solution by the Simplex method. However, there is a disadvantage in this approach: the capacity is overestimated, since the model cannot take into account dynamic interferences due to the discrete part flow, the delays introduced by the transportation system, and further constraints introduced by the limited WIP. A way to overcome the problem is to adopt a closed queueing network model to evaluate these effects in terms of machine utilization. From the optimal LP solution we can derive the routing matrix of the parts; from the solution of the network model (by the convolution algorithm or Mean Value Analysis), we obtain the machine utilizations u_j. Then the LP model is adjusted: the capacity in Eq. (6.5) is modified as $u_j C_j$, and the process is repeated. Due to the computational efficiency of queueing network models, this is a computationally feasible approach.

A further feature of queueing models is that some properties, related to concavity or convexity of performance measures with respect to decision variables, can be proven. This is obviously fundamental when gradient-like methods are used, and it cannot be obtained by simulation models. Putting such mathematical considerations aside, we may also point out the *diagnostic* value of the information provided by an performance evaluation model. A queueing network package such as MPX [1] gives (for a *stationary* scenario) information such as the WIP distribution on the workcenters and the waiting times: spotting bottlenecks is clearly useful to adjust the parameters by hand and common sense. The lead time can be disaggregated into its components. There is a waiting time for machine availability but also a waiting time for the parts of the same batch; the first part of a large batch must wait for the last one, before proceeding to the next workcenter. If this waiting time is large, we may evaluate the introduction of transfer batches. Adjusting parameters by hand requires the ability of carrying out a large number of "what if", which is possible with analytical models. After a reasonable setting has been obtained by this trial-and-error process, simulation models (possibly integrated with optimization methods) can be used.

---

[1] MPX is a registered trademark of Network Dynamics, Inc., Burlington, MA.

6.2 Optimal control of failure-prone manufacturing systems

A common issue in manufacturing systems is the need to cope with machine failures. Consider a repetitive production environment: the aim is to track a given production mix as closely as possible, i.e., to minimize inventory and backlog costs. In a deterministic case, assuming negligible setup times, the task would be trivial; the production rates would be set equal to the demand rates. If machines are subject to random failures, it is possible to keep a certain level of surplus production in order to avoid backlog problems when machines break down; clearly a compromise should be achieved between the need to avoid a backlog and the need to keep inventory costs low. Finding a suitable compromise requires explicitly modeling the stochastic part of the system, which influences the available production capacity; this results in a difficult stochastic optimization problem. A computationally viable model cannot be found within a DEDS modeling framework; it is, however, possible to rely on approximate continuous-time, continuous-flow models. In Section 3.4 we have presented a deterministic discrete-time model for a batch manufacturing system; here the same modeling framework is exploited to build a stochastic continuous-time model for a repetitive production environment. Actually, this type of continuous-flow model was the first one to be proposed in the literature. We have considered a simple continuous-time, continuous-flow model in Section 1.2. The ultimate aim is to obtain suitable parameters to implement a hierarchical scheme based on off-line optimization and on-line scheduling; on-line scheduling has the task of tracking a reference trajectory, which is obtained here by dispatching parts into the system according to part production rates computed at the higher level.

We consider here a set of parallel unrelated machines: machines are partitioned in families indexed by $j = 1, \ldots, M$. Let $\alpha_j(t)$ be the number of machines of type j which are running; $\alpha(t)$ can be considered as the *discrete* component of the system state. The continuous component is given by the production surplus $x_i(t)$ for each part type i ($i = 1, \ldots, N$). Let $\mathbf{u}(t)$ be the production rate vector; production rates are constrained by a capacity set corresponding to each state $\alpha(t)$. The capacity set is denoted by Ω_α, and it is given by

$$\sum_i \tau_{ij} u_i \leq \alpha_j(t)$$

$$u_j \geq 0,$$

where τ_{ij} is the processing time of a part of type i on a machine of type j. Note that the capacity set is polyhedral, as in LP models.

The production surplus is obviously linked to the production rates:

$$\mathbf{x}(t) = \int_0^t [\mathbf{u}(\tau) - \mathbf{d}(\tau)]\,d\tau.$$

We would like to keep the production surplus as close as possible to **0**; we may express this requirement by a convex penalty function like $\sum_i (c_i^+ x_i^+ + c_i^- x_i^-)$, where inventory and backlog costs are considered. The optimization problem can be formulated as a stochastic optimal control problem:

$$\min_{\mathbf{u}(t)} E\left[\int_0^T f(\mathbf{x}(t))\,dt\right]$$

$$\text{s.t. } \mathbf{u}(t) \in \Omega_\alpha(t)$$

$$\dot{\mathbf{x}}(t) = \mathbf{u}(t) - \mathbf{d}(t)$$

$$P\{\alpha(t+\delta t) = \alpha_n \mid \alpha(t) = \alpha_n\} = q_{mn}\delta t.$$

The last constraint specifies the failure and repair dynamics as a Markov process; q_{mn} is the infinitesimal generator of a continuous-time Markov Chain. The memoryless property of Markov processes allows tackling the problem by Dynamic Programming (DP). Here we apply the DP approach for solving an infinite-dimensional optimization problem, which is illustrated in Section D.3. Let us define the cost-to-go

$$J(x(t), \alpha(t), t) = \min_{\mathbf{u}} E\left\{\int_t^T g(x(\tau))\,d\tau \mid x(t), \alpha(t)\right\}.$$

The cost-to-go depends on both the continuous and discrete component of the state, and it satisfies an equation similar to (D.9):

$$-\frac{\partial J}{\partial t}(\mathbf{x}, \alpha, t) =$$

$$\min_{\mathbf{u}(t) \in \Omega(\alpha)} \left\{ f(\mathbf{x}) + \sum_\beta J(\mathbf{x}, \beta, t) q_{\alpha\beta} + \frac{\partial J}{\partial \mathbf{x}}(\mathbf{x}, \alpha, t)(\mathbf{u} - \mathbf{d}) \right\}$$

$$\forall \mathbf{x}, \alpha, t. \qquad (6.9)$$

The difference from (D.9) is the addition of a term taking into account the possibility of changing the current state α to β; see [83, Chapter 9] for a simple proof of this equation. At first sight, Eq. (6.9) does not look like a very pleasing result. However, consider what would happen if we knew the cost-to-go J; the resulting minimization problem is just a Linear Programming problem:

$$\min_{\mathbf{u} \in \Omega(\alpha(t))} \left[\frac{\partial J}{\partial \mathbf{x}}(\mathbf{x}, \alpha, t)\right]^T \mathbf{u}, \qquad (6.10)$$

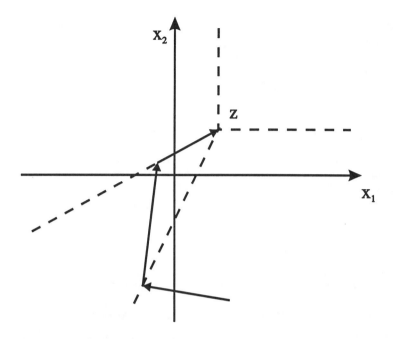

FIGURE 6.2
Optimal trajectory for the flow control problem

where $\Omega(\alpha(t))$ is a polyhedral capacity set. This happens because the minimization at time t is carried out with respect to $u(t)$, and J does not depend on this value. Actually, we do not know exactly J, but we can try to approximate it. A simple approximation is obtained by postulating a quadratic form and determining its parameters with some considerations that we cannot report here (see [83, pp. 305-307]). Furthermore, we can derive some conclusions about the *qualitative* properties of the optimal solution. The optimal production rates will be an extreme point of the capacity set. The gradient of J with respect to the continuous states is a weight on the part types; we should allocate more capacity to part types which are behind schedule and are more vulnerable to machine failures. Note that this gradient changes continuously with x, but the optimal production rates only change abruptly when the optimal solution of the LP problem (6.10) changes. Therefore, for each α we have a partition of the x space, whereby with each subregion an optimal production rate is associated.

Each state α may be feasible or not; we say that it is feasible when the corresponding capacity set is such that we can meet the demand if we stay in this state for a long time. A typical partition of the state x and an optimal trajectory are shown in Figure 6.2; the subregions are delimited by straight lines if J is approximated by a quadratic form. Consider now the derivative of J along the optimal trajectory:

$$\frac{dJ}{dt} = \frac{\partial J}{\partial t} + \left[\frac{\partial J}{\partial \mathbf{x}}\right]^T \frac{d\mathbf{x}}{dt} = \frac{\partial J}{\partial t} + \left[\frac{\partial J}{\partial \mathbf{x}}\right]^T (\mathbf{u} - \mathbf{d}).$$

The LP problem (6.10) is such that we choose $\mathbf{u}$ in order to achieve the largest reduction of J; however J is the integral of a nonnegative penalty function; hence it has a lower bound. What happens if we stay in a feasible state, and we reach the minimum of J? We reach a state (Z in figure 6.2) which is optimal in the sense that it is the optimal production surplus for α; this state is called the **hedging point**. By the way, one strategy to find an approximation of J is by estimating the hedging point and using it to find the parameters of the quadratic form.

6.3 Model management and modeling languages

A mathematical model should not be considered as something which is written once and for all. In practice, a mathematical model can be a piece of information which is created, modified, and shared; this process should be supported by suitable software systems. Another issue is how the data needed to solve the model are collected and passed to the solution procedure. This raises on the one hand the issue of interfacing the solution procedure with a database, in an easy and transparent way, and, on the other hand, the issue of generating the coefficients matrix for an LP or MILP model. Consider for instance an LP problem with constraints $\mathbf{Ax} = \mathbf{b}$. The matrix $\mathbf{A}$ must be stored in a *sparse* format to exploit the fact that only a few of its entries are nonzero in a structured model; therefore, we should load its entries by picking each coefficient in each constraint and writing it into the proper place. A common approach is to write a file (called MPS file) with a certain standard format which is accepted by most solvers. Programming all of this is not easy; matrix generators were written for this purpose. Modern model generators ease this task considerably. A recent example of model generator is EasyModeler[2], which is based on an algebraic language enabling the user to express the mathematical model in a relatively natural way.

As an example, we translate here the continuous-flow mathematical model for FFL scheduling which was described in Section 3.4.2. The first step, however, is to make sure that the model is suitable from the computational

---

[2] EasyModeler is a registered trademark of International Business Machines Corporation.

point of view; this is important, since we have to solve an MILP model by continuous relaxation. The only integer variables in the model are the variables z_{it}, which are set to 1 if the batch has not yet been completed by time bucket t. However, the model of Section 3.4.2 does not consider the fact that if $z_{ip} = 1$, then $z_{it} = 1$ for $t < p$; by the same token, if $z_{ip} = 0$, then $z_{it} = 0$ for $t > p$. Actually, these properties are implicit in the model; however, making them explicit strengthens the formulation. Alternatively, we may change the meaning of the z_{it}; as in Section 4.6.4, we may state that $z_{it} = 1$ if the batch is completed at the end of time bucket t. Then, we have to require

$$\sum_{t=1}^{T} z_{it} = 1 \quad \forall i;$$

this constraint may be suitably translated as a Special Ordered Set of type 3 (see Section C.5.1). To link z_{it} to the state variables, i.e., the cumulative production x_{ikt}, we may write

$$\sum_{p=t+1}^{T} z_{ip} \geq \frac{l_i - x_{iKt}}{l_i} \quad \forall i, t,$$

which insures that the required quantity l_i is produced on the last stage before the time bucket p for which $z_{ip} = 1$. Using this new set of decision variables, we can write the model as:

$$\min \sum_{i=1}^{N} \sum_{t=1}^{T} w_{it} z_{it} \tag{6.11}$$

$$\text{s.t.} \sum_{i=1}^{N} u_{ikt} \leq 1 \quad \forall k, t \tag{6.12}$$

$$x_{ikt} = x_{ik,t-1} + s_{ik} u_{ikt} \quad \forall i, k, t \tag{6.13}$$

$$x_{ikt} \leq x_{i,(k-1),(t-\Delta)} \quad \forall i, k, t \tag{6.14}$$

$$l_i \sum_{p=t+1}^{T} z_{ip} + x_{iKt} \geq l_i \quad \forall i, t \tag{6.15}$$

$$\sum_{t=1}^{T} z_{it} = 1 \quad \forall i \tag{6.16}$$

$$u_{ikt} \in [0, \bar{u}_{ik}] \quad \forall i, k, t \tag{6.17}$$

$$z_{it} \in \{0, 1\} \quad \forall i, t.$$

Nonnegativity and boundary conditions should be added to the model. Here u_{ikt} is the *capacity share* allocated to part type i on stage k during time bucket t, and s_{ik} is the number of parts of type i produced during a time bucket on stage k if all the available capacity is allocated to that part type ($u_{ikt} = 1$); the capacity share is restricted by $\bar{u}_{ik}$ due to constraints on additional resources such as tools and fixtures. Δ is a transportation delay, expressed as an integer number of time bucket ($\Delta \geq 1$). Note how (6.15) has been written to avoid rounding problems.

In Table 6.1 we show how this model can be translated with the Easy-Modeler language. The language is rather easy to understand; there are different sections, delimited by keywords; we have written the keywords with capital letters.

In the first line we name our model **addfun**, and we declare it of the **GENERIC** type, which corresponds to LP and MILP models; a model can also be of **QUADRATIC** type. Then we declare the subscripts of decision variables and data in the **SET** section, which is closed by the **ENDSET** keyword. We have subscripts i, k, t as in the above mathematical model; p is an auxiliary subscript for indexing time buckets. The subscript values range on the sets **batches**, **stages**, and **time_buckets**, respectively. Note that we *do not* state here how many batches, stages, and time buckets we consider in the model; EasyModeler separates the model structure from the instance data. In older systems, the data were mixed with the model structure; this was an inconvenience when the same model had to be solved with different data sets. The model description is translated to a C-language program; this program then reads some files containing the problem data and writes a **MPS** file. It is also possible to instantiate the model interactively and to call the solver directly.

Then we have the **DATA** section, in which we declare the data characterizing a model instance; **l[i]** corresponds to l_i, and so on. The **INTERNALDATA** section is used to evaluate model parameters which depend on the input data. Here **last_t** is the index corresponding to the last time bucket. This is obtained by taking the cardinality of the **time_bucket** set and subtracting one (this is due to the fact that in C-language the first location of an array is indexed by 0; the last entry of a vector of N components is indexed by $N - 1$).

Decision variables are declared in the **VARIABLE** section; here z_{it} is declared to be **LOGIC**, i.e., of the 0/1 type; variables may also be **INTEGER**; by default, they are nonnegative continuous.

Constraints are stated in the **CONSTRAINT** section; the syntax is self-explicative. A label is attached to each class of constraints. Constraints cap, dynam, link, set_z correspond to (6.12), (6.13), (6.14), (6.15), respectively, and require no explanation. Constraints dynam0 may sound odd at first. They are due to a point we overlook when writing a mathematical model: if we set $t = 1$ in (6.13), we get

$$x_{ik1} = x_{ik0} + s_{ik}u_{ik1} \qquad \forall i, k,$$

```
addfun GENERIC
SET
        i batches;
        k stages;
        t, p time_buckets;
ENDSET
DATA
        l[i];
        s[i,k];
        umax[i,k];
        Delta;
        w[i,t];
ENDDATA
INTERNALDATA
        last_t = |time_buckets| - 1;
        last_k = |stages| - 1;
ENDINTERNALDATA
VARIABLE
        x[i,k,t];
        z[i,t] LOGIC;
        u[i,k,t];
ENDVARIABLE
CONSTRAINT
    cap (FORALL k, t)
        SUM(u[i,k,t])(OVERALL i) <= 1;
    dynam (FORALL i, k; FOR t >= 1)
        x[i,k,t] = x[i,k,t-1] + s[i,k] * u[i,k,t];
    dynam0 (FORALL i)
        x[i,0,0] = s[j,0] * u[j,0,0];
    link (FORALL i; FOR k >= 1; FOR t >= Delta)
        x[i,k,t] <= x[i,k-1,t-Delta];
    setz (FORALL i, t; FOR k = last_k)
        SUM(l[i] * z[i,p])(FOR p > t) + x[i,k,t] >= l[i];
    z_sos (FORALL i)
        SOS3(z[i,t])(OVERALL t);
ENDCONSTRAINT
LIMIT
   bounds
        (FORALL i,k,t)  u[i,k,t] <= umax[i,k];
        (FORALL i) z[i,0] = 0;
        (FORALL i; FOR k=last_k; FOR t=last_t)
                                x[i,k,t] = l[i];
        (FORALL i; FOR k>=1) x[i,k,0] = 0;
ENDLIMIT
OBJFUNCTION
    MINIMIZE cost
        SUM(w[i,t] * z[i,t])(OVERALL i, t);
ENDOBJFUNCTION
```

TABLE 6.1
Sample EasyModeler program

where x_{ik0} is *not* a variable but a datum; the same issue occurs in the inventory equations for lot sizing models. In an EasyModeler model structure (where 0 corresponds to the *first* time bucket), this would lead to a decision variable x[i,k,-1], which does not exist. Therefore, we have to write (6.13) for the first time bucket apart. Constraints z_sos correspond to (6.16), which is translated as a SOS of type 3. The LIMIT section contains a simple upper bound or equality constraints. They could be added to the CONSTRAINT section, but they are dealt with in a computationally more efficient manner if set apart.

Finally, the OBJFUNCTION section contains the statement of the objective function (6.11).

For further reading

An overview of optimization approaches by simulation can be found in [81], where a good bibliography is provided. See also [119] for an application of the Response Surface Methodology to Production Planning, and [163] for a recent paper on quasi-Newton optimization methods by simulation. Many examples of the application of queueing models to the optimal design of an FMS are given in [182]. For second-order properties of performance measures, such as convexity and concavity, see [35]. Perturbation analysis is the topic of an engineering-oriented textbook [104] and a more formal one [88]. See [89] for a more contained treatment of mathematical issues. Most applications of perturbation analysis deal with queueing systems. An application to inventory management is reported in [17], and its use for the design of a flow controller (of the type described in Section 6.2) is presented in [40]. A comparison of different optimization schemes coupled with simulation for a simple queueing system can be found in [122].

Stochastic optimal control models for failure-prone manufacturing systems were originally proposed in [116]; see also [5]. A tutorial introduction to this topic can be found in [171]. Several papers have been published on this subject; we refer the reader to the bibliography presented in [83, Chapter 9].

As to model management issues in mathematical programming, there is a recent volume of the Annals of Operations Research which has been devoted to this topic [172]. More specific surveys on languages for mathematical programming can be found in [96, 139].

APPENDICES

A
Fundamentals of Mathematical Programming

In this appendix we review some fundamental concepts of Mathematical Programming, with emphasis on continuous nonlinear programming problems. This may seem a rather odd choice, since the reader will find little use of nonlinear programming in the book; most optimization models we consider are linear and discrete. Nevertheless, many *concepts* we introduce here turn out most useful to devise discrete optimization *methods*, and we believe it is better to introduce them in the most general framework.

In Section A.1 we illustrate the role of convexity in Mathematical Programming. Unconstrained optimization is dealt with in Sections A.2, where we recall basic optimality conditions, and A.3, where we outline the structure of numerical methods. Then we consider constrained problems in Section A.4, with emphasis on Lagrangian multiplier methods. Lagrangian Duality Theory, treated in section A.5, is certainly the most important concept introduced in this appendix. Lagrangian Duality is applied in decomposition strategies to tackle large-scale problems (Section A.6) and Lagrangian Relaxation based Branch and Bound algorithms (Section C.4.3); Lagrangian approaches also turn out to be most useful in developing dual heuristic methods (see e.g. Section 3.2.3). More often than not, the application of Duality Theory requires the optimization of a nondifferentiable objective function; in Section A.7 we review the basic approach, i.e., the subgradient method, for the optimization of a nondifferentiable function.

A warning to the reader is in order. Given the ultimate aims of the book, we have striven to follow the most intuitive conceptual path, and the need for mathematical rigor has been largely sacrificed. Furthermore, we have covered only the topics which are relevant for the rest of the book; for instance, primal methods for constrained optimization are completely overlooked.

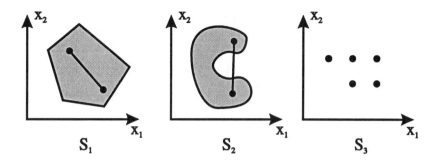

FIGURE A.1
Illustration of the convexity property

A.1 Convex sets and convex functions

Students with a Calculus background usually think that the most fundamental properties of functions are continuity and differentiability; in Mathematical Programming an even more important role is played by convexity. We first introduce the concept of convex *set*, and then we consider convex *functions*.

DEFINITION A.1 *A set $S \subseteq \mathbf{R}^n$ is a **convex set** if*

$$\mathbf{x}, \mathbf{y} \in S \Rightarrow \lambda \mathbf{x} + (1-\lambda)\mathbf{y} \in S, \quad \forall \lambda \in [0,1].$$

Example A.2
The concept of convexity can be intuitively grasped by considering that the points of the form $\lambda \mathbf{x} + (1-\lambda)\mathbf{y}$ ($0 \leq \lambda \leq 1$) are simply the points on the straight line joining $\mathbf{x}$ and $\mathbf{y}$. A set S is convex if the line joining any pair of points $\mathbf{x}, \mathbf{y} \in S$ is contained in S. Consider the three subsets of $\mathbf{R}^2$ depicted in Figure A.1. S_1 is convex, but S_2 is not. S_3 is a discrete set and it is not convex; this fact has important consequences for discrete optimization problems. ☐

PROPERTY A.1
The intersection of convex sets is a convex set.

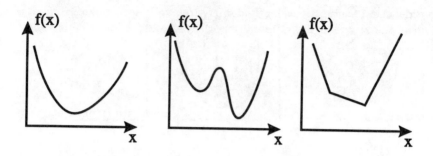

FIGURE A.2
Convex and nonconvex functions

The **convex combination** of p points $x_1, x_2, \ldots, x_p \in \mathbf{R}^n$ is defined as

$$x = \sum_{i=1}^{p} \mu_i x_i, \qquad \mu_1, \ldots, \mu_p \geq 0, \qquad \sum_{i=1}^{p} \mu_i = 1.$$

Given a set $S \subset \mathbf{R}^n$, the set of points which are the convex combinations of points in S is the **convex hull** of S (denoted by $[S]$). If S is a convex set, then $S \equiv [S]$. The convex hull of a generic set S is the smallest convex set containing S; it can also be regarded as the intersection of all the convex sets containing S.

DEFINITION A.3 *A scalar function $f: \mathbf{R}^n \to \mathbf{R}$ defined over a convex set $S \subseteq \mathbf{R}^n$ is a* **convex function** *if*

$$x, y \in S, \; \lambda \in [0,1] \Rightarrow f(\lambda x + (1-\lambda)y) \leq \lambda f(x) + (1-\lambda)f(y).$$

DEFINITION A.4 *A function f is* **concave** *if $(-f)$ is convex.*

The concept of a convex function is strictly related to the concept of convex set. Let $\text{epi}(f)$ denote the **epigraph** of f, i.e.,

$$\text{epi}(f) = \{(\mu, x) \mid f(x) \leq \mu, x \in \mathbf{R}^n, \mu \in \mathbf{R}\} \subset \mathbf{R}^{n+1}.$$

A function f is convex if and only if $\text{epi}(f)$ is a convex set.

A further link between convex sets and convex functions is that the set $S = \{x \in \mathbf{R}^n \mid g(x) \leq 0\}$ is convex if and only if g is a convex function.

Example A.5

Consider the three functions depicted in Figure A.2. The first one is convex, whereas the second is not; this can be easily checked by considering the line joining pairs of points on the respective graphs. The third function, though not everywhere differentiable, is still a convex function. □

Convexity of functions is preserved by some operations; in particular, a linear combination of convex functions f_i,

$$f(\mathbf{x}) = \sum_{i=1}^{n} \lambda_i f_i(\mathbf{x}),$$

is a convex function if $\lambda_i \geq 0$.

There are alternative characterizations of a convex function. For our purposes the most important is the following. If f is a differentiable function, then it is convex (over S) if and only if:

$$f(\mathbf{x}) \geq f(\mathbf{x}_0) + \nabla f^T(\mathbf{x}_0)(\mathbf{x} - \mathbf{x}_0) \qquad \forall \mathbf{x}, \mathbf{x}_0 \in S.$$

Note that the hyperplane

$$z = f(\mathbf{x}_0) + \nabla f^T(\mathbf{x}_0)(\mathbf{x} - \mathbf{x}_0)$$

is the usual tangent hyperplane, i.e., the first-order Taylor expansion of f at $\mathbf{x}_0$. For a differentiable function, convexity implies that the first-order approximation at a certain point $\mathbf{x}_0$ consistently underestimates the true value of the function at all the other points $\mathbf{x}$. The concept of a tangent hyperplane applies only to differentiable convex functions, but it can be generalized by the concept of a support hyperplane.

DEFINITION A.6 *Given a convex function f and a point $\mathbf{x}^0$, the hyperplane (in $\mathbf{R}^{n+1}$) given by $z = f(\mathbf{x}^0) + \gamma^T(\mathbf{x} - \mathbf{x}^0)$, which meets the epigraph of f in $(\mathbf{x}^0, f(\mathbf{x}^0))$ and lies below it is called the* **support hyperplane** *of f at $\mathbf{x}^0$.*

The concept of a support hyperplane is depicted in Figure A.3. If f is differentiable in $\mathbf{x}_0$, the support hyperplane is the usual tangent hyperplane (i.e., $\gamma = \nabla f(\mathbf{x}_0)$); if f is nondifferentiable, the support hyperplane need not be unique (see also Section A.7).

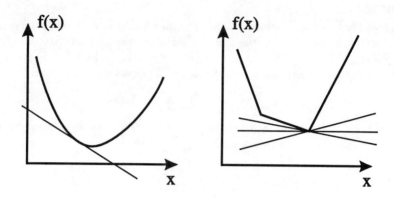

FIGURE A.3
Illustration of the support hyperplane

A.1.1 Qualitative properties of mathematical programming problems

Solving a mathematical programming problem

$$\min_{\mathbf{x} \in S} f(\mathbf{x})$$

means finding a point $\mathbf{x}^* \in S$ such that

$$f(\mathbf{x}^*) \leq f(\mathbf{x}) \qquad \forall \mathbf{x} \in S;$$

$\mathbf{x}^*$ is said a **global optimum** (the term **minimizer** is also used). Note that, in general, the global optimum is not unique. If the condition $f(\mathbf{x}^*) \leq f(\mathbf{x})$ holds only in a neighborhood of $\mathbf{x}^*$, then we speak of a **local optimum** (see Figure A.4). In this appendix we will state properties only for minimization problems; this is not a loss of generality, since a maximization problem can be readily transformed into a minimization problem

$$\max_{\mathbf{x} \in S} f(\mathbf{x}) = -\min_{\mathbf{x} \in S} (-f(\mathbf{x})).$$

DEFINITION A.7 *A mathematical programming problem*

$$\min_{\mathbf{x} \in S} f(\mathbf{x})$$

is a **convex problem** *if f is a convex function and S is a convex set.*
  A problem

$$\min_{\mathbf{x} \in S} f(\mathbf{x})$$

is a **concave problem** *if f is a concave function and S is a convex set.*

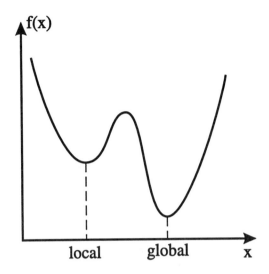

FIGURE A.4
Global and local optima

Note that a set S of the form $S = \{\mathbf{x} \in \mathbf{R}^n \mid g_i(\mathbf{x}) \leq 0,\ i \in I\}$ is convex if the functions g_i are convex. Assuming that the optimization problem has a finite solution, the following properties can be shown.

PROPERTY A.2
In a convex problem a local optimum is also a global optimum.

PROPERTY A.3
In a concave problem the global optimum lies on the boundary of the feasible region S.

A.2 Unconstrained optimization problems

We have an unconstrained problem if the feasible set $S \equiv \mathbf{R}^n$. Assuming a twice continuously differentiable function, the problem can be solved by analyzing the properties of f in $\mathbf{x}^*$.

THEOREM A.4
(**Necessary conditions for local optimality**) *The point $\mathbf{x}^*$ is a local optimum of f only if*

1. $\nabla f(\mathbf{x}^*) = \mathbf{0}$ *(stationarity condition)*;
2. *the Hessian matrix*

$$\nabla^2 f(\mathbf{x}^*) = \left[\frac{\partial^2 f}{\partial x_i \partial x_j}(\mathbf{x}^*)\right]$$

is semidifinite positive (i.e., $\mathbf{s}^T \nabla^2 f(\mathbf{x}^*)\mathbf{s} \geq 0$, $\forall \mathbf{s} \in \mathbf{R}^n$).

THEOREM A.5
(**Sufficient** conditions for local optimality) *The point $\mathbf{x}^*$ is a local optimum of f if*

1. $\nabla f(\mathbf{x}^*) = \mathbf{0}$;
2. *the Hessian matrix is definite positive (i.e., $\mathbf{s}^T \nabla^2 f(\mathbf{x}^*)\mathbf{s} > 0$, $\forall \mathbf{s} \in \mathbf{R}^n$, $\mathbf{s} \neq \mathbf{0}$).*

These two theorems are somewhat disappointing; they concern *local* rather than *global* optimality, and they not provide us with necessary *and* sufficient conditions. A better situation holds in the case of convex functions.

THEOREM A.6
(**Necessary and sufficient condition for global optimality**) *The point $\mathbf{x}^*$ is a global optimum of f if*

1. *f is a convex function;*
2. $\nabla f(\mathbf{x}^*) = \mathbf{0}$.

PROOF If f is convex and differentiable, we have

$$f(\mathbf{x}) \geq f(\mathbf{x}_0) + \nabla f^T(\mathbf{x}_0)(\mathbf{x} - \mathbf{x}_0) \qquad \forall \mathbf{x}, \mathbf{x}_0 \in S.$$

But if $\mathbf{x}^*$ is a stationarity point, we have

$$f(\mathbf{x}) \geq f(\mathbf{x}^*) + \nabla f^T(\mathbf{x}^*)(\mathbf{x}-\mathbf{x}^*) = f(\mathbf{x}^*) + \mathbf{0}^T(\mathbf{x}-\mathbf{x}^*) = f(\mathbf{x}^*) \qquad \forall \mathbf{x} \in S,$$

which simply states that $\mathbf{x}^*$ is the global optimum. ∎

The stationarity condition is a *first-order* condition, whereas the conditions on the Hessian matrix are *second-order* ones. In the following we will be interested only in first-order conditions. In Section A.7 we will see that the stationarity condition is easily extended to the case of a convex nondifferentiable function.

A.3 Numerical methods for unconstrained optimization

The stationarity condition yields a set of nonlinear equations which could be solved to spot candidate optima. Due to the difficulty of analytically solving such equations, numerical methods are adopted in practice. Numerical methods are based on an iterative search process; a sequence of points $\mathbf{x}^{(k)}$ is generated, converging to a local optimum $\mathbf{x}^*$.

A general scheme is:

1. find a search direction $\mathbf{s}^{(k)} \in \mathbf{R}^n$;
2. find a step length $\alpha^{(k)} \in \mathbf{R}_+$, e.g. by solving the one-dimensional problem
$$\min_{\alpha \geq 0} f(\mathbf{x}^{(k)} + \alpha \mathbf{s}^{(k)});$$
3. update $\mathbf{x}^{(k+1)} = \mathbf{x}^{(k)} + \alpha^{(k)} \mathbf{s}^{(k)}$.

The search direction s must be a *descent direction*. We say that s is a descent direction of f at the point x if, for a sufficiently small step α, $f(\mathbf{x} + \alpha \mathbf{s}) < f(\mathbf{x})$. If we consider the function $h(\alpha) = f(\mathbf{x} + \alpha \mathbf{s})$, a descent direction is characterized by:

$$\left.\frac{dh}{d\alpha}\right|_{\alpha=0} = \nabla f^T(\mathbf{x})\mathbf{s} < 0.$$

A possible choice is:

$$\mathbf{s}^{(k)} = -\nabla f^{(k)},$$

which yields the **steepest descent** or **gradient method**. The steepest descent method suffers from poor convergence near the minimizer (zigzagging). This is essentially due to the fact that the gradient method uses a *first-order* local approximation of f; using a second-order approximation (involving second derivatives) yields the Newton method, which has better convergence properties, as well as higher computational requirements. We will not use Newton-like methods in this book; the reader should bear in mind that in a gradient-like method, the step size should be adapted. It must be relatively large at the first iterations, and it should be reduced in order to avoid zig-zagging near the minimizer.

Another shortcoming, common to all gradient-like methods, is that they get easily trapped in local optima in the case of nonconvex objective functions. A brute-force approach to overcome the problem is to start the search process from different initial solutions. There are more sophisticated methods for continuous global optimization, but they are beyond the scope of this book; in Section C.8.4 we will consider global optimization methods for *discrete* optimization problems.

If the gradient of the objective function is difficult to compute, it is possible to approximate it with finite differences; this may be useful when the objective function is not known, but its values are obtained by evaluative models (see Chapter 6). Given a point $\hat{\mathbf{x}}$ and a vector of step sizes $\mathbf{h}$, we can approximate the first derivative of f as a finite difference,

$$\left.\frac{\partial f(\mathbf{x})}{\partial x_i}\right|_{\mathbf{x}=\hat{\mathbf{x}}} \approx \frac{f(\hat{\mathbf{x}}+h_i \mathbf{1}_i)-f(\hat{\mathbf{x}})}{h_i},$$

where $\mathbf{1}_i$ is the ith unit vector. An alternative approximation is based on **central differences**,

$$\left.\frac{\partial f(\mathbf{x})}{\partial x_i}\right|_{\mathbf{x}=\hat{\mathbf{x}}} \approx \frac{f(\hat{\mathbf{x}}+h_i \mathbf{1}_i)-f(\hat{\mathbf{x}}-h_i \mathbf{1}_i)}{2h_i}.$$

The gradient method can be extended to deal with constrained optimization problems; we will not consider such methods, but it is worth noting that for the specific case

$$\min_{\mathbf{x} \in \mathbf{R}_+^n} f(\mathbf{x})$$

it is sufficient to slightly modify the updating rule as follows:

$$\mathbf{x}^{(k+1)} = \max\left\{\mathbf{0}, \mathbf{x}^{(k)} + \alpha^{(k)} \mathbf{s}^{(k)}\right\},$$

which should be interpreted componentwise, i.e., if some component becomes negative, then set it at zero. This operation essentially amounts to *projecting* $\mathbf{x}^{(k+1)}$ onto the feasible set $\mathbf{R}_+^n$.

A.4 Constrained optimization problems

Usually, the decision variables of an optimization problem are constrained, and stationarity is no longer a necessary condition for optimality. The general form of a constrained optimization problem is

$$\min \ f(\mathbf{x})$$
$$\text{s.t.} \ \ h_i(\mathbf{x}) = 0 \quad i \in E$$
$$g_i(\mathbf{x}) \leq 0 \quad i \in I,$$

where E is the set of the **equality constraints** and I is the set of **inequality constraints**. In this section we assume that all the involved functions have suitable differentiability properties.

We discuss two classes of methods: those based on penalty functions and those based on Lagrangian multipliers. A common characteristic of the two approaches is that they transform a constrained problem into an unconstrained one. We do not describe primal (as opposed to dual) methods, i.e., methods based on the generalization of gradient search whereby the search process is kept within the feasible region. This choice is in no way dictated by algorithmic efficiency issues but by the relevance of the involved concepts for the remainder of the book.

A.4.1 The penalty function approach

Consider a problem with equality constraints

$$\min\ f(\mathbf{x})$$
$$\text{s.t.}\ h_i(\mathbf{x}) = 0 \quad i \in E.$$

It is possible to transform the constrained problem into an unconstrained one by adding a penalty term to the objective function

$$\Phi(\mathbf{x}, \sigma) = f(\mathbf{x}) + \sigma \sum_{i \in E} h_i^2(\mathbf{x}),$$

where σ is a number chosen in order to penalize constraint violations.

A possible solution algorithm is based on a sequence of unconstrained problems:

1. choose a sequence $\{\sigma^{(k)}\} \to \infty$;
2. find the minimizer $\mathbf{x}^*(\sigma^{(k)})$ of $\Phi(\mathbf{x}, \sigma)$;
3. stop if $h_i(\mathbf{x}^*)$ is sufficiently small for all i.

The case of inequality constraints

$$\min\ f(\mathbf{x})$$
$$\text{s.t.}\ g_i(\mathbf{x}) \leq 0 \quad i \in I,$$

can be tackled by two approaches: **exterior penalties** and **interior penalties**. In the exterior penalty approach we introduce a penalty function such as

$$v(y) = \begin{cases} 0 & \text{for } y \leq 0 \\ y^2 & \text{for } y > 0. \end{cases}$$

Then, we solve a sequence of problems

$$\min\ f(\mathbf{x}) + \sigma \sum_{i \in I} v(g_i(\mathbf{x})) = f(\mathbf{x}) + \sigma \sum_{i \in I} \left[g_i^+(\mathbf{x})\right]^2$$

for increasing values of σ. The name *exterior* stems from the fact that the feasible set is approached from outside, i.e., infeasible solutions may be obtained during the solution process. To overcome this problem, an interior penalty approach can be pursued, by introducing a **barrier** function

$$B(\mathbf{x}) = -\sum_{i \in I} \frac{1}{g_i(\mathbf{x})}.$$

The barrier function tends to infinity when $\mathbf{x}$ tends to the boundary of the feasible region. Then an unconstrained problem

$$\min f(\mathbf{x}) + \sigma B(\mathbf{x})$$

is solved for decreasing values of σ, until the term $\sigma B(\mathbf{x})$ is small enough.

The penalty function approach is conceptually very simple, and some convergence properties can be proved. However, severe numerical difficulties may arise, for instance, when σ gets very large in the exterior penalty approach. Nevertheless penalty functions are most useful in providing a starting point for other, more sophisticated, methods. Furthermore, they can be a valuable tool for solving production management problems with tight constraints. Consider, for instance, an inequality constraint

$$\mathbf{g}(\mathbf{x}) \le \mathbf{b},$$

where $\mathbf{b}$ is a resource availability limiting production levels modeled by $\mathbf{x}$. It might well be the case that no feasible solution exists; in practice, this does not imply that the production management problem should be dismissed. It would be useful, from a diagnostic point of view, to relax the constraints with a suitably large penalty, and solve the relaxed model in order to spot what resources are critical and when (see Section 3.1.2).

A.4.2 Multiplier methods: optimization with equality constraints

Consider the equality constrained problem (P_E)

$$\min\ f(\mathbf{x})$$
$$\text{s.t.}\ h_i(\mathbf{x}) = 0 \qquad i \in E.$$

The classical Lagrangian approach is based on the following theorem.

THEOREM A.7
Assume that the functions f, h_i in (P_E) are continuously differentiable and that $\mathbf{x}^*$ is feasible and satisfies a constraint qualification condition. Then a necessary condition for the local optimality of $\mathbf{x}^*$ is that there exists a set of numbers λ_i^*, $i \in E$, such that

$$\nabla f(\mathbf{x}^*) + \sum_{i \in E} \lambda_i^* \nabla h_i(\mathbf{x}^*) = \mathbf{0}.$$

Note that the theorem only states a *necessary* condition for *local* optimality; the same condition is necessary and sufficient for global optimality in the case of a convex problem. A sufficient condition for constraint qualification is that the gradients $\nabla h_i(\mathbf{x}^*)$ are linearly independent. The role of the constraint qualification will be clarified shortly in Example A.9. The theorem basically requires that the gradient of f at the optimum is a linear combination of the gradients of the constraints.

We can see that the theorem generalizes the stationarity condition to the equality constrained case. Consider the **Lagrangian function**

$$\mathcal{L}(\mathbf{x}, \boldsymbol{\lambda}) = f(\mathbf{x}) + \sum_{i \in E} \lambda_i h_i(\mathbf{x}) = f(\mathbf{x}) + \boldsymbol{\lambda}^T \mathbf{h}(\mathbf{x}), \qquad (A.1)$$

where the numbers λ_i are called **Lagrange multipliers**. The multipliers $\boldsymbol{\lambda}$ are also known as **dual variables**, whereas the components of $\mathbf{x}$ are called the **primal variables**. The stationarity condition of the unconstrained case must be applied to $\mathcal{L}(\mathbf{x}, \boldsymbol{\lambda})$, with respect to the primal and dual variables, i.e.,

$$\nabla_\mathbf{x} \mathcal{L}(\mathbf{x}, \boldsymbol{\lambda}) = \nabla f(\mathbf{x}) + \sum_{i \in E} \lambda_i \nabla h_i(\mathbf{x}) = \mathbf{0}, \qquad (A.2)$$

which is the condition required by the theorem above, and

$$\nabla_{\boldsymbol{\lambda}} \mathcal{L}(\mathbf{x}, \boldsymbol{\lambda}) = \begin{bmatrix} h_{i_1}(\mathbf{x}) \\ h_{i_2}(\mathbf{x}) \\ \vdots \end{bmatrix} = \mathbf{0}, \qquad (A.3)$$

which is simply the satisfaction of the equality constraints. Note that if we have m inequality constraints, Eqs. (A.2) and (A.3) yield a system of $n + m$ nonlinear equations to find the n primal variables x_i and the m dual variables λ_i.

Example A.8

Consider the quadratic program

$$\min \; x_1^2 + x_2^2$$
$$\text{s.t.} \; x_1 + x_2 = 4;$$

since this is a convex problem, the stationarity of the Lagrangian is a necessary and sufficient condition for global optimality. The Lagrangian function is

$$\mathcal{L}(x_1, x_2, \lambda) = x_1^2 + x_2^2 + \lambda(x_1 + x_2 - 4).$$

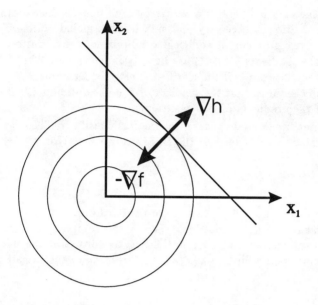

FIGURE A.5
Geometrical interpretation of the Lagrangian conditions

The stationarity conditions

$$\frac{\partial \mathcal{L}}{\partial x_1} = 2x_1 + \lambda$$

$$\frac{\partial \mathcal{L}}{\partial x_2} = 2x_2 + \lambda$$

$$\frac{\partial \mathcal{L}}{\partial \lambda} = x_1 + x_2 - 4 = 0,$$

yield $x_1^* = x_2^* = 2$ and $\lambda^* = -4$.

Figure A.5 illustrates the Lagrangian conditions geometrically: the gradient of the objective function at the optimum parallel to the gradient of the constraint. This implies that ∇f^* is orthogonal to the constraint, i.e., to improve the objective function we must get out of the feasible set. □

Example A.9
We illustrate here the importance of the constraint qualification condition. Consider the problem

$$\min\ x_1 + x_2$$

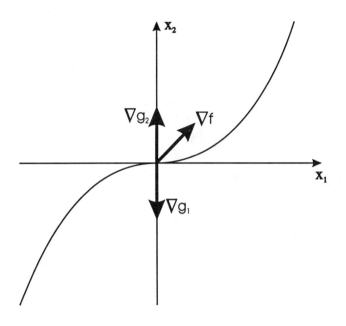

FIGURE A.6
Geometrical illustration of the constraint qualification condition

$$\text{s.t. } x_2 = x_1^3$$
$$x_2 = 0$$

From figure A.6 it is clear that the (trivial) solution of this problem is the only feasible point $(0, 0)$.

If we mechanically follow the Lagrangian approach, we build the Lagrangian function

$$\mathcal{L}(x_1, x_2, \lambda_1, \lambda_2) = x_1 + x_2 + \lambda_1(x_2 - x_1^3) + \lambda_2 x_2.$$

Stationarity yields the system

$$\frac{\partial \mathcal{L}}{\partial x_1} = 1 - 3\lambda_1 x_1^2 = 0$$
$$\frac{\partial \mathcal{L}}{\partial x_2} = 1 + \lambda_1 + \lambda_2 = 0$$

$$\frac{\partial \mathcal{L}}{\partial \lambda_1} = x_2 - x_1^3 = 0$$

$$\frac{\partial \mathcal{L}}{\partial \lambda_2} = x_2 = 0$$

which has no solution (the first equation requires that $x_1 \neq 0$, which is not compatible with the last two equations). This is due to the fact that the gradients of the two constraints are parallel at the origin and are not a basis able to express the gradient of f (see Figure A.6). □

A.4.3 Multiplier methods: optimization with inequality constraints

Consider the inequality constrained problem (P_I):

$$\min \ f(\mathbf{x})$$
$$\text{s.t.} \ g_i(\mathbf{x}) \leq 0 \qquad i \in I.$$

The Lagrangian approach is generalized by the following conditions.

THEOREM A.8
Assume that the functions f, g_i in (P_I) are continuously differentiable, and that $\mathbf{x}^*$ is feasible and satisfies a constraint qualification condition. Then a necessary condition for the local optimality of $\mathbf{x}^*$ is that there exists a set of numbers $\mu_i^* \geq 0 \ i \in I$ such that:

$$\nabla f(\mathbf{x}^*) + \sum_{i \in I} \mu_i^* \nabla g_i(\mathbf{x}^*) = \mathbf{0}; \qquad (A.4)$$

$$\mu_i^* g_i(\mathbf{x}^*) = 0. \qquad (A.5)$$

As with the equality constrained case, these conditions are necessary and sufficient for global optimality in the convex case.

Eq. (A.4) is the same as in the equality constrained case, but now the multipliers are required to be nonnegative. This nonnegativity requirements can be intuitively justified by regarding the Lagrangian function

$$f(\mathbf{x}) + \sum_{i \in I} \mu_i g_i(\mathbf{x})$$

as a sort of penalty function to be minimized. Eq. (A.5) is called the **complementary slackness** condition and can be interpreted as follows. Let us introduce first the concept of **active constraints**: an inequality constraint is active in $\mathbf{x}_0$ if $g_i(\mathbf{x}_0) = 0$; it is inactive in $\mathbf{x}_0$ if $g_i(\mathbf{x}_0) < 0$. Note that inactive constraints at the optimum could be dropped without changing the result, whereas active constraints behave as equality constraints at the optimal solution. The complementary slackness condition basically states that dual variables corresponding to inactive constraints are zero, and that nonzero dual variables are associated with active constraints (see also Section A.4.5).

A.4.4 Kuhn-Tucker conditions for the general constrained problem

We consider now the general problem (P_{EI})

$$\min \ f(\mathbf{x})$$
$$\text{s.t.} \ h_i(\mathbf{x}) = 0 \quad i \in E$$
$$g_i(\mathbf{x}) \le 0 \quad i \in I.$$

In this case the two previous theorems can be merged by considering a Lagrangian function

$$\mathcal{L}(\mathbf{x}, \boldsymbol{\lambda}, \boldsymbol{\mu}) = f(\mathbf{x}) + \sum_{i \in E} \lambda_i h_i(\mathbf{x}) + \sum_{i \in I} \mu_i g_i(\mathbf{x}). \qquad (A.6)$$

THEOREM A.9
(Kuhn-Tucker Conditions) *Assume that the functions f, h_i, g_i in (P_{EI}) are continuously differentiable, and that $\mathbf{x}^*$ is feasible and satisfies a constraint qualification condition. Then a necessary condition for the local optimality of $\mathbf{x}^*$ is that there exists numbers λ_i^* $(i \in E)$ and $\mu_i^* \ge 0$ $(i \in I)$ such that:*

$$\begin{cases} \nabla f(\mathbf{x}^*) + \sum_{i \in E} \lambda_i^* \nabla h_i(\mathbf{x}^*) + \sum_{i \in I} \mu_i^* \nabla g_i(\mathbf{x}^*) = 0 \\ \mu_i^* g_i(\mathbf{x}^*) = 0 \quad \forall j \in I. \end{cases}$$

These conditions are necessary and sufficient for global optimality in the convex case.

Example A.10
Consider the problem

$$\min \ x_1^2 + x_2^2$$
$$\text{s.t.} \ x_1 \ge 0$$
$$x_2 \ge 3$$
$$x_1 + x_2 = 4.$$

We first write the Lagrangian function
$$\mathcal{L}(\mathbf{x}, \boldsymbol{\mu}, \lambda) = x_1^2 + x_2^2 - \mu_1 x_1 - \mu_2(x_2 - 3) + \lambda(x_1 + x_2 - 4).$$
A set of numbers satisfying the Kuhn-Tucker conditions can be found by solving the following system:
$$\begin{cases} 2x_1 - \mu_1 + \lambda = 0 \\ 2x_2 - \mu_2 + \lambda = 0 \\ x_1 \geq 0, \quad x_2 \geq 3 \\ x_1 + x_2 = 4 \\ \mu_1 x_1 = 0, \quad \mu_1 \geq 0 \\ \mu_2(x_2 - 3) = 0, \quad \mu_2 \geq 0. \end{cases}$$
We have to proceed with a case-by-case analysis in order to find out which components of μ are zero.

Case 1 ($\mu_1 = \mu_2 = 0$). From the stationarity conditions we obtain the system
$$\begin{cases} 2x_1 + \lambda = 0 \\ 2x_1 + \lambda = 0 \\ x_1 + x_2 - 4 = 0, \end{cases}$$
yielding $x_1 = x_2 = 2$, which violates the second inequality constraint.

Case 2 ($\mu_1, \mu_2 \neq 0$). The complementary slackness conditions immediately yield $x_1 = 0, x_2 = 3$, violating the equality constraint.

Case 3 ($\mu_1 \neq 0, \mu_2 = 0$). We obtain
$$\begin{cases} x_1 = 0 \\ x_2 = 4 \\ \lambda = -2x_2 = -8 \\ \mu_1 = \lambda = -8. \end{cases}$$
The Kuhn-Tucker conditions are not satisfied since the value of μ_1 is negative.

Case 4 ($\mu_1 = 0, \mu_2 \neq 0$). We obtain
$$\begin{cases} x_2 = 3 \\ x_1 = 1 \\ \lambda = -2 \\ \mu_2 = 4, \end{cases}$$
which satisfy all the necessary conditions.

Since this is a convex problem, we have obtained the global optimum.

Note how nonzero multipliers correspond to the active constraints, whereas the inactive constraint $x_1 \geq 0$ has $\mu_1 = 0$. □

A.4.5 Interpreting dual variables

The complementary slackness condition (A.5) suggests that the dual variables are somehow related to the "importance" of the corresponding constraints in determining the optimal solution. If $\mu_i = 0$, then the corresponding constraint can be dropped from the formulation without changing the solution. In this section we formalize this intuition.

Consider an equality-constrained problem and apply a small perturbation to the constraints

$$h_i(\mathbf{x}) = \epsilon_i \qquad i \in E.$$

Applying the Lagrangian approach to the perturbed problem we get a new solution $\mathbf{x}^*(\epsilon)$ and a new multiplier vector $\boldsymbol{\lambda}^*(\epsilon)$, both depending on ϵ. The Lagrangian function for the perturbed problem is

$$\mathcal{L}(\mathbf{x}, \boldsymbol{\lambda}, \epsilon) = f(\mathbf{x}) + \sum_{i \in E} \lambda_i (h_i(\mathbf{x}) - \epsilon_i). \qquad (A.7)$$

Equality constraints are satisfied by the optimal solution of the perturbed problem. Hence

$$f^* = f(\mathbf{x}^*(\epsilon)) = \mathcal{L}(\mathbf{x}^*(\epsilon), \boldsymbol{\lambda}^*(\epsilon), \epsilon). \qquad (A.8)$$

We can evaluate the derivative of the optimal value with respect to each component of ϵ,

$$\frac{df^*}{d\epsilon_i} = \frac{d\mathcal{L}}{d\epsilon_i} = \underbrace{\nabla_\mathbf{x} \mathcal{L}^T \frac{\partial \mathbf{x}}{\partial \epsilon_i} + \nabla_\boldsymbol{\lambda} \mathcal{L}^T \frac{\partial \boldsymbol{\lambda}}{\partial \epsilon_i}}_{=0} + \frac{\partial \mathcal{L}}{\partial \epsilon_i} = \frac{\partial \mathcal{L}}{\partial \epsilon_i} = -\lambda_i, \qquad (A.9)$$

where we have used the stationarity condition of $\mathcal{L}$. This result shows that, apart from the sign, the dual variables express the *sensitivity* of the optimal value with respect to the constraints.

Note that the result holds under the assumption that the optimal value is differentiable with respect to the perturbation; in fact, this is not always the case, and some technical complication would be needed to cope with nondifferentiability.

Example A.11
Consider the problem

$$\min \ (x_1 - 2)^2 + (x_2 - 2)^2$$
$$\text{s.t.} \ x_1 + x_2 = b,$$

where b is a parameter. We want to investigate the variation in the optimal value as a function of b. By writing the stationarity condition for the

Lagrangian function
$$\mathcal{L}(x_1, x_2, \lambda) = (x_1 - 2)^2 + (x_2 - 2)^2 + \lambda(x_1 + x_2 - b),$$
we get the linear system
$$\frac{\partial \mathcal{L}}{\partial x_1} = 2(x_1 - 2) + \lambda = 0$$
$$\frac{\partial \mathcal{L}}{\partial x_2} = 2(x_2 - 2) + \lambda = 0$$
$$\frac{\partial \mathcal{L}}{\partial \lambda} = x_1 + x_2 - b = 0,$$
which yields
$$x_1^* = x_2^* = \frac{b}{2}, \qquad \lambda^* = 4 - b.$$
The sensitivity
$$\frac{df(\mathbf{x}^*)}{db} = -\lambda^* = b - 4$$
shows that, for $b < 4$, an increase in b leads to a decrease in the objective function; for $b > 4$, it would lead to an increase in the objective function. If $b = 4$, then the constraint plays no role, and $\lambda^* = 0$. This is illustrated in Figure A.7. □

We have only considered the equality-constrained case, but the result can be extended to inequality-constrained problems by noting that, by complementary slackness, inactive inequality constraints have zero multipliers and active constraints behave like equality constraints.

If the objective function is expressed in dollars and the constraints model the resource availability, the multipliers have the dimension of prices per unit of resource. This is why dual variables are also known as **shadow prices**. This is of particular importance in the linear programming case (see Appendix B) and plays an important role in the development of heuristics for discrete optimization problems.

A.5 Duality theory

In the preceding sections we have shown that the stationarity of the Lagrangian function plays a crucial role for constrained optimization. Stationarity is linked to some optimality condition, either for minimization or maximization. It is rather intuitive that we should minimize the Lagrangian function with respect to the primal variables, but what about the

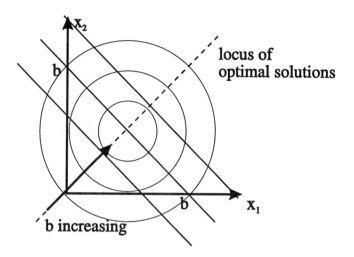

FIGURE A.7
Variation of the optimal solution with respect to a perturbation of the constraint

dual ones? In this section we show that interesting results are obtained by *maximizing* the Lagrangian function with respect to the dual variables, leading to **Duality Theory**.

Consider the inequality-constrained problem

$$(P) \quad \min f(\mathbf{x})$$
$$\text{s.t.} \quad g_i(\mathbf{x}) \leq 0 \quad i \in I \quad (A.10)$$
$$\mathbf{x} \in S \subseteq \mathbf{R}^n.$$

This problem is known as the **primal problem**. Note that the set S is any subset of $\mathbf{R}^n$, possibly a *discrete* one; furthermore, in this section we do not assume the differentiability nor the convexity of the objective function. The results obtained are therefore extremely general.

Consider the Lagrangian function obtained by *dualizing* constraints (A.10):

$$\mathcal{L}(\mathbf{x}, \boldsymbol{\mu}) = f(\mathbf{x}) + \sum_{i \in I} \mu_i g_i(\mathbf{x}).$$

For a given multiplier vector, the minimization of the Lagrangian function with respect to $\mathbf{x} \in S$ is called the **relaxed problem**; the solution of the relaxed problem defines a function $w(\boldsymbol{\mu})$, called the **dual function**:

$$w(\boldsymbol{\mu}) = \min_{\mathbf{x} \in S} \mathcal{L}(\mathbf{x}, \boldsymbol{\mu}).$$

Consider the **dual problem**:

$$(D) \qquad \max_{\mu \geq 0} w(\mu) = \max_{\mu \geq 0} \left\{ \min_{\mathbf{x} \in S} \mathcal{L}(\mathbf{x}, \mu) \right\}.$$

The following theorem holds.

THEOREM A.10
(**Weak duality theorem**) *For any $\mu \geq 0$ the dual function is a lower bound of the optimum $f(\mathbf{x}^*)$ of the primal problem (P), i.e.,*

$$w(\mu) \leq f(\mathbf{x}^*) \qquad \forall \mu \geq 0.$$

PROOF Under the hypothesis $\mu \geq 0$, it is easy to see that:

$$\nu(P) \geq \nu \begin{pmatrix} \min & f(\mathbf{x}) \\ \text{s.t.} & \mathbf{x} \in S \\ & \mu^T \mathbf{g}(\mathbf{x}) \leq 0 \end{pmatrix} \qquad (A.11)$$

$$\geq \nu \begin{pmatrix} \min & f(\mathbf{x}) + \mu^T \mathbf{g}(\mathbf{x}) \\ \text{s.t.} & \mathbf{x} \in S \\ & \mu^T \mathbf{g}(\mathbf{x}) \leq 0 \end{pmatrix} \qquad (A.12)$$

$$\geq \nu \begin{pmatrix} \min & f(\mathbf{x}) + \mu^T \mathbf{g}(\mathbf{x}) \\ \text{s.t.} & \mathbf{x} \in S \end{pmatrix}. \qquad (A.13)$$

Eq. (A.11) is justified by the fact that the points satisfying the set of constraints $g_i(\mathbf{x}) \leq 0$ also satisfy the constraint $\mu^T \mathbf{g}(\mathbf{x})$ if $\mu \geq 0$, but not vice versa. In other words, the feasible set of the first problem is a subset of the feasible set of the second one. Clearly, when we relax the feasible set, the optimal value cannot increase. Eq. (A.12) holds since the third problem involves the same feasible set as the second problem, but we have added a negative term to the objective function. Finally, Eq. (A.13) holds since the fourth problem is a relaxation of the third one (we delete a constraint).
∎

We obtain a very general, but weak condition. Under suitable conditions (essentially convexity), a stronger property holds, known as **strong duality**:

$$\nu(D) = w(\mu^*) = f(\mathbf{x}^*) = \nu(P).$$

The convexity assumption does not hold, in particular, for the case of a discrete set; therefore, in general, duality yields only a lower bound for discrete optimization problems (see section C.4.3). The following theorem is useful in establishing when the dual problem yields an optimal solution of the primal problem.

THEOREM A.11
If there is a pair $(\mathbf{x}^*, \boldsymbol{\mu}^*)$, *where* $\mathbf{x}^* \in S$ *and* $\boldsymbol{\mu}^* \geq 0$, *satisfying the following conditions:*

1. $f(\mathbf{x}^*) + \boldsymbol{\mu}^{*T}\mathbf{g}(\mathbf{x}^*) = \min_{\mathbf{x} \in S}\{f(\mathbf{x}) + \boldsymbol{\mu}^{*T}\mathbf{g}(\mathbf{x})\}$;
2. $\boldsymbol{\mu}^{*T}\mathbf{g}(\mathbf{x}^*) = 0$;
3. $\mathbf{g}(\mathbf{x}^*) \leq 0$;

then $\mathbf{x}^*$ *is a global optimum for the primal problem* (P).

In other words, the optimal solution $\mathbf{x}^*$ of the relaxed problem for a multiplier vector $\boldsymbol{\mu}^*$ is a global optimum for the primal problem if the pair $(\mathbf{x}^*, \boldsymbol{\mu}^*)$ is primal feasible, dual feasible, and it satisfies the complementary slackness conditions. Note that these are *sufficient* conditions for global optimality.

Weak duality holds in the equality constrained case, too. If we consider the optimal solution $\mathbf{x}^*$ of the primal problem

$$\min\ f(\mathbf{x})$$
$$\text{s.t.}\ h_i(\mathbf{x}) = 0 \quad i \in E$$
$$\mathbf{x} \in S,$$

and the optimal solution $\bar{\mathbf{x}}$ of the relaxed problem

$$\min_{\mathbf{x} \in S}\left\{f(\mathbf{x}) + \boldsymbol{\lambda}^T\mathbf{h}(\mathbf{x})\right\}$$

for any multiplier vector $\boldsymbol{\lambda}$ (not restricted in sign), it is easy to see that

$$f(\bar{\mathbf{x}}) + \boldsymbol{\lambda}^T\mathbf{h}(\bar{\mathbf{x}}) \leq f(\mathbf{x}^*) + \boldsymbol{\lambda}^T\mathbf{h}(\mathbf{x}^*) = f(\mathbf{x}^*).$$

Unfortunately, convexity does not hold easily for equality constraints. In fact, it holds only for linear equality constraints such as $\mathbf{a}_i^T\mathbf{x} = b_i$. Strong duality with equality constraints holds only in the Linear Programming case (see Section B.5).

Example A.12
Consider the problem

$$\min\ x_1^2 + x_2^2$$
$$\text{s.t.}\ x_1 + x_2 \geq 4$$
$$x_1, x_2 \geq 0.$$

The optimal value is 8, corrsponding to the optimal solution $(2, 2)$.

Since this is a convex problem, we can apply strong duality. The dual function is

$$w(\mu) = \min_{x_1,x_2}\{x_1^2 + x_2^2 + \mu(-x_1 - x_2 + 4); \text{ s.t. } x_1, x_2 \geq 0\} =$$

$$\min_{x_1}\{x_1^2 - \mu x_1; \text{ s.t. } x_1 \geq 0\} + \min_{x_2}\{x_2^2 - \mu x_2; \text{ s.t. } x_2 \geq 0\} + 4\mu.$$

Since $\mu \geq 0$, the optima with respect to x_1, x_2 are obtained for

$$x_1^* = x_2^* = \frac{\mu}{2}.$$

Hence,

$$w(\mu) = -\frac{1}{2}\mu^2 + 4\mu.$$

The maximum of the dual function is reached for $\mu^* = 4$, and we have $w(4) = f^* = 8$. □

In example A.12 we have found an explicit representation of the dual function. In general the maximization of the dual function must be tackled by a numerical method; in practice, an iterative procedure is adopted:

1. assign an initial value $\mu^{(0)} \geq 0$; set $k \leftarrow 0$;
2. solve the relaxed problem with multipliers $\mu^{(k)}$;
3. given the solution of the relaxed problem, compute a search direction $s^{(k)}$ and a step length $\alpha^{(k)}$, and update

$$\mu^{(k+1)} = \max\left\{0, \mu^{(k)} + \alpha^{(k)}s^{(k)}\right\}.$$

Then, set $k \leftarrow k + 1$, and go to step 2.

In order to find a search direction, one would be tempted to compute a gradient of the dual function. Unfortunately, the dual function is not everywhere differentiable, as we can see in the following example.

Example A.13
Consider the discrete optimization problem

$$\min \mathbf{c}^T\mathbf{x}$$

$$\text{s.t. } \mathbf{a}^T\mathbf{x} \geq b \qquad (A.14)$$

$$\mathbf{x} \in S = \{\mathbf{x}^1, \mathbf{x}^2, \ldots, \mathbf{x}^m\}, \qquad (A.15)$$

where $\mathbf{c}, \mathbf{a}, \mathbf{x} \in \mathbf{R}^n$, $b \in \mathbf{R}$ and S is a *discrete* set. Dualizing constraint (A.14) with a multiplier $\mu \geq 0$ we obtain the dual function

$$w(\mu) = \min_{j=1,\ldots,m}\left\{(b - \mathbf{a}^T\mathbf{x}^j)\mu + \mathbf{c}^T\mathbf{x}^j\right\}.$$

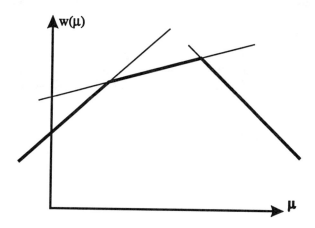

FIGURE A.8
A nondifferentiable dual function

It is easy to see that the dual function is the lower envelope of a family of affine functions, as shown in Figure A.8. We have a nondifferentiability point when the relaxed problem has multiple optimal solutions. ☐

In example A.13 we have a nondifferentiable function; still, it is a *concave function*. This property holds in general.

PROPERTY A.12

The dual function $w(\mu)$ is a concave function.

Since maximizing a concave function is equivalent to minimizing a convex function, this is a reassuring result, which we will exploit in Section A.7 in order to carry out the maximization of $w(\mu)$.

Coming back to example A.12, we can see that the dualization of the inequality constraint results in two separate problems with respect to the two decision variables. In the next section we investigate the use of dualization as a tool for decomposition.

A.6 Decomposition of large-scale optimization problems

In example A.12 we have seen how the dualization of a constraint may result in independent subproblems. The systematic application of the dualization of coupling constraints is called **dual** or **price-directed** decomposition. Consider a problem (P) characterized by an additive objective function:

$$\min \sum_{j=1}^{N} f_j(\mathbf{x}_j)$$

$$\text{s.t.} \sum_{j=1}^{N} g_j(\mathbf{x}_j) \leq b \qquad (A.16)$$

$$\mathbf{x}_j \in S_j \qquad j = 1, \ldots, N.$$

where $\mathbf{x}_j$ is a sub-vector of the vector $\mathbf{x}$. The constraint (A.16) can be interpreted as a resource availability constraint; it is a coupling constraint in the sense that deleting it yields a decomposition of the overall problem into a set of independent subproblems:

$$\nu \begin{pmatrix} \min & \sum_{j=1}^{N} f_j(\mathbf{x}_j) \\ \text{s.t.} & \mathbf{x}_j \in S_j \quad j = 1, \ldots, N \end{pmatrix} = \sum_{j=1}^{N} \nu \begin{pmatrix} \min & f_j(\mathbf{x}_j) \\ \text{s.t.} & \mathbf{x}_j \in S_j \end{pmatrix}.$$

In this sense (A.16) can be considered a "complicating constraint". Clearly, the optimal solutions of each subproblem do not necessarily satisfy (A.16). If strong duality holds, we obtain an exact decomposition by dualizing the coupling constraint with a multiplier $\mu \geq 0$. The relaxed problem, for a given μ, can be decomposed, obtaining the dual function

$$w(\mu) = \nu \begin{pmatrix} \min \sum_{j=1}^{N} \left[f_j(\mathbf{x}_j) + \mu \left(\sum_{j=1}^{N} g_j(\mathbf{x}_j) - b \right) \right] \\ \text{s.t.} \ \mathbf{x}_j \in S_j \quad j = 1, \ldots, N \end{pmatrix}$$

$$\Rightarrow \mu b + \sum_{j=1}^{N} \nu \begin{Bmatrix} \min \ [f_j(\mathbf{x}_j) + \mu g_j(\mathbf{x}_j)] \\ \text{s.t.} \ \mathbf{x}_j \in S_j \end{Bmatrix}$$

The decomposed subproblems are "coordinated" by the maximization of the dual function, as illustrated in Figure A.9. The coordinator iteratively adjusts the multiplier, which is passed down to each "local" subproblem; the solution of each subproblem is fed back to the coordinator, which updates μ according to a certain strategy. Note that, in general, the optimal solutions of the local subproblems do not satisfy the interaction constraint. One would expect that a good multiplier adjustment strategy should lead to the satisfaction of the resource availability constraint; in fact, there may be convergence problems. Nevertheless, this not a severe difficulty for the

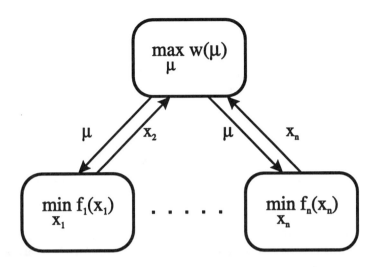

FIGURE A.9
A dual decomposition scheme

applications presented in this book. We will use dual decomposition to develop heuristics and to compute lower bounds based on weak duality (see Section C.4.3), hence, convergence is not actually a big problem.

If we want to make sure that intermediate solutions at each iteration are feasible, we may pursue a primal or **resource-directed** decomposition approach. We introduce a set of variables y_j representing the amount of resource allocated to the subsystem j and rewrite (P) as

$$\min \sum_{j=1}^{N} f_j(\mathbf{x}_j)$$

$$\text{s.t. } \sum_{j=1}^{N} y_j \leq b$$

$$g_j(\mathbf{x}_j) \leq y_j \quad j = 1, \ldots, N$$

$$\mathbf{x}_j \in S_j \quad j = 1, \ldots, N.$$

In the case of primal decomposition, the coordinator has the task of fixing and updating the allocated resource level y_j. The primal decomposition approach can be applied in a more general setting. Consider the problem:

$$\min \ f(\mathbf{x}, \mathbf{y})$$

$$\text{s.t. } g(\mathbf{x}, \mathbf{y}) \leq 0$$
$$\mathbf{x} \in S_x; \qquad \mathbf{y} \in S_y.$$

Here, we assume that **y** is the vector of "complicating variables", in the sense that for a given **y**, the resulting optimization problem with respect to **x** is easily solved. We obtain the so-called decomposition by **projection**:

$$\min_{\mathbf{y} \in S_y} \left\{ \min_{\mathbf{x} \in S_x} f(\mathbf{x}, \mathbf{y}) \quad \text{s.t.} \quad g(\mathbf{x}, \mathbf{y}) \leq 0 \right\}.$$

A.7 Nonsmooth optimization

In this section we relax the assumption of differentiability of the objective function. The optimization of a nondifferentiable function is often called **nonsmooth optimization**. For our purposes, the introduction of nonsmooth optimization concepts is motivated by the need for maximizing the dual function of discrete optimization problems, which may be a nondifferentiable function, as we have seen in Example A.13.

We consider first an unconstrained problem

$$\min_{\mathbf{x} \in \mathbf{R}^n} f(\mathbf{x}),$$

where f is continuous, convex, and almost everywhere differentiable. The last requirement means that ∇f exists almost everywhere (i.e., apart from a countable set of points), and, where it is not defined, we can compute the **directional derivative** for any direction $\mathbf{d} \in \mathbf{R}^n$, i.e.,

$$f'(\mathbf{x}; \mathbf{d}) \stackrel{\text{def}}{=} \lim_{t \downarrow 0} \frac{1}{t}[f(\mathbf{x} + t\mathbf{d}) - f(\mathbf{x})],$$

where $t \downarrow 0$ means that t approaches 0 from positive values. In practice we have stepwise continuously differentiable functions.

Example A.14
Consider the function

$$f(x) = \begin{cases} -x & \text{for } x < 0; \\ x^2 & \text{for } x \geq 0. \end{cases}$$

In this case $\nabla f(x) = -1$ for $x < 0$ and $\nabla f(x) = 2x$ for $x > 0$. In $x = 0$ the gradient is not defined, but the following limits exist:

$$\lim_{x \downarrow 0} \nabla f(x) = 0,$$

$$\lim_{x \uparrow 0} \nabla f(x) = -1. \quad \square$$

To solve a nonsmooth optimization problem, we can try to generalize the gradient method considered in section A.3. In the following, we will strongly rely on the convexity of f. We recall that, for a differentiable convex function, the gradient $\nabla f(\mathbf{x}_0)$ defines the unique support hyperplane of the graph of f at $\mathbf{x}_0$:

$$z = f(\mathbf{x}_0) + \nabla f^T(\mathbf{x}_0)(\mathbf{x} - \mathbf{x}_0).$$

If f is not differentiable at $\mathbf{x}_0$, we have a set of support hyperplanes

$$z = f(\mathbf{x}_0) + \gamma^T(\mathbf{x} - \mathbf{x}_0). \tag{A.17}$$

(see Figure A.3). If f is differentiable at $\mathbf{x}_0$, convexity implies

$$f(\mathbf{x}) \geq f(\mathbf{x}_0) + \nabla f^T(\mathbf{x}_0)(\mathbf{x} - \mathbf{x}_0) \quad \forall \mathbf{x} \in \mathbf{R}^n, \tag{A.18}$$

which in turn implies that, if $\nabla f(\mathbf{x}_0) = \mathbf{0}$, then $\mathbf{x}_0$ is a global optimum. Geometrically, optimality is associated with a horizontal support hyperplane. This property can be easily extended to nondifferentiable functions; the only difference is that at a nondifferentiability point we have more than one support hyperplane; if one of them is horizontal, then that point is a global optimum. This observation leads to the following generalization of the gradient. Given a convex function f a vector $\gamma \in \mathbf{R}^n$ such that

$$f(\mathbf{x}) \geq f(\mathbf{x}^0) + \gamma^T(\mathbf{x} - \mathbf{x}^0) \quad \forall \mathbf{x} \in \mathbf{R}^n, \tag{A.19}$$

is called a **subgradient** of f at $\mathbf{x}^0$. The name illustrates the role of γ in generalizing (A.18). In the differentiable points, the subgradient is unique and it coincides with the gradient; in the nondifferentiable points, there exists a collection of subgradients, each defining a support hyperplane of the form (A.17). The set of subgradients at $\mathbf{x}^0$ is called the **subdifferential** of f at $\mathbf{x}_0$ and is denoted by $\partial f(\mathbf{x}^0)$. A consequence of (A.19) is that a necessary and sufficient condition for global optimality is

$$\mathbf{0} \in \partial f(\mathbf{x}^0),$$

since it implies

$$f(\mathbf{x}) \geq f(\mathbf{x}^0) + \mathbf{0}^T(\mathbf{x} - \mathbf{x}^0) = f(\mathbf{x}^0) \quad \forall \mathbf{x} \in \mathbf{R}^n.$$

This discussion can be easily adapted to the maximization of a concave nondifferentiable function; we have only to reverse the inequality in (A.19).

In Section A.3 we have briefly described the gradient method for the optimization of a differentiable function. We can generalize the gradient method by using subgradients; at each step k we find $\gamma^{(k)} \in \partial f(\mathbf{x}^{(k)})$, and, having selected a suitable step size $\alpha^{(k)}$, we update

$$\mathbf{x}^{(k+1)} = \mathbf{x}^{(k)} - \alpha^{(k)}\gamma^{(k)}.$$

If the nonsmooth problem we are dealing with is the maximization of the dual function

$$\max_{\mu \geq 0} w(\mu),$$

with respect to nonnegative multipliers, we must update μ as:

$$\mu^{(k+1)} = \max\left\{0, \mathbf{x}^{(k)} + \alpha^{(k)}\gamma^{(k)}\right\}.$$

In order to apply subgradient optimization to the maximization of the dual function, we have to solve two issues: how to find a subgradient and how to select the step $\alpha^{(k)}$.

It is possible to show that the subdifferential is the convex hull of extreme subgradients given by the directional derivatives. Indeed, this link is made evident by the following result:

$$f'(\mathbf{x}; \mathbf{d}) = \max_{\gamma \in \partial f(\mathbf{x})} \gamma^T \mathbf{d}.$$

However, this is not very useful in practice, since it requires an explicit representation of the objective function, which is not reasonable in the case of the dual function $w(\mu)$. The following result allows solving the problem.

PROPERTY A.13
For $\bar{\mu} \geq 0$ let

$$Y(\bar{\mu}) = \left\{\mathbf{x} \in S \mid f(\mathbf{x}) + \bar{\mu}^T \mathbf{g}(\mathbf{x}) = w(\bar{\mu})\right\}.$$

Then, for all $\mathbf{x} \in Y(\bar{\mu})$, $\mathbf{g}(\mathbf{x})$ is a subgradient of w at $\bar{\mu}$.

In practice, this property leads to a very simple way to compute a subgradient of the dual function: given a solution $\mathbf{x}^{(k)}$ of the Lagrangian problem

$$\min_{\mathbf{x} \in S} f(\mathbf{x}) + \mu^{(k)T}\mathbf{g}(\mathbf{x}),$$

a subgradient is obtained by evaluating the dualized constraints at $\mathbf{x}^{(k)}$; then the multipliers are updated:

$$\mu^{(k+1)} = \max\left\{0, \mathbf{x}^{(k)} + \alpha^{(k)}\mathbf{g}(\mathbf{x}^{(k)})\right\}.$$

If a constraint g_i is violated (i.e., $g_i(\mathbf{x}^{(k)}) > 0$), the corresponding multiplier is increased; if it is satisfied, the corresponding multiplier is decreased. This fact leads to an interesting economic interpretation of dual decomposition, that will be pursued in Section 3.1.3.

To update the multipliers, we have to select a step length $\alpha^{(k)}$; a commonly used formula is

$$\alpha^{(k)} = \frac{l^{(k)}\left(\bar{w} - w(\mu^{(k)})\right)}{\|\,\mathbf{g}(\mathbf{x}^{(k)})\,\|^2},$$

where $l^{(k)}$ is a scalar satisfying $0 \leq l^{(k)} \leq 2$ and $\bar{w}$ is an upper bound on the optimal value of the dual function. Note that such an upper bound can be obtained by applying a heuristic method to the primal problem.

We know that the gradient method for smooth optimization is characterized by slow convergence. The same remark applies to the nonsmooth case; a further issue is that, in general, subgradients do not correspond to ascent directions, as shown in the next example.

Example A.15
Consider the nonsmooth problem

$$\max_{x_1,x_2} \min\{3 - 2x_1 + x_2, 2 + x_1 - 2x_2\},$$

which requires maximizing the lower envelope of two planes. In this case we look for an ascent direction. The two planes intersect on the line

$$3 - 2x_1 + x_2 = 2 + x_1 - 2x_2$$

i.e., $x_2 = x_1 - 1/3$. This line is the locus of nondifferentiability points (see Figure A.10). At any point on this line, the subgradients are convex combinations of the extreme subgradients $[-2, 1]^T$ (the gradient of $3 - 2x_1 + x_2$) and $[1, -2]^T$ (the gradient of $2 + x_1 - 2x_2$). From Figure A.10, checking the level lines of the two planes, it is obvious that the two extreme subgradients are not ascent directions. □

Even if the sequence of values of the objective function is not monotone, it can be shown that the subgradient method converges to the optimum if the step length is selected appropriately and the function to be minimized (maximized) is convex (concave). This last requirements is ensured in the applications we are concerned with, since the dual function is always concave.

It is possible to develop multiplier adjustment rules guaranteeing that the search directions generated are ascent directions. Such methods are not always advantageous, in comparison to plain subgradient methods, due to the increased computational requirements; however, there are specific cases in which ascent directions are easily computed by exploiting the particular problem structure.

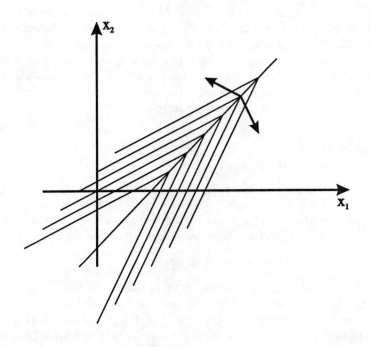

FIGURE A.10
Subgradients do not necessarily correspond to ascent direction

For further reading

A collection of excellent tutorial and survey papers on optimization can be found in [151]; topics include unconstrained and constrained optimization, global optimization, and stochastic optimization.

The reader interested in a thorough and rigorous treatment of optimization theory is referred, e.g., to [103, 144]. An excellent reference for numerical methods is [78]. Global optimization methods are illustrated in [106]. A classical reference on barrier penalty methods is [72].

Readers interested in decomposition of large-scale problems may consult [86]. A particular case of primal decomposition is Benders decomposition for mixed-integer linear programming, which is described in a readable manner in [155].

Subgradient methods for the maximization of the dual function are described in [73]. Methods to find ascent directions in subgradient optimization are described, e.g., in [155].

B

Linear Programming and Network Optimization

Linear Programming (LP in the following) is by far the most developed area within Mathematical Programming. LP software packages are available on a wide variety of hardware platforms; these packages are based on the Simplex method. From a theoretical point of view, the Simplex method is not the most efficient algorithm for LP problems, and recent LP packages provide the user with the possibility of adopting alternative approaches. Nevertheless, years of use have shown its efficiency and reliability. Since there are plenty of excellent books on LP, we do not cover here neither the alternative algorithms for LP nor the computational strategies to enhance the efficiency of the Simplex method. Our aim is only to provide the reader with a basic knowledge which will prove useful to understand some issues concerning integer linear programming.

An important subclass of LP problems consists of Graph and Network Optimization problems. Though many graph optimization problems could be solved by a generic LP algorithm, it turns out that significant computational savings are obtained by specialized algorithms. The principles of such specialized algorithms are beyond the scope of this book. For our purposes, it is sufficient that the reader gains a basic knowledge of this kind of problem in terms of models and some properties of their optimal solutions.

We start in Section B.1 with basic notions about convex polyhedra, which are the basis of the Simplex method, and graph theory. LP modeling is discussed in Section B.2, followed by some intuitive geometrical and algebraic considerations in Section B.3. The Simplex method is described in Section B.4. Section B.5 is devoted to the important topic of duality in LP. Network optimization models are described in Section B.6. Finally, in Section B.7 we outline some important results about unimodular matrices, which characterize a class of continuous LP problems whose optimal solutions have integer coordinates.

B.1 Background

B.1.1 Convex polyhedra and polytopes

Consider the hyperplane (in $\mathbf{R}^n$) $\mathbf{a}_i^T \mathbf{x} = b_i$, where $b_i \in \mathbf{R}$, $\mathbf{a}_i, \mathbf{x} \in \mathbf{R}^n$ are column vectors[1] and T denotes transposition. A hyperplane divides $\mathbf{R}^n$ into two **half-spaces** expressed by the linear inequalities $\mathbf{a}_i^T \mathbf{x} \leq b_i$ and $\mathbf{a}_i^T \mathbf{x} \geq b_i$.

DEFINITION B.1 A **polyhedron** $P \subseteq \mathbf{R}^n$ *is a set of points satisfying a finite collection of linear inequalities, i.e.,*

$$P = \{\mathbf{x} \in \mathbf{R}^n \mid \mathbf{A}\mathbf{x} \geq \mathbf{b}\}.$$

A polyhedron is therefore the intersection of a finite collection of half-spaces.

PROPERTY B.1
A polyhedron is a convex set (it is the intersection of convex sets).

DEFINITION B.2 *A polyhedron is* **bounded** *if there exists a positive number M such that*

$$P \subseteq \{\mathbf{x} \in \mathbf{R}^n \mid -M \leq x_j \leq M \quad j = 1, \ldots, n\}.$$

A bounded polyhedron is called a **polytope**.

A polytope and an unbounded polyhedron are shown in Figure B.1.

DEFINITION B.3 *A point $\mathbf{x}$ is an* **extreme point** *of a polyhedron P if $\mathbf{x} \in P$ and it is not possible to express $\mathbf{x}$ as $\mathbf{x} = \frac{1}{2}\mathbf{x}' + \frac{1}{2}\mathbf{x}''$ with $\mathbf{x}', \mathbf{x}'' \in P$ and $\mathbf{x}' \neq \mathbf{x}''$.*

A polytope P has a finite number of extreme points $\mathbf{x}^1, \ldots, \mathbf{x}^J$. Any point $\mathbf{x}$ in a polytope P can be expressed as a convex combination of its extreme points:

$$\mathbf{x} = \sum_{j=1}^{J} \lambda_j \mathbf{x}^j \qquad \sum_{j=1}^{J} \lambda_j = 1, \ \lambda_j \geq 0;$$

---

[1] Unless the contrary is stated, we assume that all vectors are columns.

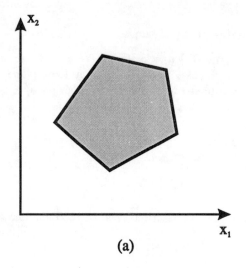

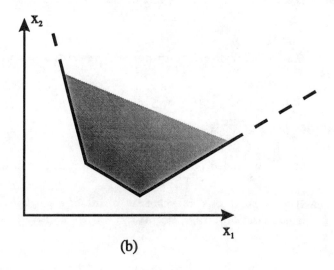

FIGURE B.1
A two-dimensional polytope (a) and an unbounded polyhedron (b)

in other words a polytope is the convex hull of its extreme points. In the case of an unbounded polyhedron this is not true, and we must introduce another concept.

DEFINITION B.4 *A vector* $\mathbf{r} \in \mathbf{R}^n$ *is called a* **ray** *of the polyhedron* $P = \{\mathbf{x} \in \mathbf{R}^n \mid \mathbf{Ax} \geq \mathbf{b}\}$ *if* $\mathbf{Ar} \geq \mathbf{0}$.

If $\mathbf{x}_0$ is a point in a polyhedron P and $\mathbf{r}$ is a ray of P, then

$$\mathbf{y} = \mathbf{x}_0 + \lambda \mathbf{r} \in P \qquad \forall \lambda \geq 0.$$

Clearly, only unbounded polyhedra have rays.

DEFINITION B.5 *A ray* $\mathbf{r}$ *of a polyhedron* P *is called an* **extreme ray** *if it cannot be expressed as* $\mathbf{r} = \frac{1}{2}\mathbf{r}_1 + \frac{1}{2}\mathbf{r}_2$ *where* $\mathbf{r}_1, \mathbf{r}_2$ *are rays of* P *such that* $\mathbf{r}_1 \neq \lambda \mathbf{r}_2$ *for any number* $\lambda > 0$.

A polyhedron P can be described in terms of its extreme rays and points, in the sense that any point $\mathbf{x} \in P$ can be expressed combining extreme rays and points:

$$\mathbf{x} = \sum_{j=1}^{J} \lambda_j \mathbf{x}^j + \sum_{k=1}^{K} \mu_k \mathbf{r}^k \qquad \sum_{j=1}^{J} \lambda_j = 1, \; \lambda_j, \mu_k \geq 0.$$

B.1.2 Graph Theory

DEFINITION B.6 *A* **graph** $G = (\mathcal{N}, \mathcal{A})$ *consists of a set* $\mathcal{N}$ *of* **nodes** *and a set* $\mathcal{A}$ *of* **arcs**. *Arcs are pairs* (i, j) *of nodes* $i, j \in \mathcal{N}$. *Arcs can be ordered or unordered pairs. If the arcs are ordered pairs we speak of a* **directed graph**; *otherwise we speak of an* **undirected graph**.

A directed graph can be pictorially represented as in Figure B.2(a). Given a directed arc (i, j), the nodes i and j are the endpoints of the arc; the arrows representing the arcs are oriented from i (the tail of the arc) to j (the head). An undirected graph is shown in Figure B.2(b). When speaking of an undirected arc, we may write it indifferently as (i, j) or (j, i). We will not consider graphs with loops, i.e., arcs of the form (i, i). Given a directed or undirected arc (i, j), we say that the arc is **incident** to the nodes i and j. When some numerical information is attached to the nodes or the arcs of a graph we obtain a **network**. The numerical information may refer to the length of an arc or to its flow capacity (see Section B.6).

DEFINITION B.7 *A graph* $G' = (\mathcal{N}', \mathcal{A}')$ *is a* **subgraph** *of* $G = (\mathcal{N}, \mathcal{A})$ *if* $\mathcal{N}' \subseteq \mathcal{N}$ *and* $\mathcal{A}' \subseteq \mathcal{A}$. G' *is called the* **subgraph induced by** $\mathcal{N}'$ *if* $\mathcal{A}'$

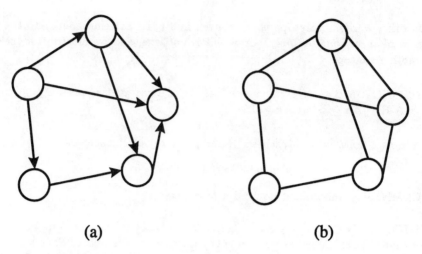

FIGURE B.2
A directed graph (a) and an undirected graph (b)

contains all the arcs in $\mathcal{A}$ with both endpoints in $\mathcal{N}'$.

*DEFINITION B.8 We say that an undirected graph is **complete** if, for any pair of distinct nodes i and j, there is an arc (i,j). A subset of nodes inducing a complete subgraph is called a **clique** (see Figure B.3).*

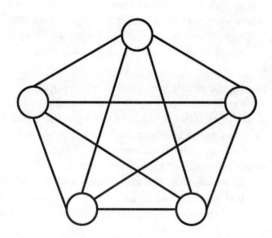

FIGURE B.3
A complete graph

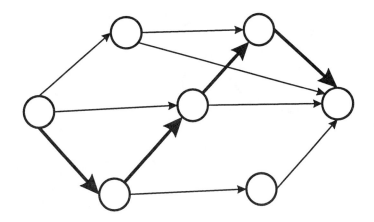

FIGURE B.4
A (directed) path

Given a directed graph, consider a sequence of nodes

$$i_1 \to i_2 \to \ldots \to i_l$$

such that no node is repeated. We say that this sequence is a **directed path** if $(i_k, i_{k+1}) \in \mathcal{A}$ for all k ranging from 1 to $l-1$ (see Figure B.4). If we take a directed path $i_1 \to i_2 \to \ldots \to i_l$ and we add the arc (i_l, i_1), we obtain a **directed cycle**, or **circuit**. In the following we will omit the term "directed" when referring to paths and cycles; furthermore we will use the terms path and cycle to refer both to node and arc sequences. A graph that does not contain any cycle is called **acyclic**.

DEFINITION B.9 A graph $G = (\mathcal{N}, \mathcal{A})$ is called **bipartite** if $\mathcal{N}$ can be partitioned into two disjoint subsets, S and T, such that any arc $(i,j) \in \mathcal{A}$ has one endpoint in S and the other in T.

A bipartite graph is shown in Figure B.6.

B.2 The standard and canonical forms of LP problems

A mathematical programming problem is an LP problem if both the objective functions and the constraints are linear.[2] An LP problem can be

---

[2] To be more precise we should speak of *affine* rather than linear functions.

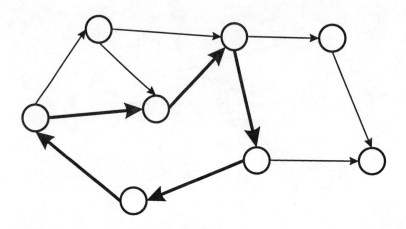

FIGURE B.5
A (directed) cycle

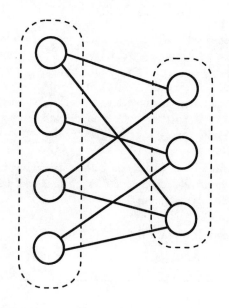

FIGURE B.6
A bipartite graph

expressed as

$$\min \ c^T x$$
$$\text{s.t.} \ a_i^T x = b_i \quad i \in E$$
$$a_i^T x \geq b_i \quad i \in I.$$

where $c, a_i, x \in \mathbf{R}^n$, $b_i \in \mathbf{R}$.

Example B.10
Here is a simple example of an LP problem:

$$\min \ x_1 + x_2 - x_3$$
$$\text{s.t.} \ x_1 + 2x_2 \geq 6$$
$$2x_1 + x_2 - x_3 \geq 6$$
$$x_1 + x_2 + x_3 = 1$$
$$x_1, x_2 \geq 0.$$

The nonnegativity constraints on x_1, x_2 could be considered just as other inequality constraints. However it is convenient to set such restrictions apart. ☐

An LP problem is said in **canonical form** if

1. all decision variables are nonnegative;
2. only inequality constraints are given.

An LP problem in canonical form is written as

$$\min \ c^T x$$
$$\text{s.t.} \ a_i^T x \geq b_i \quad i = 1, \ldots, m$$
$$x \geq 0,$$

or, in matrix form:

$$\min \ c^T x$$
$$\text{s.t.} \ Ax \geq b$$
$$x \geq 0,$$

where $c, x \in \mathbf{R}^n$, $b \in \mathbf{R}^m$, $A \in \mathbf{R}^{m,n}$. We denote the ith row (corresponding to the ith constraint) of A by a_i^T, and the jth column (corresponding to the jth variable) by A^j.

An LP problem is said in **standard form** if

1. all decision variables are nonnegative;
2. only equality constraints are given.

An LP problem in standard form is written as

$$\min\ c^T x$$
$$\text{s.t.}\ a_i^T x = b_i \quad i = 1, \ldots, m$$
$$x \geq 0,$$

or, in matrix form:

$$\min\ c^T x$$
$$\text{s.t.}\ Ax = b$$
$$x \geq 0,$$

with the same notation as in the case of the canonical form. Clearly, we must have $m < n$, so that the system of linear equations is underdetermined and there are multiple solutions. The case of a unique or no solution is not interesting for optimization.

The reader might think that the canonical and standard forms are somewhat restrictive; in fact, this is not true, since a generic LP problem can be reduced to either form using the following transformations:

- If a variable x_j is not restricted in sign, it can be rewritten as $x_j = x_j^+ - x_j^-$ where $x_j^+, x_j^- \geq 0$.
- An inequality constraint

$$a_i^T x \geq b_i$$

can be transformed into an equality constraint by introducing a slack variable $s_i \geq 0$:

$$a_i^T x - s_i = b_i.$$

- An equality constraint

$$a_i^T x = b_i$$

can be transformed into two inequality constraints

$$a_i^T x \geq b_i, \quad -a_i^T x \geq -b_i.$$

We see that the feasible set of an LP problem is a (possibly unbounded) polyhedron.

B.3 Geometric and algebraic features of Linear Programming

In this section we review some qualitative properties of the optimal solution of an LP problem; such properties are the foundation of the Simplex method. LP problems are convex problems, since both the objective function and the feasible set are convex; this implies that a local optimum is also a global one (see Section A.1.1). Furthermore, a linear objective function is also a concave function; this implies that the optimum lies on the boundary of the feasible region. This can be also understood by geometrical intuition.

Example B.11
Consider the LP problem

$$\max\ x_1 + x_2$$
$$\text{s.t.}\ x_1 + 2x_2 \leq 6$$
$$2x_1 + x_2 \leq 6$$
$$x_1, x_2 \geq 0.$$

The feasible region and some level curves of the objective function are shown in Figure B.7. The feasible region is a bidimensional polytope characterized by the extreme points, (0,0), (0,3), (2,2), and (3,0). Geometrical intuition shows that the optimal solution is the extreme point (2,2). □

Given an LP problem, one of the three following cases occurs:

1. the feasible set is empty, and the problem has no solution;
2. the optimal solution is unbounded; this case may occur only if the feasible set is an unbounded polyhedron;
3. the problem has a finite optimal solution, corresponding to an extreme point of the feasible set; note that we have an infinite set of optimal solutions if the level curves of the objective function are parallel to an edge of the polyhedron.

Since there is a finite number of extreme points in a polyhedron, it is possible to devise an algorithm able to explore the set of extreme points, without considering the other points of the feasible set. Furthermore, due to the convexity of LP, we can develop a solution approach iteratively exploring the extreme points of the feasible set. Given a current extreme point, we should look for an adjacent extreme point improving the objective

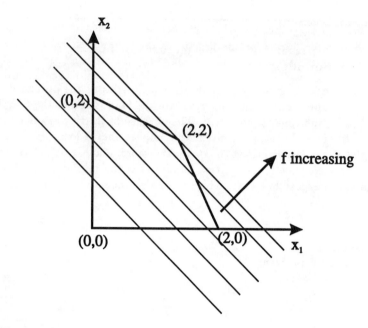

FIGURE B.7
Feasible region and level curves for a simple LP problem

function; when no improving adjacent extreme point can be found, we are sure that a globally optimal solution has been obtained.

To actually implement this idea, the geometrical intuition must be translated into algebraic terms. To this aim, it is convenient to work on the standard LP form. For convenience, assume that the system of linear equations

$$\mathbf{Ax} = \mathbf{b} \qquad (B.1)$$

is such that $\mathbf{A}$ has full rank; this assumption is not necessary, as the Simplex method is able to spot redundant equations and to eliminate them. A feasible solution of the LP problem is any solution $\mathbf{x}$ of (B.1) such that $\mathbf{x} \geq \mathbf{0}$. It is useful to consider a solution $\mathbf{x}$ as a way to express the vector $\mathbf{b}$ as a linear combination of the columns of $\mathbf{A}$:

$$\mathbf{Ax} = \begin{bmatrix} \mathbf{A}^1 \mathbf{A}^2 \cdots \mathbf{A}^n \end{bmatrix} \begin{bmatrix} x_1 \\ x_2 \\ \vdots \\ x_n \end{bmatrix} = \sum_{j=1}^{n} x_j \mathbf{A}^j = \mathbf{b}.$$

This underdetermined system (recall that $m < n$) has infinite solutions, but we consider only a subset, i.e., those such that at least $n - m$ components x_j are null. Such solutions are called **basic solutions**; the name derives from the fact that the m column vectors associated with the m possibly nonnull variables are sufficient to express the m-dimensional vector **b**. A basic solution is associated with a basis of $\mathbf{R}^m$ consisting of m columns of **A**. The m variables corresponding to the selected columns are called basic variables; the others are nonbasic variables. A basic solution with non-negative components is called **basic feasible**. Basic feasible solutions are fundamental because it can be shown that they correspond to the extreme points of the feasible set. Therefore, we can look for the optimal solution considering only basic feasible solutions, or, in other words, by considering bases consisting of subsets of columns of **A**.

Example B.12
Consider the following system of linear equations:

$$\begin{bmatrix} -1 & 1 & 1 & -1 & 0 \\ 0 & 1 & 0 & 4 & 0 \\ 0 & 0 & 2 & 2 & 1 \end{bmatrix} \begin{bmatrix} x_1 \\ x_2 \\ x_3 \\ x_4 \\ x_5 \end{bmatrix} = \begin{bmatrix} 1 \\ 3 \\ 1 \end{bmatrix}$$

A basic solution is

$$x_1 = 2, \ x_2 = 3, \ x_3 = x_4 = 0, \ x_5 = 1,$$

which corresponds to the basis formed by the columns $\mathbf{A}^1, \mathbf{A}^2, \mathbf{A}^5$. This solution is also feasible. If we take the basis formed by $\mathbf{A}^2, \mathbf{A}^3, \mathbf{A}^5$, we obtain the basic solution

$$x_1 = 0, \ x_2 = 3, \ x_3 = -2, \ x_4 = 0, \ x_5 = 5,$$

which is not feasible since $x_3 < 0$. ☐

B.4 The Simplex method

We have seen in the previous section that we need only to consider basic feasible solutions when solving an LP problem. The Simplex method is an iterative algorithm; given a current extreme point (or basic feasible solution, or basis), it looks for an adjacent extreme point such that the objective function is improved, and it stops when no improving adjacent extreme point is found.

Consider a basic feasible solution **x**; we will consider later how to obtain an initial basic feasible solution. We can partition the vector **x** into two subvectors: the subvector $\mathbf{x}_B \in \mathbf{R}^m$ of the basic variables and the subvector

$\mathbf{x}_N \in \mathbf{R}^{n-m}$ of the nonbasic variables. Using a suitable permutation of the variable indexes, we may rewrite the system of linear equations

$$\mathbf{A}\mathbf{x} = \mathbf{b}$$

as

$$[\mathbf{A}_B \mathbf{A}_N] \begin{bmatrix} \mathbf{x}_B \\ \mathbf{x}_N \end{bmatrix} = \mathbf{A}_B \mathbf{x}_B + \mathbf{A}_N \mathbf{x}_N = \mathbf{b}, \qquad (B.2)$$

where $\mathbf{A}_B \in \mathbf{R}^{m,m}$ is nonsingular and $\mathbf{A}_N \in \mathbf{R}^{m,n-m}$. If $\mathbf{x}$ is basic feasible,

$$\mathbf{x} = \begin{bmatrix} \mathbf{x}_B \\ \mathbf{x}_N \end{bmatrix} = \begin{bmatrix} \hat{\mathbf{b}} \\ \mathbf{0} \end{bmatrix},$$

where

$$\hat{\mathbf{b}} = \mathbf{A}_B^{-1}\mathbf{b} \geq \mathbf{0}.$$

The objective function value corresponding to $\mathbf{x}$ is

$$\hat{f} = [\mathbf{c}_B^T \ \mathbf{c}_N^T] \begin{bmatrix} \hat{\mathbf{b}} \\ \mathbf{0} \end{bmatrix} = \mathbf{c}_B^T \hat{\mathbf{b}}.$$

Now we must find out if it is possible to improve the current solution. This can be obtained by moving to an adjacent extreme point of the feasible set; in algebraic terms this means changing the basis, i.e., replacing a basic variable with a nonbasic one. Therefore, we need to evaluate the benefit of introducing a nonbasic variable into the basis. This can be obtained by eliminating the basic variables from the expression of the current value of the objective function. Using (B.2) we may express the basic variables as

$$\mathbf{x}_B = \mathbf{A}_B^{-1}(\mathbf{b} - \mathbf{A}_N \mathbf{x}_N) = \hat{\mathbf{b}} - \mathbf{A}_B^{-1}\mathbf{A}_N \mathbf{x}_N; \qquad (B.3)$$

then we rewrite the objective function value

$$\mathbf{c}^T \mathbf{x} = \mathbf{c}_B^T \mathbf{x}_B + \mathbf{c}_N^T \mathbf{x}_N = \mathbf{c}_B^T(\hat{\mathbf{b}} - \mathbf{A}_B^{-1}\mathbf{A}_N \mathbf{x}_N) + \mathbf{c}_N^T = \hat{f} + \hat{\mathbf{c}}_N^T \mathbf{x}_N,$$

where

$$\hat{\mathbf{c}}_N^T = \mathbf{c}_N^T - \mathbf{c}_B^T \mathbf{A}_B^{-1} \mathbf{A}_N. \qquad (B.4)$$

The **reduced costs** $\hat{\mathbf{c}}_N$ measure the marginal variation of the objective function with respect to the nonbasic variables. If $\hat{\mathbf{c}}_N \geq \mathbf{0}$, it is not possible to improve the objective function; in this case, bringing a nonbasic variable into the basis can only increase the cost, since a variable can assume only nonnegative values. Therefore, the current basis is optimal if $\hat{\mathbf{c}}_N \geq \mathbf{0}$. If there exists a $q \in N$ such that $\hat{c}_q < 0$, it is possible to improve the objective function by bringing x_q into the basis. A simple strategy is to choose q such that:

$$\hat{c}_q = \min_{j \in N} \hat{c}_j.$$

Note that this selection does not necessarily result in the best performance of the algorithm; we should consider not only the rate of change in the objective function, but also the value attained by the new basic variable. When x_q is brought into the basis, a basic variable must leave the basis in order to maintain $\mathbf{Ax} = \mathbf{b}$; the new solution must be basic feasible too. To spot the leaving variable we can reason as follows. Given the current basis, we can use it to express both $\mathbf{b}$ and the column $\mathbf{A}^q$ corresponding to the entering variable,

$$\mathbf{b} = \sum_{i=1}^{m} x_{B(i)} \mathbf{A}^{B(i)} \qquad (B.5)$$

$$\mathbf{A}^q = \sum_{i=1}^{m} d_i \mathbf{A}^{B(i)}, \qquad (B.6)$$

where $B(i)$ is the index of the ith basic variable ($i = 1, \ldots, m$), and

$$\mathbf{d} = \mathbf{A}_B^{-1} \mathbf{A}^q.$$

If we multiply (B.6) by a number θ and subtract it from (B.5) we obtain:

$$\mathbf{b} = \sum_{i=1}^{m} \left(x_{B(i)} - \theta d_i \right) \mathbf{A}^{B(i)} + \theta \mathbf{A}^q. \qquad (B.7)$$

From Eq. (B.7) we see that θ is the value of the entering variable in the new solution, and that the value of the current basic is affected in a way depending on the sign of d_i. If $d_i \leq 0$, $x_{B(i)}$ remains nonnegative when x_q increases, but, if there is an index i such that $d_i > 0$, then we cannot increase x_q at will, since there is a limit value for which a currently basic variable becomes zero. This limit value is attained by the entering variable x_q, and the first current basic variable which gets zero leaves the basis:

$$x_q = \min_{\substack{i = 1, \ldots, m \\ d_i > 0}} \frac{\hat{b}_i}{d_i}.$$

If $\mathbf{d} \leq \mathbf{0}$, there is no limit on the increase of x_q, and the optimal solution is unbounded.

Example B.13
We show here very briefly how the Simplex method would work on the simple problem illustrated in Example B.11. Our aim is just to provide the reader with a "feeling" about the implementation of the method. For the book, a deeper understanding of the involved issues is not needed; the interested reader is referred to any textbook on LP for a thorough explanation.

First, we must rewrite the problem in minimization form:

$$\min \; -x_1 - x_2$$
$$\text{s.t.} \; x_1 + 2x_2 \leq 6$$
$$2x_1 + x_2 \leq 6$$
$$x_1, x_2 \geq 0.$$

It is useful to intepret the steps of the method geometrically by referring to Figure B.7. To transform the problem into standard form, we introduce the slack variables s_1 and s_2:

$$\min \; -x_1 - x_2$$
$$\text{s.t.} \; x_1 + 2x_2 + s_1 = 6$$
$$2x_1 + x_2 + s_2 = 6$$
$$x_1, x_2, s_1, s_2 \geq 0.$$

To start the Simplex method, we need an initial basis. We show in the next section how this can be accomplished in general, but in this case it is easy to see that a basic feasible solution is

$$x_1 = x_2 = 0, \; s_1 = s_2 = 6.$$

This solution corresponds with the extreme point $(0,0)$ of the feasible set. The Simplex method involves computing the inverse of the basis matrix $\mathbf{A}_B$. To accomplish this task easily, we must write the problem according to the following *tableau* format:

-1	-1	0	0	0
1	2	1	0	6
2	1	0	1	6

The tableau is separated into four parts.

- The upper left part consists of the cost coefficients of the variables; during each step of the method we read the reduced costs for the nonbasic variables; in *this* particular case, the cost coefficients of the objective function are the reduced costs of the nonbasic variables x_1 and x_2. We see that, since the reduced costs are negative, it is convenient to bring either x_1 or x_2 into the basis.
- The upper right part is the value of the objective function, with the sign changed; now it is 0, since x_1 and x_2 are nonbasic.

- The lower left part contains the matrix **A**, which will be transformed during the computation. Here we can read the numbers d_i which allow us to determine which basic variable should leave the basis.
- The lower right part contains the vector **b**. During the steps of the method, we can read the vector $\hat{\mathbf{b}}$.

We emphasize the fact that writing the initial tableau is very easy in this case, since the initial basis contains only slack variables and the corresponding matrix is just the identity matrix. In general some calculations are required.

Now, since the reduced costs are equal, we arbitrarily decide to bring x_1 into the basis; this corresponds to moving to the extreme point $(3,0)$ of the feasible set. To determine which slack variable must leave the basis, we must compute

$$\min_{\substack{i=1,2 \\ d_i > 0}} \frac{\hat{b}_i}{d_i} = \min\left\{\frac{6}{1}, \frac{6}{2}\right\} = 3,$$

from which we see that s_2 leaves the basis (the second inequality constraint becomes active). To update the tableau, we apply the well-known transformations of Gaussian elimination for solving systems of linear equations. The transformations are required to tranform the first column of the tableau as follows:

$$\begin{array}{|c|} \hline -1 \\ \hline 1 \\ 2 \\ \hline \end{array} \Rightarrow \begin{array}{|c|} \hline 0 \\ \hline 0 \\ 1 \\ \hline \end{array}$$

The transformed tableau is the following:

0	$-\frac{1}{2}$	0	$\frac{1}{2}$	3
0	$\frac{3}{2}$	1	$-\frac{1}{2}$	3
1	$\frac{1}{2}$	0	$\frac{1}{2}$	3

We see that the new solution is

$$x_1 = s_1 = 3$$

with a corresponding value -3. Since the reduced cost of x_2 is negative, we bring it into the basis. To find out which variable leaves the basis, we compute:

$$\min_{\substack{i=1,2 \\ d_i > 0}} \frac{\hat{b}_i}{d_i} = \min\left\{3\frac{2}{3}, 3\frac{2}{1}\right\} = 2,$$

from which we see that s_1 becomes nonbasic. Now both constraints are active and we have moved to the extreme point $(2,2)$. The transformed tableau is

0	0	$\frac{1}{3}$	$\frac{1}{3}$	4
0	1	$\frac{2}{3}$	$-\frac{1}{3}$	2
1	0	$-\frac{1}{3}$	$\frac{2}{3}$	2

Since the reduced costs of the slack variables are positive, this is indeed the optimal tableau.   □

B.4.1 Getting a starting basis

To start the Simplex iteration, we must be able to obtain a starting basic feasible solution. This is easily accomplished by introducing a vector $\mathbf{y} \in \mathbf{R}^m$ of artificial variables and solving:

$$\min \sum_{i=1}^{m} y_i$$
$$\text{s.t. } \mathbf{Ax} + \mathbf{y} = \mathbf{b}$$
$$\mathbf{x}, \mathbf{y} \geq \mathbf{0}.$$

If the LP formulation is transformed in such a way that $\mathbf{b} \geq \mathbf{0}$, a starting basic feasible solution is easily found for this artificial problem:

$$\mathbf{x} = \mathbf{0}, \quad \mathbf{y} = \mathbf{b}.$$

If there is a basic feasible solution to the original problem, the application of the Simplex algorithm to the artificial problem yields an optimal solution such that $\mathbf{y}^* = \mathbf{0}$, and $\mathbf{x}^*$ is a basic feasible solution for the original problem. This preliminary phase is called the *Phase I* of the Simplex method; apart from generating a starting basic feasible solution for the Phase II, its purpose is also to spot redundant constraints which are eliminated.

B.4.2 Computational issues and interior point methods

In the previous section we have illustrated an example of the application of the Simplex method. In practice, several improvements are needed from the computational point of view. Such improvements are implemented in good commercial packages for LP, and here we just point out the related issues, without going in any detail.

Efficient ways to update the tableau. In the above example we have updated the whole tableau at each step; this is rather time consuming for large problems. The method can be streamlined by noting that we actually need only the inverse of the matrix $\mathbf{A}_B$ to compute what we need. This idea is implemented in the **revised simplex method**.

Numerical stability. It is well known that Gaussian elimination is not a numerically stable method for solving systems of linear equations. Numerical stability is a problem for the Simplex method too. Numerically stable matrix factorization techniques are adopted in practice; such techniques are also important for efficiency. Another point is that numerical stability depends on the order of magnitude of the problem data; scaling techniques are used to avoid problems due to large differences in the numerical values.

Pricing and cycling. When we evaluate the possibility of bringing a nonbasic variable into the basis we compute its reduced price; the variable with the smallest negative price is selected. This is not always the best approach, and alternative strategies are available. Apart from efficiency issues, another issue must be considered. The Simplex method is actually a local optimization method, and we have said that, due to the convexity of LP problems, this is not a problem. Actually, the following difficulty may arise: it is possible that a variable enters the basis at a value zero. In this case the value of the objective function does not change; it may happen that a sequence of such changes of basis is found, without affecting the value of the objective function. If we come to a previously visited basis, **cycling** occurs and the method gets stuck. This problem can be overcome by better rules for selecting the nonbasic variable entering the basis.

Interior point methods. A last point is that the Simplex method is, from a theoretical point of view, a "bad" algorithm. In the worst case, an exponential number of steps is required for its convergence; alternative algorithms have been proposed, in the sense that they require only a polynomial number of steps (see Supplement S2.1 for more details on computational complexity issues). These more recent algorithms are **interior point** algorithms, in the sense that, unlike the Simplex method, they do not explore only extreme points on the boundary of the feasible set, but they move in its interior. These algorithms are commercially available and can outperform the Simplex method; nevertheless, at present there is no definite evidence concerning the overall superiority of one approach over the other in the average case.

B.5 Duality in Linear Programming

We have dealt with duality in nonlinear programming in Section A.5. Duality in LP can be developed without considering the more general nonlinear case, but we prefer to put it in a more general framework. Note that, due to the convexity of LP problems, strong duality holds.

Let us start with an LP problem (P_1) in the following form:

$$(P_1) \quad \min \ c^T x$$
$$\text{s.t. } Ax \geq b.$$

If we dualize the inequality constraints with a vector $\mu \in \mathbf{R}_+^m$ of dual variables, we get the dual problem

$$\max_{\mu \geq 0} \min_x \left\{ c^T x + \mu^T (b - Ax) \right\} = \max_{\mu \geq 0} \left\{ \mu^T b + \min_x \left(c^T - \mu^T A \right) x \right\}.$$

Since x is unrestricted in sign, the inner minimization problem has a finite value if and only if

$$c^T - \mu^T A = 0;$$

otherwise, each component of x is set to $\pm \infty$ depending on the sign of the corresponding cost coefficient, and this results in a value $-\infty$ for the dual function. Since we want to maximize the dual function, we may enforce the above condition, and the dual problem (D_1) turns out to be:

$$(D_1) \quad \max \ \mu^T b$$
$$\text{s.t. } A^T \mu = c$$
$$\mu \geq 0.$$

The dual problem is still an LP problem, resulting from exchanging b with c and by transposing A. Note that the dual problem need not be feasible. In fact, if the primal problem is unbounded, the dual one is unfeasible and vice versa; it may also happen that neither the primal nor the dual problem is feasible. If both problems have a finite value solution, the two optimal values are equal. The duality relation between (P_1) and (D_1) can be interpreted the other way round too:

$$\begin{pmatrix} \max & x^T c \\ \text{s.t.} & Ax = b \\ & x \geq 0 \end{pmatrix} \iff \begin{pmatrix} \min & b^T \nu \\ \text{s.t.} & A^T \nu \geq c \end{pmatrix}$$

Given an LP problem (P_2) in standard form

$$(P_2) \quad \min \ c^T x$$

Duality in Linear Programming

Primal	Dual
min $c^T x$	max $\mu^T b$
$a_i^T x = b_i$	μ_i unrestricted
$a_i^T x \geq b_i$	$\mu_i \geq 0$
$x_j \geq 0$	$\mu^T A^j \leq c_j$
x_j unrestricted	$\mu^T A^j = c_j$

TABLE B.1
Duality relationships

B.5.1 Applications of LP duality

Consider the LP problem in standard form

$$\min c^T x$$
$$\text{s.t.} \quad Ax = b$$
$$x \geq 0,$$

we can use the above relationship to find out its dual:

$$\begin{pmatrix} \min c^T x \\ \text{s.t.} \quad Ax = b \\ x \geq 0 \end{pmatrix} \iff \begin{pmatrix} \max x^T(-c) \\ \text{s.t.} \quad (A^T)^T x = b \\ x \geq 0 \end{pmatrix} \iff \begin{pmatrix} \min b^T \nu \\ \text{s.t.} \quad A^T \nu \geq -c \end{pmatrix}$$

$$\iff \begin{pmatrix} \min -b^T \mu \\ \text{s.t.} \quad -A^T \mu \geq -c \end{pmatrix} \iff \begin{pmatrix} \max b^T \mu \\ \text{s.t.} \quad A^T \mu \leq c \end{pmatrix}$$

where we have introduced $\mu = -\nu$; we obtain the dual (D_2) of (P_2).

Note the similarities and the differences between the two dual pairs. The dual variables are restricted in sign when the constraints of the primal problem are inequalities, and are unrestricted in sign in the other case. When the variables are restricted in sign in the primal we have inequality constraints in the dual, whereas in the case of unrestricted variables we have equality constraints in the dual. In Table B.1 we summarize the "recipe" for building the dual of a generic LP.

B.5.1 Applications of LP duality

Consider the LP problem in standard form

$$\min c^T x$$
$$\text{s.t.} \quad Ax = b$$
$$x \geq 0,$$

and its dual

$$\max \mathbf{b}^T \boldsymbol{\mu}$$
$$\text{s.t.} \quad \mathbf{A}^T \boldsymbol{\mu} \leq \mathbf{c}. \qquad (B.8)$$

We have seen in Section A.4.5 that the dual variables can be interpreted as *sensitivities* of the optimal value with respect to perturbations in the constraints. This is evident here, since if both problems have a finite solution we have:

$$f^* = \mathbf{c}^T \mathbf{x}^* = \boldsymbol{\mu}^T \mathbf{b} \Rightarrow \frac{\partial f^*}{\partial b_i} = \mu_i^*.$$

This seems intuitively true, but we should note that some difficulties may occur when the dual problem has multiple optimal solutions. In this case, the function $f^*(\mathbf{b})$ is nondifferentiable; this is related to what we have seen in Section A.5 about the nondifferentiability of the dual function. Apart from this technical difficulty, it remains true that LP duality theory is fundamental in developing efficient ways to analyze how the optimal solution of an LP problem changes when the problem is perturbed. The types of perturbation that can be considered are not limited to $\mathbf{b}$; we may consider variations in $\mathbf{A}$ and $\mathbf{c}$ as well. Here we point out how we can cope with row and column additions; as shown in Appendix C, this is important for algorithms dealing with MILP problems.

The starting point is to realize the intimate relationship between the dual feasibility condition (B.8) and the optimality conditions for the primal problem. The Simplex method stops when the reduced costs are nonnegative:

$$\hat{\mathbf{c}}^T = \mathbf{c}^T - \mathbf{c}_B^T \mathbf{A}_B^{-1} \mathbf{A} \geq \mathbf{0}.$$

If we set

$$\boldsymbol{\mu}^T = \mathbf{c}_B^T \mathbf{A}_B^{-1},$$

dual feasibility is satisfied, and the optimal values are equal, since

$$\boldsymbol{\mu}^T \mathbf{b} = \mathbf{c}_B^T \mathbf{A}_B^{-1} \mathbf{b} = \mathbf{c}_B^T \hat{\mathbf{b}}.$$

The Simplex method starts with a primal feasible, and proceeds to attain dual feasibility; we can devise an algorithm starting with a dual feasible solution aimed at obtaining a primal feasible solution. By exploiting the apparent symmetry of the primal-dual pair, a dual Simplex method can be built, exploiting the same information as the tableau we have seen. The point is that using the same data structures we obtain both the primal and the dual variables.

Now consider the optimal solution $\mathbf{x}^*$ of an LP problem in standard form and suppose we add an inequality constraint

$$bfa_{m+1}^T x \geq b_{m+1}.$$

If x^* satisfies the additional constraint, it remains optimal. Otherwise, there is no need to restart the Simplex method from scratch. In fact, we can easily obtain a dual feasible solution. Let x_{n+1} be the slack variable of the additional constraint; then the augmented problem is

$$\min \ c^T x + 0 \cdot x_{n+1}$$

$$\text{s.t.} \ \left[\begin{array}{c|c} A & 0 \\ \hline a_{m+1}^T & 1 \end{array}\right] \left[\begin{array}{c} x \\ \hline x_{n+1} \end{array}\right] = \left[\begin{array}{c} b \\ \hline b_{m+1} \end{array}\right]$$

$$x, x_{n+1} \geq 0.$$

If we add x_{n+1}, whose cost coefficient is 0, to the optimal basis of the original problem, it is easy to see that dual feasibility, i.e., nonnegativity of the reduced costs, is preserved:

$$\tilde{c}^T = [c^T \mid 0] - [c_B^T \mid 0] \left[\begin{array}{c|c} A_B & 0 \\ \hline 0 & 1 \end{array}\right]^{-1} \left[\begin{array}{c|c} A & 0 \\ \hline a_{m+1}^T & 1 \end{array}\right]$$

$$= [c^T \mid 0] - [c_B^T \mid 0] \left[\begin{array}{c|c} A_B^{-1} & 0 \\ \hline 0 & 1 \end{array}\right] \left[\begin{array}{c|c} A & 0 \\ \hline a_{m+1}^T & 1 \end{array}\right]$$

$$= [c^T \mid 0] - [c_B^T \mid 0] \left[\begin{array}{c} A_B^{-1} A \\ \hline 1 \end{array}\right]$$

$$= [\hat{c}^T \mid 0].$$

In a similar way we can cope with the addition of an equality constraint.

The addition of a variable x_{n+1} is dealt with in a dual way. In this case we add a column A^{n+1} and a cost coefficient c_{n+1}. In this case primal feasibility is maintained, but dual feasibility need not. We have simply to compute the reduced price of the new variable, with respect to the current optimal basis, and to check the dual feasibility.

B.6 Network Optimization

A network is a graph with some numerical information attached to the nodes or the arcs. Unless the contrary is stated, we consider here only

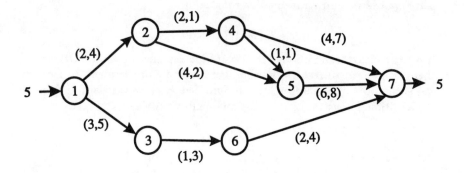

FIGURE B.8
A sample flow graph; there are a single source and a single demand node; for each arc there is a (cost, capacity) pair

directed graphs and networks. We review some basic problems along with their models. Most of them can be formulated as LP problems and solved by the Simplex method; however, specialized algorithms allow solving them much more efficiently. We will not consider such algorithms in detail; our aim is only to make the reader aware of this important class of models and the possibility of solving some of them quite efficiently.

B.6.1 The minimum cost flow problem

Consider a directed network modeling a distribution system; nodes correspond to routing points and arcs to transportation links. A single commodity flows in the network. Some nodes i are **source nodes**, in the sense that they send a known amount s_i of the commodity into the network; this amount is called a **supply**. Other nodes act as **demand nodes** (or sinks), in the sense that they are the final destination of the commodity; demand nodes are characterized by a demand which can be modeled by a negative supply. For transit nodes we have $s_i = 0$. If flows are conserved, we must have

$$\sum_{i \in \mathcal{N}} s_i = 0.$$

Let x_{ij} be the amount of commodity flowing on arc (i,j), from node i to node j. Any arc (i,j) has an associated capacity u_{ij}, i.e., an upper bound on x_{ij}. Each arc has a cost c_{ij}: in the standard minimum cost flow problem a linearity assumption is made, i.e., the cost of sending an amount x_{ij} of commodity from i to j is $c_{ij}x_{ij}$. In Figure B.8 a sample flow graph is shown. We want to find a flow routing such that the required amount of commodity is shipped at minimum cost. The following LP problem is obtained:

$$\min \sum_{(i,j) \in \mathcal{A}} c_{ij} x_{ij}$$

$$\text{s.t.} \sum_{j | (i,j) \in \mathcal{A}} x_{ij} - \sum_{j | (j,i) \in \mathcal{A}} x_{ji} = s_i \quad \forall i \in \mathcal{N} \quad (B.9)$$

$$x_{ij} \in [0, u_{ij}] \quad \forall (i,j) \in \mathcal{A}.$$

Eq. (B.9) represents the conservation of flow; the difference between the flow leaving node i and the flow entering it is the supply s_i.

More general cost models may be considered, possibly resulting in much more complex problems. For instance, fixed-charge arcs may be given. When a positive amount of commodity flows on a fixed-charge arc, a fixed cost must be paid, independent of the value of the flow that arc (see e.g. the fixed-charge problem in Section C.1). Another possible complication arises when different commodities flow in the network, competing for the available transportation capacity. This multi-commodity flow problem can be easily modeled as an LP problem, but the specialized algorithms for the single-commodity flow problem do not apply directly to this case.

B.6.2 The transportation problem

The transportation problem can be considered as a special case of the minimum cost flow problem, which arises when the network is bipartite, i.e., there is a set of source nodes and a set of demand nodes, with no intermediate transit node (see Figure B.9).

We obtain the following LP model:

$$\min \sum_{(i,j) \in \mathcal{A}} c_{ij} x_{ij}$$

$$\text{s.t.} \sum_{j | (i,j) \in \mathcal{A}} x_{ij} = a_i \quad \forall i \in \mathcal{S}$$

$$\sum_{i | (i,j) \in \mathcal{A}} x_{ij} = b_j \quad \forall j \in \mathcal{T}$$

$$x_{ij} \geq 0 \quad \forall (i,j) \in \mathcal{A},$$

where a_i is the amount of commodity available at the supply node i, b_j is the required amount of commodity at the demand node j, and

$$\sum_{i \in \mathcal{S}} a_i = \sum_{j \in \mathcal{T}} b_j.$$

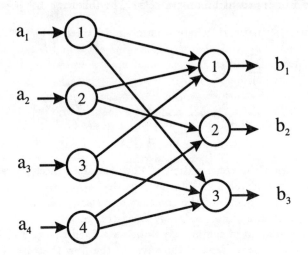

FIGURE B.9
A bipartite flow graph corresponding to a transportation problem

Actually, this last restriction is not necessary; by introducing dummy nodes it is possible to deal with unbalanced supplies and demands. The transportation problem can be solved by any algorithm for the minimum cost flow problem; however, specialized algorithms are available exploiting its structure.

B.6.3 The maximum flow problem

In the maximum flow problem we have a source node s and a single demand node t. The amount of commodity flowing in the network is not known: we want to find out the maximum amount of commodity which can be routed in the network given the arc capacities. Denoting the unknown flow by v, we obtain

$$\max v$$
$$\text{s.t.} \sum_{j|(i,j)\in\mathcal{A}} x_{ij} - \sum_{j|(j,i)\in\mathcal{A}} x_{ji} = \begin{cases} v & \text{for } i = s \\ -v & \text{for } i = t \\ 0 & \text{for } i \in \mathcal{N} - \{s,t\} \end{cases}$$
$$x_{ij} \in [0, u_{ij}] \quad \forall (i,j) \in \mathcal{A}.$$

Specific algorithms are available for this problem, too.

B.6.4 The Shortest and Longest Path problems

Consider a directed network where a number is attached to the arcs, expressing its length (in practical applications, the arc "length" does not need to refer to a physical length). Given a start node s and a destination node t, the shortest path problem consists of finding a path leading from s to t such that the total distance traveled is minimal. In the longest path problem we are interested in maximizing the total distance traveled; the longest path in a graph is also known as the **critical path**. We will not consider here the specialized algorithms for such problems; in Appendix D we show how they can be coped with by using dynamic programming. However, it is worth mentioning that these problems can be modeled with LP. In the case of the shortest path, we can think of sending a unit flow from s to t. It is easy to see that the following minimum cost flow formulation, where c_{ij} is the arc length, allows finding the shortest path.

$$\min \sum_{(i,j)\in \mathcal{A}} c_{ij} x_{ij}$$

$$\text{s.t.} \sum_{j|(i,j)\in \mathcal{A}} x_{ij} - \sum_{j|(j,i)\in \mathcal{A}} x_{ji} = \begin{cases} 1 & \text{for } i = s \\ -1 & \text{for } i = t \\ 0 & \text{for } i \in \mathcal{N} - \{s,t\} \end{cases}$$

$$x_{ij} \in [0,1] \quad \forall (i,j) \in \mathcal{A}.$$

According to this model, there is no apparent guarantee that the unit flow we send from s will not split, resulting in a fractional solution. In Section B.7 we show that minimum cost flow problems have an integer optimal solution if all the data are integers.

B.6.5 The Traveling Salesman Problem

In the Traveling Salesman Problem (TSP for short), we are given a set of "cities" represented by nodes of a graph; the distance between each pair of cities is represented by an arc of given length. Given a start node s, we want to find a cycle starting from s, visiting each node exactly once and returning to s, such that the total distance covered is minimal. This cycle is known as the **Hamiltonian cycle**. If the graph is directed, we have the **asymmetric** TSP (ATSP); if the graph is undirected (i.e., the distance from i to j is the same as from j to i), we have the **symmetric** TSP.

Unlike the previous problems, TSP cannot be modeled with LP and no efficient algorithm is known for it; in Section C.1 we show how it can be modeled with Mixed-Integer LP.

B.7 Total unimodularity

The optimal solution of an LP problem is, in general, a vector with noninteger components. In most production management problems, it is necessary to enforce an integrality requirement on some decision variable; in this case the Simplex method does not yield a feasible solution to the problem. Nevertheless, in some cases, solving the problem as a continuous optimization problem yields an integer optimal solution. This certainly happens when the extreme points of the feasible set have integer coordinates. An important class of problems for which this holds is characterized by a property of the matrix **A** called **unimodularity**.

Example B.14
We consider here the so-called **assignment problem**. We are given two disjoint sets of objects which should be paired. The sets have the same cardinality, and we require that each element of the first set has exactly one mate in the second set. Let $i, j = 1, \ldots, N$ be the indexes of the two sets, and c_{ij} the cost of pairing object i with object j. By introducing binary variables x_{ij} modeling the decision of pairing i and j, we obtain the following model:

$$\min \sum_{i=1}^{N} \sum_{j=1}^{N} c_{ij} x_{ij}$$

$$\text{s.t.} \quad \sum_{j=1}^{N} x_{ij} = 1 \quad i = 1, \ldots, N$$

$$\sum_{i=1}^{N} x_{ij} = 1 \quad j = 1, \ldots, N$$

$$x_{ij} \in \{0, 1\} \quad i, j = 1, \ldots, N.$$

If we relax the integrality constraints, i.e., we only require $x_{ij} \in [0, 1]$ and solve the problem as a continuous LP problem, we obtain an integer solution. □

The fact that the assignment problem has a "naturally" integer optimal solution is no coincidence.

DEFINITION B.15 *A full-rank integer matrix* **A** *is* **unimodular** *if the determinant of any basis matrix* $\mathbf{A}_B$ *is* ± 1.

DEFINITION B.16 *An integer matrix* **A** *is* **totally unimodular** *if the determinant of any square submatrix is* ±1 *or* 0.

Clearly a totally unimodular matrix is also unimodular, but not vice versa. An important property of unimodular matrices is that every solution of a system $\mathbf{Ax} = \mathbf{b}$ is integer if **b** is integer; this is easy to see by considering the Cramer rule for solving a system of linear equations.

THEOREM B.2
If a $(0, \pm 1)$ *matrix* **A** *has no more than two nonzero entries in each column, and if* $\sum_i a_{ij} = 0$ *when column j has two nonzero entries, then* **A** *is totally unimodular.*

A consequence of this theorem is that the matrix **A** of a network optimization problem is totally unimodular, since each column contains exactly one 1 and one −1 element.

Example B.17
Consider the network of Figure B.8, and write the flow conservation equations for nodes N_1 and N_2:

$$x_{12} + x_{13} = 5$$

$$x_{24} + x_{25} - x_{12} = 0.$$

We note that the variable x_{12} relative to the flow on arc $(1,2)$ occurs twice, with a coefficient 1 in the equation of node N_1, and with a coefficient -1 in the equation of node N_2. In fact, the network can be represented by the following node-arc **incidence** matrix:

	(1,2)	(1,3)	(2,4)	(2,5)	(3,6)	(4,5)	(4,7)	(5,7)	(6,7)
N_1	1	1	0	0	0	0	0	0	0
N_2	−1	0	1	1	0	0	0	0	0
N_3	0	−1	0	0	1	0	0	0	0
N_4	0	0	−1	0	0	1	1	0	0
N_5	0	0	0	−1	0	−1	0	1	0
N_6	0	0	0	0	−1	0	0	0	1
N_7	0	0	0	0	0	0	−1	−1	−1

This matrix is totally unimodular. Actually, the matrix of the overall flow problem includes entries due to the upper bounds, but this does not change the property. □

This is why the LP representations of minimum cost flow, maximum flow, and shortest and longest path problems have integer solutions even when

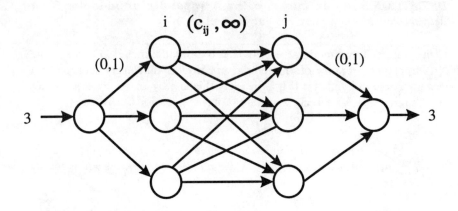

FIGURE B.10
A minimum cost flow graph equivalent to the assignment problem

solved as continuous problems, provided that the data are expressed by integers. This happens in the assignment case, too, since it is easily seen that it is a particular case of a minimum cost flow problem (see Figure B.10).

For further reading

Readers interested in a deeper understanding of Linear Programming and the Simplex method in all its forms are referred to the many textbooks available, among which we mention [148]. More information on polyhedral theory can be found in [150, Chapter I.4].

An introductory textbook on graph and network optimization is [71]. A most comprehensive reference is [4], where the most recent algorithmic developments are covered; additional material on advanced algorithms can be found in [22].

C
Enumerative and Heuristic Methods for Discrete Optimization

In Appendix A and B we have considered optimization problems involving continuous decision variables. In many production management problems some decision variables can only take integer values; we obtain Integer Programming (IP) models. This is often due to the need to choose one among a set of discrete alternatives, which leads to Combinatorial Optimization problems. Note, however, that IP is only one approach to solving Combinatorial Optimization problems; for instance, Dynamic Programming is an alternative method, that does not require the formulation of an IP model. We will use the term Discrete Optimization to refer both to Combinatorial Optimization and Integer Programming.

Discrete optimization problems can be solved either exactly or approximately; in most cases, the exact solution requires some form of partial enumeration. The need for enumeration is due to the fact that, in general, there is no characterization of the optimal solutions of a discrete optimization problem, and the only way to solve it is, in principle, to enumerate all the feasible solutions and to pick the best one. This is clearly possible when the set of feasible solutions is finite; partial, or implicit, enumeration can be considered as a way to reduce the computational burden by limiting the number of enumerated solutions. In this Appendix we deal with Branch and Bound and heuristic methods; in Appendix D we deal with Dynamic Programming.

We start in Section C.1 by listing some classical discrete optimization problems and showing how they can be modeled as integer programming problems. Modeling with binary variables is the topic of Section C.2. We describe the general structure of Branch and Bound methods in Section C.3. A successful Branch and Bound algorithm stems from a combination of factors: the ability of computing lower bounds on the optimal value of a subproblem (Section C.4); a good strategy to generate subproblems (Section C.5); a way to explore the tree of subproblems (Section C.6);

the use of additional conditions to reduce the search effort (Section C.7). Many practical problems cannot be solved at optimality with a reasonable computational effort; in this case, heuristic strategies must be adopted. Due to the variety of heuristics that have been proposed, in Section C.8 we describe only some *general* principles which can be used to design heuristic strategies; it should be emphasized that for a specific problem it can be advantageous to devise a specific solution approach.

C.1 Classical discrete optimization problems

We consider here some prototypical discrete optimization problems, along with possible integer programming formulations. These problems are fundamental, since they often appear as subproblems or substructures of more complex problems.

The Knapsack problem

In the knapsack problem we are given a set of items characterized by a weight and a value. We must find a subset of items in order to maximize the total value of the items selected, taking into account the maximum total weight the knapsack can carry. As a possible manufacturing oriented application, consider the problem of selecting a set of jobs to be executed on a single machine; the jobs have a priority and a processing time, and the available machining time is limited.

To model the problem, we introduce a set of binary variables x_i ($i = 1, \ldots, N$) such that

$$x_i = \begin{cases} 1 & \text{if object } i \text{ is selected} \\ 0 & \text{otherwise.} \end{cases}$$

If we denote by w_i and v_i, the weight and the value of object i, respectively, and by C the available capacity, we can build the following 0/1 programming model:

$$\max \sum_{i=1}^{N} v_i x_i$$

$$\text{s.t.} \sum_{i=1}^{N} w_i x_i \leq C$$

$$x_i \in \{0, 1\} \quad \forall i.$$

One would be tempted to sort the items in ascending order of the ratio v_i/w_i and to select the items until the capacity is saturated; unfortunately this yields only a suboptimal solution. However, quite efficient Branch and Bound algorithms are available for the knapsack problem.

The Traveling Salesman Problem

We have introduced the TSP problem in Section B.6.5. Given a (directed or undirected) graph with weighted edges, we want to find a cycle visiting each vertex exactly once and minimizing the total weight of the selected arcs (i.e., the length of the cycle). If the arcs are directed, we have the Asymmetric TSP; the case of undirected arcs is called Symmetric TSP. As an application, consider the problem of sequencing jobs on a single machine characterized by sequence dependent setup times (see Section 4.6.3). Our aim is to sequence the jobs in order to minimize the time needed to process all the jobs. This problem can be modeled as an Asymmetric TSP by introducing a node for each job ($i = 1, \ldots, N$) and a dummy node 0, representing the current state of the machine. The arc lengths are set to the setup times, the arcs from node 0 have a length corresponding to the setup time from the current machine state, and the arcs to node 0 have length 0, if we do not care about the last machine state. Let $\mathcal{A}$ denote the set of arcs.

To model the problem, we introduce a set of binary variables x_{ij} defined as

$$x_{ij} = \begin{cases} 1 & \text{if the arc } (i,j) \text{ is used} \\ 0 & \text{otherwise.} \end{cases}$$

The objective function is

$$\sum_{(i,j)\in\mathcal{A}} c_{ij} x_{ij}.$$

Clearly, each node must have exactly one predecessor and one successor:

$$\sum_{\substack{j=0 \\ i \neq j}}^{N} x_{ij} = 1 \quad \forall i$$

$$\sum_{\substack{i=0 \\ i \neq j}}^{N} x_{ij} = 1 \quad \forall j.$$

Note that these constraints are not sufficient to guarantee a feasible solution: in fact they do not rule out the possibility of subtours, i.e., two

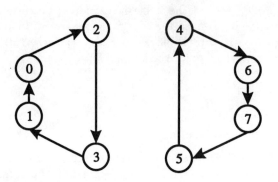

FIGURE C.1
Assignment constraints do not rule out subtours in the TSP problem.

(or more) disjoint cycles (see figure C.1). One way to avoid subtours is to introduce a set of continuous variables u_i (one for each node apart from the start node 0), subject to the constraints

$$u_i - u_j + Nx_{ij} \leq N - 1 \qquad \forall i,j = 1, 2, \ldots, N;\ i \neq j. \qquad (C.1)$$

Note that the constraints are not written for $i, j = 0$, and that u_0 is not defined. Consider a subtour not including node 0. Let the subtour consist of k nodes and let $\sigma(m)$ denote the name of the mth node in the cycle. Now write (C.1) for each arc in the cycle:

$$u_{\sigma(1)} - u_{\sigma(2)} + Nx_{\sigma(1),\sigma(2)} \leq N - 1$$
$$u_{\sigma(2)} - u_{\sigma(3)} + Nx_{\sigma(2),\sigma(3)} \leq N - 1$$
$$\ldots$$
$$u_{\sigma(k)} - u_{\sigma(1)} + Nx_{\sigma(k),\sigma(1)} \leq N - 1.$$

If we sum up these constraints, we get

$$Nk \leq (N-1)k.$$

Then, these constraints rule out subtours not including node 0; the key point is that node 0 is not considered in (C.1). It is easy to see that no undesired restriction is introduced by these constraints. Consider for instance an arc on a path, and write the corresponding constraint, where $x_{\sigma(k),\sigma(k+1)} = 1$,

$$u_{\sigma(k+1)} - u_{\sigma(k)} \geq 1$$

In general, if we sum the constraints on a path of m nodes not including node 0, we get

$$u_{\sigma(k+m)} - u_{\sigma(k)} \geq m.$$

Thus, the value taken by u_i is simply the sequence number of node i in the Hamiltonian cycle.

The Bin Packing problem

The Bin Packing problem is somehow related to the knapsack problem. We are given a set of N items (indexed by i) of weight w_i and a set of M bins (indexed by j) of capacity C_j. We are required to allocate the items in the bins in order to minimize the number of bins used (M is an upper bound on this number). As an application, consider the problem of scheduling the activity of a batch processor, i.e., a machine able to process a set of jobs together (see Section 4.2.5). Each job requires an amount of space w_i, and the available space is C; we want to minimize the number of batches required to process the set of jobs. Note that in this case the "bins" are the batches and their capacities are equal.

We introduce two sets of binary decision variables:

$$y_j = \begin{cases} 1 & \text{if bin } j \text{ is used} \\ 0 & \text{otherwise} \end{cases}$$

$$x_{ij} = \begin{cases} 1 & \text{if item } i \text{ is loaded on bin } j \\ 0 & \text{otherwise.} \end{cases}$$

The model can be formulated as follows:

$$\min \sum_{j=1}^{M} y_j$$

$$\text{s.t.} \sum_{i=1}^{N} w_i x_{ij} \leq C_j \quad \forall j \tag{C.2}$$

$$\sum_{j=1}^{M} x_{ij} = 1 \quad \forall i \tag{C.3}$$

$$x_{ij} \leq y_j \quad \forall i, j \tag{C.4}$$

$$y_j, x_{ij} \in \{0, 1\} \quad \forall i, j.$$

M is an upper bound on the number of bins that will be used. Eq. (C.2) insures that no bin capacity is exceeded. Eq. (C.3) states that each item must be loaded in exactly one bin. Eq. (C.4) relates the two sets of decision variables; item i can be allocated to bin j only if bin j is activated.

The Fixed-Charge problem

We are given a set of activities indexed by $i = 1, \ldots, N$. The level of activity i is measured by a nonnegative continuous variable x_i; the activity levels are subject to a set of constraints, formally expressed as $\mathbf{x} \in S$. Each activity has a cost proportional to the level x_i, and a fixed cost f_i which is paid whenever $x_i > 0$. The fixed cost does not depend on the activity level. We want to carry out the activities at minimum cost. A common case of such a problem is the minimum cost flow with fixed charges, whereby a fixed cost is incurred whenever an arc is used (the minimum cost flow problem has been introduced in Section B.6.1). In manufacturing, fixed costs are usually associated with setup issues (see Section 3.2).

If we introduce binary variables y_i such that

$$y_i = \begin{cases} 1 & \text{if } x_i > 0 \\ 0 & \text{otherwise,} \end{cases}$$

we can build the following model:

$$\min \sum_{i=1}^{N} (c_i x_i + f_i y_i)$$
$$\text{s.t.} \quad x_i \leq M_i y_i \quad \forall i \qquad (C.5)$$
$$\mathbf{x} \in S$$
$$y_i \in \{0, 1\} \quad \forall i,$$

where M_i is an upper bound on the level of activity i. Eq. (C.5) is a common way to model fixed charge costs. If $y_i = 0$, then necessarily $x_i = 0$; if $y_i = 1$, then we obtain $x_i \leq M_i$ which is a nonbinding constraint if M_i is large enough. Note that we are able to model a *discontinuous* objective function by this technique.

The Set Covering, Set Packing and Set Partitioning problems

Consider a finite set S of N items, indexed by i, and a family of M subsets $S_j \subseteq S$; each subset has an associated cost c_j. The Set Covering problem consists of selecting a minimum cost collection of subsets, such that each item in S is included *at least* once.

Let $\mathbf{A}$ be an incidence matrix, whose entries c_{ij} are 1 if item i belongs to subset S_j, 0 otherwise. The Set Covering problem can be modeled as

$$\min \sum_{j=1}^{M} c_j x_j$$

$$\text{s.t.} \sum_{i=1}^{N} \overline{c}_{ij} x_j \geq 1 \quad \forall i$$

$$x_j \in \{0,1\} \quad \forall j.$$

where x_j is a binary variable modeling the selection of subset j. We can put the Set Covering model in the compact form:

$$\min \ \mathbf{c}^T \mathbf{x}$$

$$\text{s.t.} \ \mathbf{Ax} \geq \mathbf{1}$$

$$\mathbf{x} \in \{0,1\}^N.$$

Note that we allow overlapping subsets. If we forbid overlapping subsets, i.e., we require $S_{j_1} \cap S_{j_2} = \emptyset$, we obtain the Set Partitioning problem

$$\min \ \mathbf{c}^T \mathbf{x}$$

$$\text{s.t.} \ \mathbf{Ax} = \mathbf{1}$$

$$x_j \in \{0,1\} \quad \forall j.$$

The Set Packing problem is obtained when we do not require that every item is included, but that each item may belong to, *at most*, one of the selected subsets; in this case it is more natural to think of maximizing the total profit.

$$\max \ \mathbf{c}^T \mathbf{x}$$

$$\text{s.t.} \ \mathbf{Ax} \leq \mathbf{1}$$

$$x_j \in \{0,1\} \quad \forall j.$$

The Set Covering and the Set Partitioning problems can be useful for formulating complex problems and obtaining **modal formulations** (see Section C.2.3). Furthermore, they can be effectively solved by good heuristics.

The Generalized Assignment problem

We are given a set of N jobs (indexed by i) and M machines (indexed by j). For each pair (i,j) we have a processing time p_{ij} and a cost c_{ij}. Each machine has a capacity B_j. We want to assign the jobs to the machines in order to minimize the total cost, subject to the capacity constraints.

We can model the problem as

$$\min \sum_{i=1}^{N} \sum_{j=1}^{M} c_{ij} x_{ij}$$

$$\text{s.t.} \sum_{i=1}^{N} p_{ij} x_{ij} \leq B_j \quad \forall j$$

$$\sum_{j=1}^{M} x_{ij} = 1 \quad \forall i$$

$$x_{ij} \in \{0, 1\} \quad \forall i, j,$$

where x_{ij} models the decision of assigning job i to machine j.

The Generalized Assignment problem can be considered a variation of the Assignment problem introduced in Section B.7. Unlike the Assignment problem, if we relax the integrality constraints, i.e., we only require $x_{ij} \in [0, 1]$, the optimal solution is not necessarily integer.

C.2 Modeling with binary variables

From the above list of examples, the reader may have noticed that the integer variables are all binary. We will see little use of general integer variables. In practice, most integer variables are binary and are used to model logical conditions or to express nonlinear constraints or objective functions. It is important to be aware of all the possible modeling "tricks" based on the use of binary variables.

C.2.1 Expressing logical constraints

Consider a set of binary decision variables $y, x_1, x_2, \ldots, x_N$. Each variable may model, for instance, the decision to start a certain activity. We can model the following relations or constraints:

- at least one activity x_i is started (the *inclusive* OR is true):

$$\sum_{i=1}^{N} x_i \geq 1;$$

- at most one activity x_i is started:
$$\sum_{i=1}^{N} x_i \leq 1;$$

- exactly *one* activity x_i is started (the *exclusive* OR is true):
$$\sum_{i=1}^{N} x_i = 1;$$

- activity y is necessarily started if *any* activity x_i is started (y is the inclusive OR of the x's):
$$x_i \leq y \quad i = 1, \ldots, N;$$

we can look at this the other way round; if y is not started, no x_i can;

- activity y is *necessarily started* if *all* the activities x_i are started:
$$y \geq \sum_{i=1}^{N} x_i - N + 1;$$

- activity y *may be started* only if all activities x_i are started:
$$y \leq x_i \quad i = 1, \ldots, N.$$

C.2.2 Modeling piecewise linear objective functions

The fixed-charge model is an example of how IP models can be used to cope with nonlinear or even discontinuous objective functions. A common class of nonlinear objective functions is that of piecewise linear objective functions, like the following:

$$f(x) = \begin{cases} c_1 x & 0 \leq x \leq x^{(1)} \\ c_2(x - x^{(1)}) + c_1 x^{(1)} & x^{(1)} \leq x \leq x^{(2)} \\ c_3(x - x^{(2)}) + c_1 x^{(1)} + c_2(x^{(2)} - x^{(1)}) & x^{(2)} \leq x \leq x^{(3)} \end{cases}$$

If $c_1 < c_2 < c_3$, then $f(x)$ is convex (see Figure C.2) and it can be coped with by continuous LP models. The function $f(x)$ can be converted to a linear form by introducing three auxiliary variables y_1, y_2, y_3 and substituting:

$$x = y_1 + y_2 + y_3$$
$$0 \leq y_1 \leq x^{(1)}$$
$$0 \leq y_2 \leq (x^{(2)} - x^{(1)})$$
$$0 \leq y_3 \leq (x^{(3)} - x^{(2)}).$$

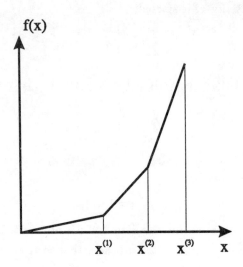

FIGURE C.2
A convex piecewise linear function.

Then we can express
$$f(x) = c_1 y_1 + c_2 y_2 + c_3 y_3,$$
since $c_1 < c_2$, y_2 is positive in the optimal solution only if y_1 is set to its upper bound. Similarly, y_3 is activated only if both y_1 and y_2 are saturated. When $c_1 > c_2 > c_3$ (as in Figure C.3), i.e., the function is concave, this is no longer the case, and IP must be used. Note that concave piecewise linear functions are useful to model economies of scale. To model a general piecewise linear function, we can note that any point x in the interval $[x^{(i)}, x^{(i+1)}]$ can be expressed as a convex combination
$$x = \lambda_i x^{(i)} + \lambda_{i+1} x^{(i+1)},$$
where $\lambda_i + \lambda_{i+1} = 1$ and $\lambda_i, \lambda_{i+1} \geq 0$. By the same token, $f(x)$ in the same interval can be expressed as
$$f(x) = \lambda_i f(x^{(i)}) + \lambda_{i+1} f(x^{(i+1)}).$$
Now we must rule out convex combinations of nonadjacent points; this is accomplished by introducing a binary decision variable s_i for each interval $[x^{(i)}, x^{(i+1)}]$. In the example above we obtain:
$$f(x) = \lambda_0 f(0) + \lambda_1 f(x^{(1)}) + \lambda_2 f(x^{(2)}) + \lambda_3 f(x^{(3)})$$

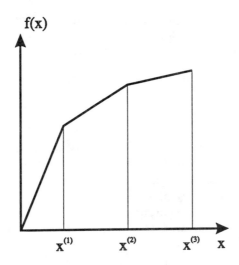

FIGURE C.3
A concave piecewise linear function.

$$x = \lambda_0 0 + \lambda_1 x^{(1)} + \lambda_2 x^{(2)} + \lambda_3 x^{(3)}$$

$$\lambda_0 + \lambda_1 + \lambda_2 + \lambda_3 = 1$$

$$0 \leq \lambda_0 \leq s_0$$

$$0 \leq \lambda_1 \leq s_0 + s_1$$

$$0 \leq \lambda_2 \leq s_1 + s_2$$

$$0 \leq \lambda_3 \leq s_2$$

$$s_0 + s_1 + s_2 = 1$$

$$s_0, s_1, s_2 \in \{0, 1\}.$$

A much more direct, and computationally convenient, way to express this kind of relation exploits Special Ordered Sets, as illustrated in Section C.5.

C.2.3 Modal formulations

Modal formulations are obtained when a point of the feasible set is expressed as a convex combination of other points [210, pp. 37-38]. In other words, a way to carry out certain activities is expressed as a combination of "extreme" ways, which can be referred to as *modes*.

In this book we will use the term modal formulation in a slightly more general way, when the feasible solutions of a certain problem are expressed as combinations of "elementary" solutions, or "recipes". Then the problem of optimally combining the recipes can be basically formulated as a Set Partitioning or Set Covering problem (possibly with additional constraints).

Example C.1
Consider the Generalized Assignment Problem; we outline here a modal formulation similar to the one developed in [46].

The basic idea is to build a set of recipes, i.e., a set of possible assignments of jobs to each machine; for each machine j we consider K_j possible assignments, and for each assignment we compute the cost $d_j^{(k)}$. Note that we do not consider every possible assignment, but only a subset. Let $\alpha_{ij}^{(k)} = 1$ if job i is assigned to machine j in recipe k, 0 otherwise. Then we may reformulate the Generalized Assignment Problem as a Set Partitioning Problem

$$\min \sum_{j=1}^{M} \sum_{k=1}^{K_j} d_j^{(k)} y_j^{(k)}$$

$$\text{s.t.} \sum_{k=1}^{K_j} y_j^{(k)} = 1 \quad \forall j \tag{C.6}$$

$$\sum_{j=1}^{M} \sum_{k=1}^{K_j} \alpha_{ij}^{(k)} y_j^{(k)} = 1 \quad \forall i \tag{C.7}$$

$$y_j^{(k)} \in \{0,1\} \quad \forall j,\ k=1,\ldots,K_j.$$

Here the binary variables $y_j^{(k)}$ model the decision of adopting the recipe k for machine j. Eq. (C.6) states that exactly one recipe must be selected for each machine. Eq. (C.7) ensures that each job is assigned to exactly one machine. ☐

In the above formulation we have neglected some practical issues: for instance, what should be done if no feasible solution is found? This may happen if the number of recipes is too small. On the other hand, we should avoid generating too many recipes, since this would result in a huge problem. In practice, the problem is tackled by *dynamically* generating the recipes. Generating a recipe implies adding a decision variable together with its column in the constraints matrix; this is why the overall approach is known as **column generation**. The way columns are generated is highly problem dependent, and it will not be discussed here. It may be useful to exploit heuristic procedures to generate a good initial set of columns. As to the Set Partitioning problem, it may be too large for an exact solution approach; however, heuristic strategies can be devised. One possibility is to solve it as a continuous LP problem and to round the solution; this is a viable approach if the number of fractional variables in the optimal solution is small. We will see an application of these ideas in Section 3.2.5.

C.2.4 Linearizing nonlinear 0/1 problems

When we must consider the joint effect of decisions, we cannot build a linear model. We consider here nonlinear 0/1 programming; this is a peculiar case since the product of binary variables is itself binary. Clearly, it is useless to consider terms of the form x_j^n, since $x_j^n = x_j$. It is sufficient to cope with product terms like

$$\prod_{j \in P} x_j.$$

It is possible to linearize the problem by introducing an auxiliary variable y, subject to the following constraints:

$$\sum_{j \in P} x_j - y \leq |P| - 1$$

$$y \leq x_j \quad \forall j \in P,$$

where $|P|$ is the cardinality of the set P. The first constraint ensures that $y = 1$ whenever all the x_j are 1; the second set of constraints ensures that $y = 0$ if there is a j such that $x_j = 0$.

A typical example of nonlinear 0/1 problem is the **Quadratic assignment problem**, which can be used to model layout problems. Consider two production departments i and j, and let f_{ij} be the amount of material flow between them. Let k and l be two possible locations for the two departments, and let c_{kl} be the cost per unit flow between these locations. We can model the decision of placing a department in a location by binary variables δ_{ik} and δ_{jl}; however, the objective of minimizing the flow cost contains quadratic terms such as $f_{ij}c_{kl}\delta_{ik}\delta_{jl}$.

C.3 Branch and Bound methods

In the previous sections we have considered model formulation issues; here we turn our attention to solution methods. Consider a generic optimization problem

$$P(S) \qquad \min_{\mathbf{x} \in S} f(\mathbf{x}),$$

and assume that it cannot be tackled directly since there is no convenient way to characterize an optimal solution. A possible solution strategy is to decompose the problem by splitting the feasible set S into a collection of subsets $S_1, \ldots, S_q$ such that

$$S = S_1 \cup S_2 \cup \ldots \cup S_q;$$

then, we have

$$\min_{\mathbf{x} \in S} f(\mathbf{x}) = \min_{i=1,\ldots,q} \left\{ \min_{\mathbf{x} \in S_i} f(\mathbf{x}) \right\}.$$

The rationale behind this decomposition of the feasible set is that we may expect that solving the problems over smaller sets is easier. For efficiency reasons it is advisable, but not strictly necessary, to partition the set S in such a way that:

$$S_i \cap S_j = \emptyset \qquad i \neq j.$$

This type of decomposition is called **branching**.

Example C.2
Consider the binary programming problem:

$$\min \ \mathbf{c}^T \mathbf{x}$$
$$\text{s.t.} \ \mathbf{x} \in S = \{\mathbf{x} \mid \mathbf{A}\mathbf{x} \geq \mathbf{b}; x_j \in \{0,1\}\}.$$

The problem is decomposed by picking a variable x_p and fixing it, respectively, to 1 and 0:

$$S_1 = \{\mathbf{x} \in S; \ x_p = 0\}$$
$$S_2 = \{\mathbf{x} \in S; \ x_p = 1\}.$$

The resulting problems $P(S_1)$ and $P(S_2)$ can be decomposed in turn, until eventually all the variables have been fixed. The branching process can be pictorially represented as a search tree, as shown in figure C.4. □

The branching process leads to easier problems. In the example, the leaves of the search tree are trivial problems, since all the variables are fixed to a value; actually, the search tree is, in this case, just a way to enumerate the possible solutions. Unfortunately, there is a large number of leaves; if $\mathbf{x} \in \{0,1\}^N$, there are 2^N possible solutions. Actually, the constraints $\mathbf{A}\mathbf{x} \geq \mathbf{b}$ rule out many of them, but a brute-force enumeration is not viable but for the smallest problems.

To reduce the computational burden, one can try to eliminate a subproblem $P(S_k)$, or equivalently a node of the tree, by showing that it cannot

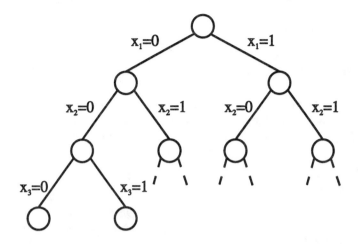

FIGURE C.4
Search tree for a binary programming problem.

lead to the optimal solution of $P(S)$. This can be accomplished if it is possible to compute a **lower bound** on the optimum of each subproblem. Let $\nu[P(S_k)]$ denote the optimal value of problem $P(S_k)$; a lower bound is a number $\beta[P(S_k)]$ such that

$$\beta[P(S_k)] \leq \nu[P(S_k)].$$

Now assume that we know a feasible solution $\hat{x}$ of $P(S)$; clearly, there is no point in solving $P(S_k)$ if

$$\beta[P(S_k)] \geq f(\hat{x}). \qquad (C.8)$$

In this case we can eliminate $P(S_k)$ from further consideration; this elimination is called **fathoming** and it corresponds to pruning a branch of the search tree. Note that $P(S_k)$ can be fathomed only by comparing $\beta[P(S_k)]$ to an upper bound ν^* on $\nu[P(S)]$. It is *not* correct to fathom $P(S_k)$ on the basis of a comparison with a subproblem $P(S_i)$ such that

$$\beta[P(S_i)] < \beta[P(S_k)].$$

Fathoming by comparing two subproblems is sometimes possible, but it requires the application of **dominance conditions** (see Section C.7).

The branching and fathoming mechanism is the foundation of a wide class of algorithms known as **Branch and Bound**. The fundamental Branch and Bound algorithm can be outlined as follows. At each step of the algorithm we have an **incumbent solution**, which is the best solution found so far.

Fundamental Branch and Bound algorithm.

1. *Initialization.* The list of candidate subproblems is initialized to $P(S)$; the value of the incumbent solution ν^* is set to $+\infty$.
2. *Selecting a candidate subproblem.* A subproblem $P(S_k)$ is selected from the list of candidate subproblems.
3. *Bounding.* Compute a lower bound $\beta(S_k)$ on $\nu[P(S_k)]$. If $\beta(S_k) \geq \nu^*$, go to step 7.
4. *Generate a feasible solution.* If possible generate a feasible solution $\hat{x}$. If $f(\hat{x}) \geq \nu^*$, or it is not possible to generate any feasible solution, go to step 6.
5. *Saving the new incumbent solution.* If at step 5 a solution $\hat{x}$ is found improving the incumbent, set $\nu^* \leftarrow f(\hat{x})$ and $x^* \leftarrow \hat{x}$. If it is possible to show that $\hat{x}$ is the optimal solution of $P(S_k)$, i.e. $f(\hat{x}) = \beta[P(S_k)]$, go to step 7.
6. *Branching.* Replace $P(S_k)$ in the list of candidate subproblems with a list of child subproblems $P(S_{k1}), P(S_{k2}),\ldots, P(S_{kq})$, obtained by partitioning S_k; go to step 2.
7. *Fathoming.* Problem $P(S_k)$ is eliminated from the list of candidate subproblems. If the list is empty, return the current incumbent solution as the optimal solution x^* and stop; otherwise go to step 2.

In Step 4 it is necessary to try generating a feasible solution; we will discuss how this can be accomplished at the end of Section C.5.

To successfully apply this algorithm we must cope with the following issues:

- computing a strong lower bound efficiently; by *strong* we mean, for a minimization problem, a large bound, as close as possible to the true optimal value for the related subproblem;
- generating subproblems;
- selecting the right candidate from the list of open subproblems.

The next sections treat such issues in more detail.

It is worth noting that the Branch and Bound idea is not limited to discrete optimization problems; in fact it can also be used for global (nonconvex) optimization.

C.4 Relaxation methods for bounding

To compute a lower bound on the optimal solution of problem $P(S)$, a common strategy is to solve a *relaxed* version $P(\overline{S})$,

$$\min_{\mathbf{x} \in \overline{S}} f(\mathbf{x}),$$

where $S \subset \overline{S}$. By enlarging the feasible set, the optimal value cannot increase. In many cases a much easier problem is obtained by relaxing some constraints. Examples of this approach are: constraints elimination, continuous relaxation, and surrogate relaxation. Note that if the optimal solution of $P(\overline{S})$ belongs to S, it is also optimal for $P(S)$.

An alternative approach, Lagrangian relaxation, based on weak duality theory, also involves a modification of the objective function.

C.4.1 Constraints elimination

The constraints elimination approach is simply based on the recognition of "complicating" constraints. We can write the feasible set S as

$$S = S_d \cap S_e,$$

where S_d and S_e are the difficult and easy constraints, respectively, in the sense that the problem

$$\min_{\mathbf{x} \in S_e} f(\mathbf{x})$$

is much easier to solve than the original one.

Then a lower bound is obtained by solving the easy problem

$$\beta[P(S)] = \nu \left(\min_{\mathbf{x} \in S_e} f(\mathbf{x}) \right).$$

This approach, unlike the following ones, does not need a mathematical statement of the problem. An application of this approach to a scheduling problem is described in Section 4.7.3.

C.4.2 Continuous relaxation

Continuous relaxation is most useful for MILP problems, and it is used in commercial software packages implementing Branch and Bound. Given a MILP problem:

$$P(S) \quad \min \ \mathbf{c}^T \mathbf{x} + \mathbf{d}^T \mathbf{y}$$
$$\text{s.t.} \ \mathbf{A}\mathbf{x} + \mathbf{E}\mathbf{y} \leq \mathbf{b}$$

$$\mathbf{x} \in \mathbf{R}_+^{n_1} \quad \mathbf{y} \in \mathbf{Z}_+^{n_2},$$

the continuous relaxation is obtained by relaxing the integrality constraints, i.e.,

$$P(\overline{S}) \quad \min \ \mathbf{c}^T \mathbf{x} + \mathbf{d}^T \mathbf{y}$$
$$\text{s.t.} \ \mathbf{Ax} + \mathbf{Ey} \leq \mathbf{b}$$
$$\begin{bmatrix} \mathbf{x} \\ \mathbf{y} \end{bmatrix} \in \mathbf{R}_+^{n_1+n_2}.$$

Example C.3
Given the binary programming problem $P(S)$,

$$\min \ \mathbf{c}^T \mathbf{x}$$
$$\text{s.t.} \ \mathbf{x} \in S = \{\mathbf{x} \mid \mathbf{Ax} \geq \mathbf{b}; \ x_j \in \{0,1\}\},$$

the continuous relaxation is problem $P(\overline{S})$,

$$\min \ \mathbf{c}^T \mathbf{x}$$
$$\text{s.t.} \ \mathbf{x} \in S = \{\mathbf{x} \mid \mathbf{Ax} \geq \mathbf{b}; \ 0 \leq x_j \leq 1\}. \quad \square$$

Since the relaxed problem is a continuous LP problem, it can be solved by the Simplex method in order to obtain a lower bound.

C.4.3 Lagrangian relaxation

We have shown in Section (A.5) that, for a generic problem (P)

$$\min \ f(\mathbf{x})$$
$$\text{s.t.} \ g(\mathbf{x}) \leq 0$$
$$\mathbf{x} \in S,$$

the following weak duality property holds:

$$\nu \left(\max_{\mu \geq 0} \left\{ \min_{\mathbf{x} \in S} f(\mathbf{x}) + \mu^T g(\mathbf{x}) \right\} \right) \leq \nu \left(\begin{array}{l} \min \ f(\mathbf{x}) \\ \text{s.t.} \ g(\mathbf{x}) \leq 0 \\ \phantom{\text{s.t.}} \ \mathbf{x} \in S \end{array} \right).$$

Weak duality lends itself to the computation of lower bounds when the primal problem (P) can be thought of as an easy problem, $\min_{\mathbf{x} \in S} f(\mathbf{x})$, complicated by the "nasty" constraints $g(\mathbf{x}) \leq 0$. A similar property holds when the complicating constraints are equality constraints; the only modification is that Lagrangian multipliers need not be restricted in sign (see Section A.5).

To obtain a strong lower bound, we have to choose a good vector of multipliers; this requires solving the dual problem, which, for the case of inequality constraints, is

$$(D_L) \quad \max_{\mu \geq 0} \nu(P\mu).$$

The maximization of the dual function is complicated by the fact that it is, in general, a nondifferentiable function. For the MILP case, it can be shown that it is a piecewise linear concave function (see Example A.13). In this case (P) is

$$\min \ c^T x$$
$$\text{s.t.} \ Ax \geq b$$
$$x \in S.$$

The relaxed problem $(P\mu)$ is

$$\min_{x \in S} \left\{ c^T x + \mu^T (b - Ax) \right\}.$$

If S is a closed and bounded set, $\nu(P\mu)$ is the lower envelope of a finite collection of linear functions, one for each $x \in S$. Then $\nu(P\mu)$ is piecewise linear; since it is the minimum of a collection of concave functions, it is itself concave (see Property A.12).

A possible strategy to compute good multipliers is the subgradient method introduced in section A.7. We know that a subgradient is easily computed as

$$b - A\hat{x}$$

where $\hat{x}$ is the optimal solution of the relaxed problem. The relaxed problem may have multiple solutions; indeed, the nondifferentiability points correspond to such multiple solutions. All the equivalent solutions yield a subgradient. Actually, the set of subgradients of $\nu(P\mu)$ (the subdifferential) in any μ is the convex hull of the subgradients corresponding to the optimal solutions:

$$\partial w(\mu) = \left\{ \sum_{\hat{x} \in T(\mu)} \rho^{(\hat{x})} (b - A\hat{x}) \, \Big| \, \sum_{\hat{x} \in T(\mu)} \rho^{(\hat{x})} = 1, \, \rho^{(\hat{x})} \geq 0, \, \forall \hat{x} \in T(\mu) \right\}$$

where $T(\mu) \stackrel{\text{def}}{=} \{\hat{x} \in S : \hat{x} \text{ solves } (P\mu)\}$, i.e., the (finite) set of optimal solutions of the relaxed problem.

The subgradient method is quite easy to implement, but its convergence properties are rather poor. This may not be a real difficulty, since all we are searching for is a good lower bound and only a few iterations are carried out for each subproblem. A simple way to improve its performance may be to initialize the multipliers for a new subproblem with the values obtained for its parent subproblem in the search tree.

The subgradient method is the simplest and most general approach to obtaining good bounds by Lagrangian relaxation, but there are alternative strategies:

- it is possible to select a subgradient in order to insure that it is an ascent direction; this may seem obviously appealing, but the advantage comes at the price of an increased computational effort;
- special purpose strategies may be devised, depending on the problem at hand.

C.4.4 Surrogate relaxation

The surrogate relaxation is obtained by forming a linear combination of constraints:

$$\left\{ \begin{array}{ll} \min & f(\mathbf{x}) \\ \text{s.t.} & \mathbf{g}(\mathbf{x}) \leq \mathbf{0} \\ & \mathbf{x} \in S \end{array} \right\} \Rightarrow \left\{ \begin{array}{ll} \min & f(\mathbf{x}) \\ \text{s.t.} & \boldsymbol{\mu}^T \mathbf{g}(\mathbf{x}) \leq 0 \\ & \mathbf{x} \in S \end{array} \right\}$$

for any vector $\boldsymbol{\mu} \geq \mathbf{0}$. Note that one scalar constraint is obtained. The case of equality constraints is handled by multipliers which are not restricted in sign.

As with the Lagrangian relaxation, we have to solve a dual maximization problem, the surrogate dual D_S, in order to obtain a good lower bound. Surrogate relaxation is usually more difficult to apply than Lagrangian relaxation.

C.4.5 Comparing alternative relaxations

From the proof of theorem A.10 it is obvious that

$$\nu(D_L) \leq \nu(D_S). \tag{C.9}$$

In other words, the best lower bound obtained by Lagrangian relaxation of a set of constraints cannot be better than the best lower bound obtained by surrogating the same constraints. Nevertheless, Lagrangian bounds are usually easier to compute: therefore (C.9) does not imply that surrogate relaxation should be preferred to Lagrangian relaxation.

We consider now, for the MILP case, the relation between continuous and Lagrangian relaxation. The following theorem can be shown:

THEOREM C.1
Let (P) and (D_L) be defined as above. Then:

$$\nu(D_L) = \nu \left(\begin{array}{ll} \min & \mathbf{c}^T \mathbf{x} \\ s.t. & \mathbf{Ax} \geq \mathbf{b} \\ & \mathbf{x} \in [S] \end{array} \right)$$

where $[S]$ is the convex hull of the points in S.

The continuous relaxation of (P) is:

$$(\overline{P}) \qquad \min \ \mathbf{c}^T \mathbf{x}$$
$$\text{s.t. } \mathbf{Ax} \geq \mathbf{b}$$
$$\mathbf{x} \in \overline{S},$$

where $\overline{S}$ is the continuous relaxation of S. However, since the convex hull is the smallest convex set containing S, we have $[S] \subseteq \overline{S}$, which readily implies:

$$\nu(\overline{P}) \leq \nu(D_L).$$

It is also interesting to note that $\nu(\overline{P}) = \nu(D_L)$ whenever S has the **integrality property**, i.e., when its continuous relaxation yields an integer solution. This happens when $\overline{S} \equiv [S]$, a typical example being the case of a unimodular matrix $\mathbf{A}$.

We can conclude that continuous relaxation yields the weakest lower bound, and that surrogate relaxation yields the strongest one. Another useful consequence of the last theorem is that it may be enable us to compare alternative Lagrangian relaxations.

Example C.4
Consider the Generalized Assignment problem:

$$\min \ \sum_{i=1}^{N} \sum_{j=1}^{M} c_{ij} x_{ij}$$

$$\text{s.t. } \sum_{i=1}^{N} p_{ij} x_{ij} \leq B_j \qquad \forall j \tag{C.10}$$

$$\sum_{j=1}^{M} x_{ij} = 1 \qquad \forall i \tag{C.11}$$

$$x_{ij} \in \{0, 1\} \qquad \forall i, j.$$

We may dualize either constraints (C.10) or (C.11).

In the first case we obtain a trivial problem, solved by assigning each job to the cheapest machine. In the second one we obtain a Knapsack problem for each machine. Clearly, the second lower bound is more difficult to compute, but it is stronger, since the first one yields problems with a totally unimodular matrix: thus, in this case, the integrality property holds and the lower bound is the same as the continuous relaxation.

The last remark could lead us to conclude that when the integrality property holds, there is little point in adopting Lagrangian relaxation: after all, we get the same lower bound as with continuous relaxation, with the added complexity of searching for an optimal set of multipliers. Nevertheless, Lagrangian relaxation may be preferable for two reasons:

1. the relaxed problem may be trivial to solve, whereas continuous relaxation requires some iteration of the Simplex algorithm;
2. it may be easier to obtain feasible solutions from the Lagrangian problem since the integrality of the decision variables is maintained.

The last point is important both for exact and heuristic methods; we will see such an example in Section 4.8.2. ▯

C.5 Branching strategies

Branching is the mechanism by which subproblems are generated within a Branch and Bound algorithm; such a mechanism is generally problem-dependent, and it may be a critical point for the efficiency of the method. A related issue is the generation of feasible solutions: branching may be not necessary if it is easy (or trivial) to derive an optimal solution for the current node of the search tree. First we describe the standard branching mechanisms adopted in commercial codes for the solution of MILP models with continuous relaxation. They include branching on binary and integer variables, and other mechanisms such as semicontinuous variables and Special Ordered Sets that may ease both model building and model solving. Then we give an example of a special purpose strategy for a problem not requiring any mathematical model. Other important issues we just outline are the selection of the branching variable and the generation of feasible solutions.

C.5.1 Branching for MILP models

Consider first a MILP model and a subproblem $P(S_k)$ within the search tree: clearly, if the optimal solution of the continuous relaxation $P(\overline{S_k})$ is all-integer (i.e., the integer variables satisfy the integrality requirements), we may fathom the corresponding node and possibly improve the current upper bound (thus fathoming other nodes). Now assume that a binary variable x_j takes a fractional value (say 0.45) in the optimal solution of the relaxed problem. It is natural to generate two subproblems (and two child nodes of the current node) by fixing $x_j = 0$ for one subproblem and $x_j = 1$ for the other. If, however, we have general integer variables taking fractional values, we cannot generate a node for each possible value $x_j = 0, 1, 2, \ldots$. Let $\bar{x}_j$ be the fractional value of x_j in the optimal solution of $P(\overline{S_k})$. Then two subproblems are generated; in the **down-child** we add the constraint

$$x_j \leq \lfloor \bar{x}_j \rfloor;$$

in the **up-child** we add

$$x_j \geq \lfloor \bar{x}_j \rfloor + 1.$$

For instance, if $\bar{x}_j = 4.2$ we generate two subproblems with the addition of constraints $x_j \leq 4$ (for the down-child) and $x_j \geq 5$ (for the up-child).

Actually, branching on binary variables may be seen as a subcase of branching on integer variables; a Branch and Bound method for MILP models just requires branching for integer variables. Commercial codes however, include modeling mechanisms that may be useful both for model building and model solving[1]. A first example are **semicontinuous variables**. A semicontinuous variable is constrained to take values of the (nonconvex) set $\{0\} \cup [c, +\infty)$, i.e., it can be zero or a value larger than a minimum threshold value. A typical trick to model such a case is to introduce a binary variable δ_j and to enforce the following constraints:

$$x_j \geq \delta_j c$$
$$x_j \leq \delta_j M,$$

where M is an upper bound on the value of x_j. Some commercial codes offer the possibility of directly treating semicontinuous variables, without the need to add the constraints to do the trick. Branching on semicontinuous variables is automatically dealt with by the software.

Another useful tool is the different types of **Special Ordered Sets**. A Special Ordered Set (SOS) is an ordered collection of variables $(x_1, \ldots, x_m)$ with some additional condition influencing the branching mechanism:

- in an SOS of type 1 (SOS1), *at most* one variable may take a nonzero value;
- in an SOS of type 2 (SOS2), *at most two adjacent* variables may take nonzero values.

---

[1] We give here a general description of the approaches; the actual implementation details may depend on the software package at hand.

A typical application of an SOS1 is for the case when mutually exclusive decisions must be taken. We may write a constraint

$$\sum_{j=1}^{m} x_j = 1$$

and state that the variables are an SOS1 (note that there is no need to further require that the variables are binary). Actually, this case is so common that another type of SOS (type 3) has been introduced to deal with it. An SOS2 is typically used to cope with piecewise linear functions without adopting the modeling approach of Section C.2.2. Instead of introducing the binary variables s_0, s_1, s_2, we may simply require that $\lambda_0, \lambda_1, \lambda_2, \lambda_3$ are an SOS2.

There are special rules for branching on an SOS. Consider for instance an SOS1. In this case two nodes are created: in the down-child the variables $\{x_1, \ldots, x_r\}$ are set to zero; in the up-child the variables $\{x_{r+1}, \ldots, x_m\}$ are set to zero. This branching strategy is more powerful in the sense that fewer nodes are created than with the plain modeling approach.

C.5.2 Branching in special purpose Branch and Bound algorithms

Branch and Bound algorithms do not necessarily require the formulation of a mathematical model: sometimes it is advisable to work on a direct representation of the solution (see Sections 4.7.2 and 4.7.3). In such a case, special branching strategies are adopted. Consider, for instance, a problem whereby the solutions are represented by permutations or sequences of objects. Branching in this case means the selection of the next item in the permutation; sequences are built step by step. In this case, we obtain the search tree depicted in Figure C.5.

All the branching mechanisms we have considered work with partial or unfeasible solutions. For specific problems, it is possible to devise branching mechanisms working on *feasible* solutions; in such cases surprisingly efficient algorithms may be obtained. We give an example of this approach in Section 4.7.2.

C.5.3 Selecting the branching variable

Another issue to be coped with is the selection of the variable (or the SOS) to be branched on. A common strategy used in commercial packages is to estimate the degradation in the objective function when a fractional variable is forced up or down. It may be sensible to branch on the variable yielding the largest degradation, since this may help to fathom the node (or its descendants) as early as possible. A further criterion for selecting the

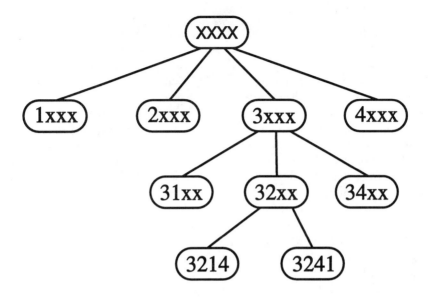

FIGURE C.5
Search tree for an optimal permutation problem.

branching variable is to aim at an integer solution. If we have an almost integer variable, we could argue that it is not interesting to branch on that variable, since it will probably be forced to an integer value when enforcing constraints on other variables.

Alternatively, priorities can be explicitly enforced. For instance, consider a triangular constraint matrix

$$\mathbf{A} = \begin{bmatrix} c_{11} & c_{12} & c_{13} & \cdots & c_{1n} \\ 0 & c_{22} & c_{23} & \cdots & c_{2n} \\ 0 & 0 & c_{33} & \cdots & c_{3n} \\ \vdots & \vdots & \vdots & \ddots & \vdots \\ 0 & 0 & 0 & \cdots & c_{nn} \end{bmatrix}.$$

Then, if we have a subset of constraints of the form

$$\mathbf{Ax} \geq \mathbf{b},$$

where the coefficients of the matrix $\mathbf{A}$ are more or less of the same order of magnitude, it may be advisable to branch on x_n first, since it has the largest impact on the rest of the variables.

C.5.4 Obtaining feasible solutions

Feasible solutions may be obtained within a Branch and Bound algorithm in different ways. In a commercial code based on continuous relaxation, a feasible solution is obtained when the optimal solution of a relaxed problem satisfies the integrality requirements. We cannot foresee at which level in the search tree this will happen, but we would like this to happen at the lowest level possible. The selection of the branching variable plays a major role from this point of view. In principle, there is no guarantee that we will actually end up with a feasible solution: it is possible to show that in some (rather pathological) cases the Branch and Bound tree grows indefinitely. This cannot happen when the feasible region is bounded; in this case, adding constraints of the form $x_j \leq c$ or $x_j \geq c+1$ leads either to an unfeasible problem (the feasible set is empty) or to an integer solution.

In the search tree for an optimal permutation problem, feasible solutions can only be obtained when the sequence has been completely determined: this corresponds to *leaves* in the tree.

When a feasible solution is obtained, we fathom the corresponding node and possibly update the current upper bound on the optimal value (the value of the incumbent solution); the larger the upper bound, the more nodes will be fathomed. The ability to generate feasible solutions at *any level* of the search tree may be useful for significantly reducing the computational effort. Note that this means being able to take an *infeasible* solution of the relaxed problem and to alter it in order to obtain a feasible one. Unfortunately, this requires special purpose approaches depending on the problem at hand. As a general rule, we may say that a Lagrangian relaxation approach makes this task easier than continuous relaxation, since integer variables are integer in the solution of the relaxed problem.

C.6 Search strategies

During the execution of a Branch and Bound procedure, we have to select the next subproblem to be examined in the list of open subproblems. The selection of the next subproblem may have a large impact on the computation times and memory requirements. We consider here two basic strategies, corresponding to two different ways of visiting nodes in the search tree.

Depth-first strategy. According to this strategy, the next subproblem to tackle is the last generated one. In other words, the list of the open subproblems is managed according to a LIFO (last in, first out) strategy. The name of the strategy derives from the way the nodes of the search tree are visited: given a subproblem $P(S_i)$, we do not consider other problems in the candidate list until all the children of $P(S_i)$ have been fathomed.

Best-first strategy. In this strategy, the next subproblem is the "most promising" one, i.e., the problem with the best lower bound. The rationale behind this approach is that by finding a good feasible solution, more open subproblems can be fathomed.

Depth-first and best-first are two "prototypical" approaches; in practice hybrid approaches may be used. For instance, it is natural to introduce a best-first element within a depth-first strategy by sorting the children of the current subproblem according to their lower subproblems.

A further consideration is that it is not fair to compare two nodes at different tree depths; since lower bounds increase when adding constraints, the lower bound of deeper nodes are more reliable. Moreover, in an MILP problem, we should also consider how many variables are already integer. In some cases aiming at integer solutions allows pruning a node earlier (by forcing degradation of the objective function) or getting good upper bounds.

In general, it is difficult to predict which strategy yields the best results, but some general considerations are worth mentioning. The depth-first strategy is much simpler to implement and requires less storage space, since only a few subproblems are kept open; on the contrary, the best-first strategy may require a large amount of memory when a "jumping" behavior occurs, i.e., when the search tree is visited jumping from one branch to the other one without reaching a leaf or a feasible solution. This jumping behavior is not expected to occur when lower bounds are sharp.

An advisable strategy could be to use depth-first until a feasible solution is obtained, then to switch to best-first. The depth-first phase may be skipped if an upper bound is obtained by running a good heuristic algorithm before starting the Branch and Bound search.

C.7 Dominance conditions

As we pointed out earlier, it is not correct to fathom a subproblem by comparing its lower bound with the lower bound of another subproblem. Still, it may be possible to develop **dominance conditions** allowing further pruning of the search tree.

Example C.5
Consider a knapsack problem

$$\max \sum_i v_i x_i$$
$$\text{s.t.} \sum_i w_i x_i \leq C$$
$$x_i \in \{0, 1\},$$

and its search tree. Let $\mathcal{I} \subset \{1, 2, \ldots, N\}$ represent a subset of the items, and consider two partial assignments $\hat{x}$ and $\tilde{x}$, i.e., two different ways of fixing x_i for $i \in \mathcal{I}$. Assume that

$$\sum_{i \in \mathcal{I}} v_i \tilde{x}_i \leq \sum_{i \in \mathcal{I}} v_i \hat{x}_i$$
$$\sum_{i \in \mathcal{I}} w_i \tilde{x}_i \geq \sum_{i \in \mathcal{I}} w_i \hat{x}_i.$$

Clearly the node corresponding to the partial assignment $\tilde{x}$ can be discarded. □

The dominance conditions of the example are strictly related to Dynamic Programming (see Example D.2). Dominance conditions are problem-specific, and are not available in commercially available codes. Furthermore they require additional amounts of storage and computation, and it is difficult to evaluate a priori if they are useful or not.

C.8 Heuristic methods for discrete optimization

Despite considerable progress both in hardware technology and in Branch and Bound methods, many practical problems defy any attempts to obtain optimal solutions with reasonable computational efforts. In such cases, one has to resort to heuristic methods, with the aim of getting a near-optimal solution.

Heuristic strategies can be classified according to different attributes. In **one-shot** methods one solution is built step by step; in **iterative** methods, many solutions are obtained during the execution of the algorithms. Usually, one-shot methods are faster but less effective. Heuristic methods may be derived from mathematical statements of the problems, or can be purely based on commonsense. Another fundamental distinction is between **general purpose** and **special purpose** methods. Special purpose methods solve specific problems by exploiting their peculiarities. An obvious drawback is that a slight change in the problem, either in the objective function or the constraints, may make the method useless. Here we describe only rather *general* principles that can be applied to the design of heuristic algorithms.

C.8.1 Heuristics from mathematical models

Most large-scale MILP models cannot be solved at optimality. Nevertheless it is often possible to derive a good heuristic from a mathematical statement of the problem.

In **LP-based heuristics** the continuous relaxation of the model is solved, and its optimal solution is rounded to an integer feasible solution by more or less sophisticated methods. Note that, in general, rounding a fractional LP solution is no easy task: it may even be difficult to obtain a *feasible* solution, let alone an optimal one; still, the structure of the problem at hand can be exploited to devise a sensible rounding strategy.

In **dual heuristics**, the optimization problem is dealt with as an easy problem with complicating constraints; by dualizing the nasty constraints and maximizing the dual function we get a lower bound and a generally primal infeasible solution. However, it is often possible to derive a good primal feasible solution from the solution of the relaxed problem.

Note that with LP-based and dual heuristics we obtain both a lower and an upper bound on the optimal value; hence, we have an *a posteriori* "certificate" on the quality of the heuristic solution.

Finally, it is also possible to decompose a complex MILP model into submodels, which can be solved sequentially; the solution of the higher level subproblem is used to derive constraints on the solution of the lower level ones. Such an approach is called **hierarchical decomposition**.

C.8.2 Greedy algorithms

Greedy algorithms are one-shot methods, whereby the discrete optimization problem is seen as a sequential decision problem: at each step the locally best decision is taken, disregarding the future consequences of the choice. In general, this results in a suboptimal solution.

Example C.6
Consider the graph depicted in Figure C.6. We want to obtain the shortest path from node 1 to node 6. Actually, this is an easy optimization problem, but let us apply a greedy strategy: starting from node 1, we build a shortest path step-by-step by selecting the closest node. This approach yields the path

$$1 \to 3 \to 4 \to 6$$

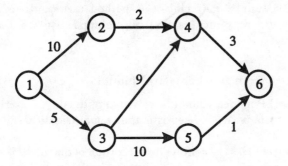

FIGURE C.6
A greedy approach does not yield the shortest path.

with a length 17. However, the optimal solution is clearly

$$1 \to 2 \to 4 \to 6$$

with a length 15. □

A greedy approach is similar to a Branch and Bound method in which we generate subproblems from the current node and we select the one with the lowest bound, forgetting the others. To reduce the myopia of this strategy, one can consider, in the set of decisions that can be taken at each stage, not only the most promising one, which is a subset of cardinality one, but a subset of greater cardinality. One could, for example, explore the consequences of the three locally best decisions, forgetting the remaining ones. The larger the cardinality of the subset considered at each stage, the larger the probability of ending up with a globally optimal set of decisions, with a corresponding increase in computational cost. This idea leads to **beam search** algorithms. In Figure C.7 we illustrate the evolution of a beam search tree if the beam width is 2. A greedy approach is a beam search with beam width 1.

To evaluate the suitability of a node of the search tree, we may use the cumulated cost of the decisions already taken, a lower bound on the optimal solution of each subproblem, or the value of a completion of the partial solution obtained by a one-shot heuristic method.

C.8.3 Truncated exponential schemes

In an exact Branch and Bound scheme for a minimization problem $P(S)$, one is allowed to eliminate a subproblem $P(S_k)$ from further consideration

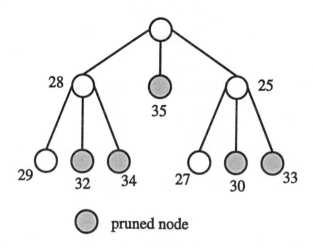

FIGURE C.7
Search tree for a beam search method

only if a lower bound $\beta[P(S_k)]$ for its optimal value is found which is not less than the value of a known upper bound $\hat{\nu}[P(S)]$ for the optimal solution of $P(S)$. In a truncated Branch and Bound one relaxes the condition

$$\beta[P(S_k)] \geq \hat{\nu}[P(S)]$$

by requiring only

$$\beta[P(S_k)] \geq (1-\epsilon)\hat{\nu}[P(S)]$$

where $0 < \epsilon < 1$. By choosing the value of ϵ we trade off solution quality with execution speed.

Unlike other heuristic strategies, truncated exponential schemes provide an *a priori* guarantee on the quality of the solution found, i.e., one is sure to find a solution within a given ϵ% from the optimal one. However, truncated exponential schemes require good lower bounding procedures like their exact counterparts, which is not always possible with reasonable computational efforts. We mention here the possibility of truncating other exact optimization methods, such as Dynamic Programming.

C.8.4 Local Search

Local search algorithms are similar to the gradient method for nonlinear programming. The basic idea is to improve a known solution by applying a set of local perturbations. Consider a generic optimization problem:

$$\min_{x \in S} f(x),$$

defined over a discrete set S. Given a feasible solution x, a neighborhood $\mathcal{N}(x)$ is defined as the set of solutions obtained by applying a set of simple perturbations to x. For instance, if our problem consists in finding an optimal permutation of items, the neighborhood of a given solution could consist of the set of permutations obtained by swapping pairs of items. Different perturbations yield different **neighborhood structures**.

The simplest local search algorithm is known as **local improvement**. Given a current solution $\bar{x}$, an alternative (candidate) solution x° is searched for, such that:

$$f(x^\circ) = \min_{x \in \mathcal{N}(\bar{x})} f(x).$$

If the neighborhood structure $\mathcal{N}(\cdot)$ is simple enough, the above minimization can be performed by an exhaustive search. Clearly, there is a tradeoff between the effectiveness of the neighborhood structure (the larger the better) and the efficiency of the algorithm. If $f(x^\circ) < f(\bar{x})$, then x° is set as the new current solution, and the process is iterated. If $f(x^\circ) \geq f(\bar{x})$, then the algorithm is stopped. A possible variation is to *partially* explore the neighborhood of the current solution until an improving solution is found; this approach is known as **first-improving**, since we do not explore the whole neighborhood before committing to a new current solution.

This basic idea is generally easy to apply, but it has one major drawback: usually the algorithm stops in a *locally* (with respect to the neighborhood structure) optimal solution; this is illustrated in Figure C.8. This is the difficulty we face when applying the gradient method to a nonconvex objective function; the reason behind all of this is that only improving perturbations (i.e., such that $\Delta f = f(x^\circ) - f(\bar{x}) < 0$) are accepted. To avoid getting stuck in a local optimum, we must relax this assumption. In the following we describe two local search approaches that have been proposed to overcome the limitations of local improvement: simulated annealing and tabu search. Due to the generality of these approaches, they are often referred to as **meta-heuristics**.

Applications of simulated annealing and tabu search to scheduling problems are described in Sections 4.8.3 and 4.8.4 respectively.

Simulated Annealing

It has been pointed out that in order to overcome the problem of local minima, we have to accept, in some disciplined way, nonimproving pertur-

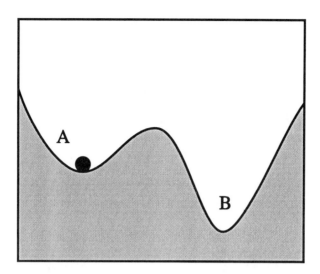

FIGURE C.8
A local improvement method can get stuck in a locally optimal solution

bations, i.e., perturbations for which $\Delta f > 0$. Simulated annealing is based on an analogy between cost minimization in discrete optimization and energy minimization in physical systems. The local improvement strategy behaves much like physical systems do according to classical mechanics; it is impossible for a system to have a certain energy, at a certain time, and to increase it without external input. In other words, if we place a ball in the hole A shown in Figure C.8, it is impossible for the ball to shift to hole B, since this would imply tunneling through a potential barrier. This is not true in thermodynamics and statistical mechanics: according to these physical models, at a temperature above absolute zero, thermal noise makes an increase in the energy of a system possible. An increase in energy is more likely to occur at high temperatures. The probability P of this 'upward jump' depends on the amount of energy ΔE acquired and the temperature T, according to the Boltzmann distribution

$$P(\Delta E, T) = \exp(-\frac{\Delta E}{KT})$$

where K is the Boltzmann constant.

Annealing is a metallurgical process by which a melted material is slowly cooled in order to obtain good (low energy) solid state configurations. If the temperature is decreased too fast, the system gets trapped in a local energy minimum, and a glass is produced. But if the process is slow enough,

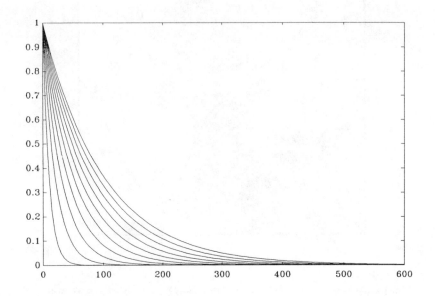

FIGURE C.9
Acceptance probabilities as a function of cost increase for different temperatures

random kinetic fluctuations due to thermal noise allow the system to escape from local minima, reaching a point very close to the global optimum.

In strict analogy with statistical mechanics, in the simulated annealing method a perturbation of the current solution yielding $\Delta f < 0$ is always accepted; a perturbation with $\Delta f > 0$ is accepted with a probability given by a Boltzmann-like probability distribution

$$P(\Delta f, T) = \exp(-\frac{\Delta f}{T})$$

This probability distribution is a decreasing exponential in Δf, whose shape depends on the parameter T, acting as a 'temperature' (see Figure C.9). The probability of accepting a nonimproving perturbation decreases as the deterioration of the solution increases. For a given Δf, the acceptance probability is higher at high temperatures. For $T \to 0$ the probability collapses into a step function, and the method behaves like local improvement. For $T \to +\infty$ the probability is one everywhere, and we have a random exploration of the solutions space. The parameter T allows balancing the need to *exploit* the solution at hand by improving it and the need to *explore* the solution space.

The simulated annealing method simply substitutes the deterministic acceptance rule of local improvement with a probabilistic one. The temperature is set to a relatively high initial value T_1, and the algorithm is iterated using at step k a temperature T_k, until some termination criterion is satisfied. The strategy by which the temperature is decreased is called the **cooling schedule**. The simplest cooling schedule is

$$T_k = \alpha T_{k-1}, \qquad 0 < \alpha < 1$$

In practice, it is advisable to keep the temperature constant for a certain number of steps, in order to reach a "thermodynamic equilibrium" before changing the control parameter. More sophisticated adaptive cooling strategies have been proposed, but the increase in complexity does not always seem justified. A very simple implementation of the annealing algorithm could be the following:

Step 1: choose an initial solution x_{old}, an initial temperature T_1 and a decrease parameter α; let $k = 1$, $f_{\text{old}} = f(x_{\text{old}})$; let $\hat{f} = f_{\text{old}}$ and $\hat{x} = x_{\text{old}}$ be the current optimal value and solution.

Step 2: randomly choose a candidate solution x_{new} from the neighborhood of x_{old}, and compute its value f_{new}.

Step 3: set the acceptance probability

$$P = \min\{1, \exp(-\frac{f_{\text{new}} - f_{\text{old}}}{T_k})\}.$$

Step 4: accept the new solution with probability P; if accepted, set $x_{\text{old}} = x_{\text{new}}$ and $f_{\text{old}} = f_{\text{new}}$; if necessary update $\hat{f}$ and $\hat{x}$.

Step 5: if some termination condition is met, stop; otherwise set $k = k + 1$, set the new temperature according to the cooling schedule, and go to step 2.

The probabilistic acceptance is easy to implement. P is evaluated according to the Boltzmann distribution; then a pseudo-random number r, uniformly distributed between 0 and 1, is computed and the move is accepted if $r \leq P$.

The termination condition could be related to a maximum iteration number, to a minimum temperature, or to a maximum number of steps in which the current solution remains unchanged. Note that here we do not explore the whole neighborhood of the current solution; the method is of the first-improving type. If a candidate solution is rejected, we select another candidate in the neighborhood of the current solution. In principle, it is possible to visit the same solution twice; if the neighborhood structure is rich enough, this is unlikely. It is necessary to save the best solution found, since the freezing point (the last current solution), need not be the best solution visited.

An implementation of the annealing algorithm is therefore characterized by the solution space, the neighborhood structure, the rule by which the neighborhood is explored, and the cooling schedule. It can be shown, under some conditions, that the method asymptotically converges (in a probabilistic sense) to the global optimum. The convergence property is a reassuring one, but it is usually considered of little practical value, since its conditions would require impractical running times. However, the experience suggests that, in many practical settings, very good solutions (often optimal) are actually found.

The running time of the algorithm to obtain high quality solutions is problem dependent. For simple scheduling problems, the running times are very short (a few minutes); they are obviously expected to increase in more complex problems. For difficult integrated circuit design problems (for which the algorithm was originally developed), running times in the order of some hours are reported.

Tabu Search

Tabu search, like simulated annealing, is a neighborhood search based meta-heuristic aimed at escaping local minima. Unlike simulated annealing, tabu search tries to keep the search biased towards good solutions.

The basic idea of tabu search is that the best solution in the neighborhood $\mathcal{N}$ of the current solution should be chosen as the new current solution, even if this implies increasing the cost. If we are in a local minimum, this means accepting a nonimproving perturbation. The problem with this basic idea is that the possibility of cycling arises. If we try to escape from a local minimum by choosing the best solution in its neighborhood, it might well be the case that, at the next iteration, we fall back into the local minimum, since this could be the best solution in the new neighborhood.

In order to prevent cycling, we must prevent revisiting solutions. One way would be to keep a record of the already visited solutions; however, this would be both memory- and time-consuming, since checking a candidate solution against the list of visited ones would require a substantial effort. A better idea could be to record only the most recent solutions. A practical alternative is to keep in memory only some *attributes* of the solutions or of the applied perturbations; such attributes are called tabu. For instance, the reverse of the selected perturbation at each step can be marked as tabu, restricting the neighborhood to be considered. Consider the problem of determining an optimal permutation of items; if we swap items c_i and c_j, in the next steps we might forbid any move changing their new positions. A tabu attribute of a solution could be the value of the objective function. In practice, it is necessary to keep only a record of the most recent tabu attributes to avoid cycling; the data structure implementing this function is the tabu list.

The basic tabu navigation algorithm can be described as follows:

Step 1: choose an initial current solution x_cur, a tabu list size; let $k = 1$, $\hat{f} = f(x_\text{cur})$, $\hat{x} = x_\text{cur}$.

Step 2: evaluate the neighborhood $\mathcal{N}(x_\text{cur})$; update the current solution with the best non tabu solution in the neighborhood; if necessary update $\hat{x}$ and $\hat{f}$.

Step 3: add some attribute of the new solution or of the applied perturbation to the tabu list.

Step 4: if the maximum iteration number has been reached, stop; otherwise set $k = k + 1$, and go to step 2.

Note that, unlike simulated annealing, tabu search explores the *whole* neighborhood of the current solution; tabu search is a strategy of the **best-improving** rather than first-improving type.

There are several issues and refinements to consider in order to implement an effective and efficient algorithm. They are rather problem specific; this shows that, although local search meta-heuristics are general purpose, a certain degree of "customization" is necessary.

- **Tabu relaxation by aspiration criteria.** The tabu attributes are a cheap way to avoid revisiting previously considered sequences, without keeping track of these solutions explicitly. It may well be the case that an interesting solution, never considered before, is considered tabu because of a tabu recorded in the tabu list. To avoid restricting the search process too much, the tabu may be relaxed according to relaxations criteria: for example, a tabu should be relaxed if the corresponding perturbation yields a new optimal solution.

- **How long should the tabu list be?** A short tabu list may not prevent cycling; a long tabu list may restrict the search process too much. Usually one can adopt a long tabu list if the neighborhood structure is sufficiently rich. It is also possible to dynamically adapt the length of the tabu list.

- **How should tabu attributes be selected?** It is not always obvious what attribute must be chosen; tabu attributes may be more or less restrictive. If we have an optimal permutation problem and we swap c_i and c_j, we may forbid any swap move involving either c_i or c_j. This rules out a relatively large number of moves. Alternatively, we may just forbid swapping c_i and c_j; this tabu is clearly less restrictive. If the attribute is not restrictive enough, cycling may still occur; if it is too restrictive, this may adversely affect the search process. A good balance must be obtained between the tabu attributes and the neighborhood structure.

- **How to restrict the neighborhood?** A very large and rich neighborhood structure has some disadvantages; the most obvious is that it takes too much time to evaluate. It is possible to restrict the neighborhood by randomly sampling it, or by selecting a subset according to a certain strategy. If the neighborhood consists of swap moves in a permutation, we may consider swapping only items in positions i and j such that $|i - j| \leq k$. A subtler issue is feasibility. In general, there is no guarantee that by applying a perturbation to a feasible current solution we obtain feasible candidates. Since spotting infeasibility may be time-consuming, it may be advisable to restrict the neighborhood in such a way as to guarantee feasibility a priori.

- **Long-term memory and diversification.** Tabu search was introduced by an analogy with human problem solving, which is characterized by a hierarchical structure of memory levels, e.g. short and long-term memory levels. The tabu list can be considered as the short-term memory component of the algorithm. A long-term memory component can be used to diversify search; we may keep memory of recurring features of the visited solutions in order to restart the search process from a new solution when no progress is being done; the new initial solution is selected in such a way as to maximize its "difference" with respect to the already visited solutions.

- **Strategic oscillation.** When constraints must be enforced on the solutions, it is easier to get trapped in local optima. A possible way to overcome the difficulty is to relax the constraints by a penalty function, characterized by a penalty coefficient. When no process is being done, it is possible to lower the penalty coefficient, allowing the search process to get out of the feasible set. Then the penalty coefficient is gradually increased, so that the search process enters again into the feasible set. The hope is that, oscillating in and out of the feasible set, a diversification behavior is introduced.

For further reading

The most comprehensive text on discrete optimization methods is [150]; other useful references are [154, 155]. As to *modeling* with integer programming we recommend [210].

More information on the heuristic aspects of Branch and Bound methods based on continuous relaxations can be found in [155, pp. 194-204]. Another recommended reference for the computational implementation of Branch and Bound is [211, Chapter 7]. Special Ordered Sets were introduced by Beale and Tomlin [18].

Classical references for Lagrangian relaxation are [73, 87]. See [155, pp. 210-211] for a proof of theorem C.1 concerning the relationship between continuous and Lagrangian relaxation bounds. The problem of selecting a subgradient in order to insure that it is an ascent direction is treated in [97]; an example of dual ascent strategies can be found in [75], where the idea is applied to the Generalized Assignment Problem. Multiplier adjustment methods for surrogate relaxation be found in [155].

Branch and Bound for continuous global optimization are described in [106].

There are many references for special purpose methods for particular problems, such as Set Partitioning and Set Covering; surveys can be found in [107, Chapter 3] and [150, Chapter II.6]. See [138] for the Knapsack problem, [47] for the Generalized Assignment Problem and [129] for the Traveling Salesman Problem. An overview of nonlinear 0/1 programming methods can be found in [100].

Local search meta-heuristics are a growing research area. Simulated annealing was originally proposed in [118]; a comprehensive text on this method is [197], where convergence results and cooling schedules are treated in detail; a convergence proof can also be found in [135]; see [66] for a survey on applications. As to tabu search, it was originally proposed in [90]; a useful reference on implementation details is [92], and a collection of applications can be found in [91]. We have completely neglected another class of local search meta-heuristics, Genetic Algorithms, we refer the interested reader to [93].

D
Dynamic Programming

Dynamic Programming (DP), unlike the Simplex method, is *not* a standard, ready-to-use, algorithm; it is rather a *general principle* which can be applied to virtually all optimization problems, including discrete, stochastic, and infinite-dimensional problems. Such a remarkable generality comes with two drawbacks: first, the difficulty of tailoring the generic principle to specific problems; second, the difficulty of solving the equations obtained by applying the DP principle. In some cases there is no way to solve them analytically; in other cases, the solution is conceptually trivial but computationally too expensive for a practical size problem.

This is why DP is not available in standard software packages. Nevertheless, DP yields efficient algorithms in some cases, and in others it yields valuable insights into the structure of the optimal solution to a seemingly intractable problem, paving the way for the development of effective heuristic approaches.

We first introduce DP through its application to the shortest path problem in a network (Section D.1). Then we generalize the idea to discrete-time sequential decision problems (Section D.2), obtaining Bellman's Optimality Principle (Section D.2.1). Finally we show how DP can be applied to continuous-time optimal control problem (Section D.3).

D.1 The shortest path problem

The best way of introducing DP is by considering one of its most natural applications, i.e., the problem of finding the shortest path in a network (see Section B.6.4). Let $\mathcal{N} = \{0, 1, 2, \ldots, N\}$ and $\mathcal{A}$ be the node and arc sets; let the start and final nodes be 0 and N respectively. For simplicity, we assume that the network is acyclic and that the arc lengths c_{ij} are non-negative. This section has a didactic purpose: more efficient algorithms

The shortest path problem

are available for the shortest path problem.

The starting point is to find a *characterization* of the optimal solution, that can be translated into a constructive algorithm. Let $F(j)$ be the length of the shortest path from node 0 to a node $j \in \mathcal{N}$ (denoted by $0 \xrightarrow{*} j$). Assume that, for a specific $j \in \mathcal{N}$, a node i lies on $0 \xrightarrow{*} j$. Then the following property holds: $0 \xrightarrow{*} i$ is a subpath of $0 \xrightarrow{*} j$. In other words, the optimal solution for a problem is obtained by assembling optimal solutions for subproblems. To understand why, consider the decomposition of $0 \xrightarrow{*} j$ into the subpaths $0 \to i$ and $i \to j$. The length of $0 \xrightarrow{*} j$ is the sum of the lengths of the two subpaths:

$$F(j) = L(0 \to i) + L(i \to j). \tag{D.1}$$

Note that the second subpath is not affected by *how* we go from 0 to i; this is strongly related to the concept of state in dynamic systems. Now, assume that $0 \to i$ is not the optimal path from 0 to i. Then we could improve the first term of (D.1) by considering the path consisting of $0 \xrightarrow{*} i$ followed by $i \to j$. The length of this new path would be

$$L(0 \xrightarrow{*} i) + L(i \to j) < L(0 \to i) + L(i \to j) = F(j),$$

which is a contradiction.

This observation leads to the following recursive equation for the shortest path from 0 to a generic node j:

$$F(j) = \min_{(i,j) \in \mathcal{A}} \{F(i) + c_{ij}\} \qquad \forall j \in \mathcal{N}. \tag{D.2}$$

This kind of recursive equation, whose exact form depends on the problem at hand, is the heart of DP and is called the **functional equation**. The overall problem requires computing $F(N)$; the functional equation allows solving an overall problem by solving a set of smaller and simpler subproblems. In the shortest path problem, the shortest path from 0 to N is computed by finding the shortest paths from 0 to the predecessors of N; these in turn are computed considering the predecessors of the predecessors of N. By unfolding the recursion, we end up solving the trivial problem of finding the shortest paths from 0 to its immediate successors, as shown in the following example.

Example D.1

Consider the network depicted in Figure D.1. To find the shortest path, we proceed by labeling each node $j = 1, \ldots, 7$ with $F(j)$ and its predecessor in the shortest path $0 \xrightarrow{*} j$. Note that we must label a node only after all of its predecessors have been labeled. We have numbered the nodes of the network in such a way that this is accomplished if we label the nodes

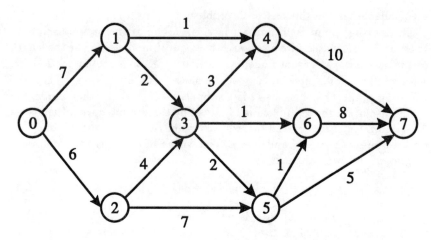

FIGURE D.1
A shortest path problem

following the order of their attached numbers. Such numbering can always be found in *acyclic* networks.

The initial condition is $F(0)$. To label node 1 we solve

$$F(1) = \min_{(i,1)\in\mathcal{A}} \{F(i) + c_{i1}\} \qquad \forall j \in \mathcal{N}.$$

In this case this is trivial, since node 0 is the only predecessor of node 1:

$$F(1) = \min\{7 + 0\} = 7.$$

Similarly, for node 2 we have:

$$F(2) = \min\{6 + 0\} = 6.$$

For node 3, we must consider the two predecessor nodes 1 and 2:

$$F(3) = \min\left\{\begin{array}{l} F(1) + c_{13} \\ F(2) + c_{23} \end{array}\right\} = \min\left\{\begin{array}{l} 7+2 \\ 6+4 \end{array}\right\} = 9.$$

Going on this way yields:

$$F(4) = \min\left\{\begin{array}{l} F(1) + c_{14} \\ F(3) + c_{34} \end{array}\right\} = \min\left\{\begin{array}{l} 7+1 \\ 9+3 \end{array}\right\} = 8$$

$$F(5) = \min\left\{\begin{array}{l} F(2) + c_{25} \\ F(3) + c_{35} \end{array}\right\} = \min\left\{\begin{array}{l} 6+7 \\ 9+2 \end{array}\right\} = 11$$

$$F(6) = \min\left\{\begin{array}{l} F(3) + c_{36} \\ F(5) + c_{56} \end{array}\right\} = \min\left\{\begin{array}{l} 9+1 \\ 11+1 \end{array}\right\} = 10$$

$$F(7) = \min \left\{ \begin{array}{l} F(4) + c_{47} \\ F(5) + c_{57} \\ F(6) + c_{67} \end{array} \right\} = \min \left\{ \begin{array}{l} 8 + 10 \\ 11 + 5 \\ 10 + 8 \end{array} \right\} = 16.$$

Therefore, the shortest path is

$$0 \to 1 \to 3 \to 5 \to 7,$$

with a length 16. □

In the previous example we have used a **forward functional equation**, since we proceed from the start node to the final one. We can repeat the above argument the other way round, i.e., considering paths from a node i to node N, and obtaining a **backward functional equation**,

$$B(i) = \min_{(i,j) \in \mathcal{A}} \{c_{ij} + B(j)\} \qquad \forall i \in \mathcal{N}, \qquad (D.3)$$

where $B(i)$ is the length of the shortest path from a node $i \in \mathcal{N}$ to N. The cost $B(i)$ is called, for obvious reasons, the **cost-to-go**; another term, which is more meaningful for the forward case, is **value function**. Depending on the problem at hand, it may be better to adopt a backward or a forward approach.

A point worth mentioning is that the above approach is also applicable to problem of finding the longest path in a network. As previously mentioned, things get more complicated in the case of cyclic networks or networks with negative length arcs.

D.2 Sequential decision processes

In this section we generalize the functional equation approach developed in the previous section for the shortest path problem. Consider a discrete-time dynamic system modeled by the state equation

$$\mathbf{x}_t = \mathbf{h}_t(\mathbf{x}_{t-1}, \mathbf{u}_t) \qquad t = 1, 2, \ldots \qquad (D.4)$$

where $\mathbf{x}_t$ is the vector of the state variables *at the end* of time interval t and $\mathbf{u}_t$ is the vector of the control variables applied *during* time interval t. Note that we assume here that the dynamics are time-varying.

A sequential decision process consists of selecting a set of controls $\mathbf{u}_t$ over a finite or infinite time horizon in order to minimize a certain objective function, possibly subject to constraints on the state and the control variables. We consider here only finite horizon problems. The objective function depends, in general, on the state trajectory $(x_1, \ldots, x_T)$ and the **control policy** $(u_1, \ldots, u_T)$. Our optimization problem is

$$\min_{u_1,\ldots,u_T} f(x_0, x_1, \ldots, x_T; u_1, \ldots, u_T) \qquad (D.5)$$

subject to dynamic constraints (D.4). Actually, we can write the objective function in a more compact form, since, given (D.4):

$$x_1 = h_1(x_0, u_1)$$
$$x_2 = h_2(x_1, u_2) = h_2\{h_1(x_0, u_1), u_2\}$$
$$\ldots$$

The optimal value depends only on x_0 and $(u_1, \ldots, u_T)$. Since the optimal value of the objective function depends essentially on the initial state x_0 and the horizon length, i.e., the number T of steps of the decision process, we can denote its value as the cost-to-go $F_T(x_0)$. In general we denote by $F_N(x)$ the optimal value of the N-step optimization problem with initial state x.

The optimal control (or optimal policy)

$$(u_1^*, \ldots, u_T^*),$$

and the optimal trajectory

$$(x_0, x_1^*, \ldots, x_T^*)$$

must comply with different types of constraints. Eq. (D.4) is only one of them; we could have constraints on the initial and/or the final state and on the admissible controls. A typical discrete-time optimal control problem is:

$$\min \sum_{t=1}^{T} f_t(x_{t-1}, u_t) \qquad (D.6)$$
$$\text{s.t.} \quad x_t = h_t(x_{t-1}, u_t) \qquad t = 1, 2, \ldots, T$$
$$g_t(x_{t-1}, u_t) \leq 0 \qquad t = 1, 2, \ldots, T,$$

where usually x_0 is given. We could also add conditions or cost penalties for the final state x_T. The functions g_t express constraints on the state and control variables. Note that we have assumed an *additive* objective function; this is fundamental in order to derive the functional equation, which is a form of decomposition. Note, however, that decomposition is possible in more general settings.

D.2.1 The optimality principle and solving the functional equation

The objective function (D.6) is *separable*, in the sense that, for a given r, the contribution of the last r steps depends only on the state $\mathbf{x}_{T-r}$ and the r controls $\mathbf{u}_{T-r+1}, \ldots, \mathbf{u}_T$. Furthermore, a similar separation property (known as *Markovian state property*) holds for the trajectory, in the sense that the state $\mathbf{x}_{t+1}$ reached from $\mathbf{x}_t$ by applying the control $\mathbf{u}_{t+1}$ depends only on $\mathbf{x}_t$ and $\mathbf{u}_{t+1}$, and not on $\mathbf{x}_0, \ldots, \mathbf{x}_{t-1}$. A consequence of such separation properties is the **optimality principle**.

> *An optimal policy* $(\mathbf{u}_1^*, \mathbf{u}_2^*, \ldots, \mathbf{u}_T^*)$ *is such that, whatever the initial state* $\mathbf{x}_0$ *and the first control* $\mathbf{u}_1^*$, *the next controls* $(\mathbf{u}_2^*, \ldots, \mathbf{u}_T^*)$ *are an optimal policy for the* $(T-1)$-*stage problem with initial state* $\mathbf{x}_1$, *obtained by applying the first control* $\mathbf{u}_1^*$.

Therefore, we may write a recursive functional equation to obtain the optimal policy:

$$F_T(\mathbf{x}_0) = \min_{\mathbf{u}_1} \left\{ f_1(\mathbf{x}_0, \mathbf{u}_1) + F_{T-1}(\mathbf{h}_1(\mathbf{x}_0, \mathbf{u}_1)) \right\}, \qquad (D.7)$$

where the minimization is carried out taking into account constraints on the control variable. This is a *backward* functional equation. The functional equation has a boundary condition that helps to start unfolding the recursion. If the final state $\mathbf{x}_T$ is given, $F_1(\mathbf{x}_{T-1})$ must be simply computed for each state $\mathbf{x}_{T-1}$ from which we can reach $\mathbf{x}_T$ in one step.

The T-stage problem can be dealt with by solving the $(T-1)$-stage problem, finding the optimal policies $(\mathbf{u}_2^*(\mathbf{x}_1), \ldots, \mathbf{u}_T^*(\mathbf{x}_1))$ for each possible initial state $\mathbf{x}_1$ reachable from x_0. The $(T-1)$-stage problem can be dealt with by solving a set of $(T-2)$-stage problems, and so on, until trivial problems are obtained. When the $(T-1)$-stage problems are solved, we must find the first optimal control $\mathbf{u}_1^*$ by considering, for each pair $(\mathbf{u}_1, \mathbf{x}_1)$, only the optimal policy starting from $\mathbf{x}_1$; the alternative trajectories starting from $\mathbf{x}_1$ are discarded when solving the T-stage problem. This is why computational savings are obtained, with respect to the complete enumeration of all the possible trajectories. Nevertheless, high requirements from the point of view of storage and computation may hinder the practical application of Dynamic Programming in many cases.

It is worth noting that a sequential decision process with a finite state space and a finite time horizon is equivalent to a shortest path problem in a *layered* network (see Figure D.2): the arc lengths are equal to the cost of the corresponding state transition. Due to the structure of the network, we have no difficulty in guaranteeing that all the successors (or predecessors) of a node have been labeled; we have only to label nodes layer after layer. Obviously, we can go forward or backward in the network; the best choice depends on the problem at hand.

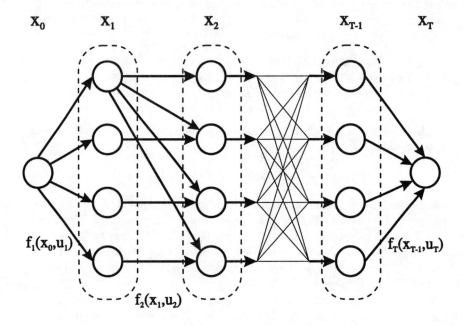

FIGURE D.2
A shortest path representation of a finite sequential decision process (for clarity, not all transitions are shown)

So far, we have considered sequential decision processes whose explicit dynamic structure makes the use of DP relatively easy. However, DP can be also applied to problems lacking any apparent dynamic structure. This requires some "art", as we will see in Section 4.4, where we apply DP to scheduling problems. For now, we outline a DP approach to a simple resource allocation problem, i.e., the Knapsack problem described in Section C.1. This is a good example of transformation of a static problem to a sequential decision problem.

Example D.2
Consider the Knapsack problem

$$\max \sum_{i=1}^{n} v_i x_i$$

$$\text{s.t.} \sum_{i=1}^{n} w_i x_i \leq C$$

$$x_i \in \{0, 1\} \quad \forall i.$$

Assuming that all the problem data are integers, we may write the functional equation as follows. For each integer m ($1 \leq m \leq n$) and for each integer z ($0 \leq z \leq C$) we define the value function

$$f_m(z) = \max \left\{ \sum_{i=1}^{m} v_i x_i \mid \sum_{i=1}^{m} w_i x_i \leq z, \ x_i \in \{0,1\} \ i = 1, \ldots, m \right\}.$$

The meaning of $f_m(z)$ is the value obtained by optimally allocating a capacity z to items 1 through m. The overall problem requires computing $f_n(C)$. We obtain a forward dynamic programming formulation, for $m = 2, \ldots, n$,

$$f_m(z) = \begin{cases} f_{m-1}(z) & \text{for } 0 \leq z \leq w_m \\ \max\{f_{m-1}(z), f_{m-1}(z - w_m) + v_m\} & \text{for } w_m \leq z \leq C \end{cases},$$

where the state z is the remaining capacity.
The initial condition is

$$f_1(z) = \begin{cases} 0 & \text{for } 0 \leq z \leq w_1 \\ c_1 & \text{for } w_1 \leq z \leq C. \end{cases}$$

We have to compute $f_m(z)$ for $m = 1, \ldots, n$, for each possible value $z = 1, \ldots, C$; therefore, the computational complexity of the method is $O(nC)$ (computational complexity is dealt with in Section 2.3 and Supplement S2.1). □

D.3 Dynamic Programming for infinite-dimensional problems

In this section we apply the DP principles to an infinite-dimensional problem, i.e., a continuous-time optimal control problem of the form[1]:

$$\min \int_0^T f(\mathbf{x}(t), \mathbf{u}(t))dt$$
$$\text{s.t.} \quad \dot{\mathbf{x}}(t) = \mathbf{h}(\mathbf{x}(t), \mathbf{u}(t), t)$$
$$\mathbf{x}(0) = \mathbf{x}_0.$$

This is the continuous-time equivalent of the discrete-time sequential decision problem tackled in the previous section. As customary in elementary calculus and physics, we bridge the gap between discrete and continuous time by considering a small time increment δt, and then letting $\delta t \to 0$. As usual in this book, the treatment is only valid from an intuitive point of view.

To derive a functional equation, we first define the cost-to-go

$$J(\mathbf{x}, t) = \min \int_t^T f(\mathbf{x}(\tau), \mathbf{u}(\tau))d\tau,$$

i.e., the value of the optimal control over the interval $[t, T]$ starting from state $\mathbf{x}$. The overall problem is

$$J(\mathbf{x}_0, 0) = \min \int_0^T f(\mathbf{x}(t), \mathbf{u}(t))dt,$$

which can be decomposed in a way similar to the shortest path case (see Eq. D.1):

$$J(\mathbf{x}_0, 0) = \min_{\substack{\mathbf{u}(t) \\ t \in [0,T]}} \left\{ \int_0^{t_1} f(\mathbf{x}(t), \mathbf{u}(t))dt + \int_{t_1}^T f(\mathbf{x}(t), \mathbf{u}(t))dt \right\}$$

$$= \min_{\substack{\mathbf{u}(t) \\ t \in [0,t_1]}} \left\{ \int_0^{t_1} f(\mathbf{x}(t), \mathbf{u}(t))dt + \min_{\substack{\mathbf{u}(t) \\ t \in [t_1,T]}} \left[\int_{t_1}^T f(\mathbf{x}(t), \mathbf{u}(t))dt \right] \right\}$$

$$= \min_{\substack{\mathbf{u}(t) \\ t \in [0,t_1]}} \left\{ \int_0^{t_1} f(\mathbf{x}(t), \mathbf{u}(t))dt + J(\mathbf{x}(t_1), t_1) \right\}.$$

---

[1] We will apply the ideas illustrated here only in Section 6.2: therefore the reader may wish to skip this section until later.

We obtain a recursive equation, which can be rewritten for an interval $[t_1, T]$ and a small time increment δt, considering a decomposition into the subintervals $[t_1, t_1 + \delta t]$ and $[t_1 + \delta t, T]$:

$$J(\mathbf{x}(t_1), t_1) = \min_{\substack{\mathbf{u}(t) \\ t \in [t_1, t_1 + \delta t]}} \left\{ \int_{t_1}^{t_1 + \delta t} f(\mathbf{x}(t), \mathbf{u}(t)) dt + J(\mathbf{x}(t_1 + \delta t), t_1 + \delta t) \right\}. \tag{D.8}$$

For a sufficiently small δt we can approximate the integral in (D.8) as

$$\int_{t_1}^{t_1 + \delta t} f(\mathbf{x}(t), \mathbf{u}(t)) dt = f(\mathbf{x}(t_1), \mathbf{u}(t_1)) \delta t + o(\delta t).$$

Assuming that the cost-to-go J is differentiable, we can write the first-order Taylor expansion

$$J(\mathbf{x}(t_1 + \delta t), t_1 + \delta t) = J(\mathbf{x}(t_1), t_1) + \frac{dJ}{dt}(t_1, \mathbf{x}(t_1)) \delta t + o(\delta t),$$

where

$$\frac{dJ}{dt} = \frac{\partial J}{\partial t} + \left(\frac{\partial J}{\partial \mathbf{x}}\right)^T \dot{\mathbf{x}}(t_1) = \frac{\partial J}{\partial t} + \left(\frac{\partial J}{\partial \mathbf{x}}\right)^T \mathbf{h}(\mathbf{x}(t_1), \mathbf{u}(t_1), t_1).$$

Therefore we have

$$J(\mathbf{x}(t_1 + \delta t), t_1 + \delta t) = J(\mathbf{x}(t_1), t_1) + \left[\frac{\partial J}{\partial t} + \left(\frac{\partial J}{\partial \mathbf{x}}\right)^T \mathbf{h}(\mathbf{x}(t_1), \mathbf{u}(t_1), t_1)\right] \delta t + o(\delta t).$$

Putting it all together we obtain

$$J(\mathbf{x}(t_1), t_1) = \min_{\substack{\mathbf{u}(t) \\ t \in [t_1, t_1 + \delta t]}} \left\{ J(\mathbf{x}(t_1), t_1) \right.$$
$$\left. + \left[f(\mathbf{x}(t_1), \mathbf{u}(t_1)) + \frac{\partial J}{\partial t} + \left(\frac{\partial J}{\partial \mathbf{x}}\right)^T \mathbf{h}(\mathbf{x}(t_1), \mathbf{u}(t_1), t_1)\right] \delta t + o(\delta t) \right\}.$$

Note that neither $J(\mathbf{x}(t_1), t_1)$ nor $\frac{\partial J}{\partial t}$ depend on $\mathbf{u}$, and can be taken outside the minimization; then, dividing by δt and taking the limit for $\delta t \to 0$, we get

$$\frac{\partial J}{\partial t}(\mathbf{x}(t_1), t_1) + \min_{\mathbf{u}(t_1)} \left\{ f(\mathbf{x}(t_1), \mathbf{u}(t_1)) + \left(\frac{\partial J}{\partial \mathbf{x}}\right)^T \mathbf{h}(\mathbf{x}(t_1), \mathbf{u}(t_1), t_1) \right\} = 0.$$

By rewriting the equation for a generic state $\mathbf{x}$ and time t, we get the Bellmann equation

$$\frac{\partial J}{\partial t}(\mathbf{x}, t) + \min_{\mathbf{u}} \left\{ f(\mathbf{x}, \mathbf{u}) + \left(\frac{\partial J}{\partial \mathbf{x}}\right)^T \mathbf{h}(\mathbf{x}, \mathbf{u}, t) \right\} = 0. \qquad (D.9)$$

This equation cannot be solved in general, but it is useful in deriving structural insights into the optimal policy, as shown in Section 6.2. We just note here that in practice, one looks for a solution of (D.9) yielding a *feedback control law*, i.e., $\mathbf{u} = \mathbf{u}(\mathbf{x})$.

For further reading

A good reference on DP is [58]; the reader may also consult [21] for stochastic DP and feedback control policies. In Section D.1 we have presented a basic method for the shortest path problem; advanced methods are described in [4, 22]. The DP formulation for the Knapsack problem that we have presented in Example D.2 has been taken from [187]. The derivation of the Bellman equation for continuous-time optimal control has been taken from [83], where stochastic optimal control is also dealt with.

E

Stochastic modeling

In most optimization models one assumes that all the relevant data are known with certainty. However, this is hardly the case in practice. Typical sources of uncertainty are machine breakdowns, unpredictable arrival patterns of jobs and random processing times. In some cases it is necessary to explicitly take into account the randomness of the manufacturing environment. In the case of generative models, adding random elements makes the problem much harder. Luckily, in the case of evaluative models, a stochastic modeling approach may lead to quite manageable and convenient models, such as queueing network models. Queueing models, along with their advantages and disadvantages, are dealt with in Chapter 5; in this Appendix we review the background of the "easy" stochastic models (essentially, the steady-state analysis of Markov chains).

We assume a basic understanding of probability theory and random variables; a suitable reference is [162].

E.1 Motivation

The best way to introduce stochastic modeling is by a simple queueing example.

Example E.1
Consider a single machine to which parts arrive for processing at random times. If a part arrives when the machine is busy, it is queued in a buffer. We assume that the service time is also a random variable (this is particularly true in the case of manual operations).

Let $X(t)$ denote the WIP level at the machine at time t, including the parts waiting in the queue and the one under service. Obviously $X(t)$ is,

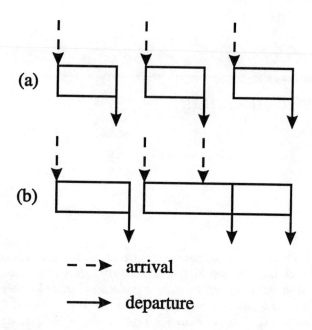

FIGURE E.1
Comparison of deterministic (a) and random (b) arrivals and services

for each t, a random variable; such a collection of time-indexed random variables is called a **stochastic process**.

It is instructive to compare the evolution of this stochastic system with the evolution of a deterministic system whose arrival and service parameters are the average values of the stochastic model. The issue is illustrated in Figure E.1. In the deterministic system arrivals and departures are periodically repeated. It is easy to see that in the deterministic system there will never be a buildup of WIP, unless the average service time is larger than the average interarrival time, in which case the WIP level goes to infinity. On the contrary, in the stochastic system we have time intervals in which $X(t) > 0$. This shows that we cannot cope with uncertainty by building a deterministic model based on average values. ☐

The stochastic process of the last example can be represented as shown in Figure E.2: we have a graph whose nodes are the number of jobs in the system, and the arcs are related, in a way we will clarify later, to the probability of passing from one state to another. It is interesting to note that the graph of Figure E.2 can represent different systems, like the one in the next example.

Motivation

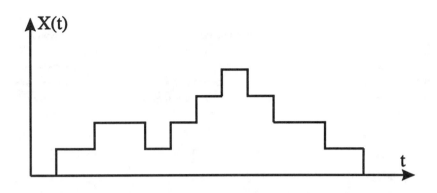

FIGURE E.2
A representation of the evolution of a single-machine queueing system

Example E.2
Consider a manufacturing cell consisting of N failure-prone machines; we say that the machine is **up** when it is working or available to process a part, and it is **down** when it is unavailable because of a breakdown. We start at time $t = 0$ with all the machines in the up state. Machine failures occur randomly. When a machine goes down, a repair activity is started; the duration of the repair is random. Let $X(t)$ be the number of up machines at time t. It is easy to see that this is essentially the same stochastic process as in Example E.1; the only difference is that here we have a *finite* state space. ▫

The nodes of the graph are called the **states** of the stochastic process: in Example E.1 we have an infinite but countable state space. In a stochastic system we cannot control all the inputs, and we cannot predict exactly the evolution of the system; we can only study the time evolution of the probability of being in a certain state. When speaking of the state of a stochastic process, some care must be taken; we have defined the state of a dynamic system as an interface between the past and the future, in the sense that the knowledge of the state is all we need to predict the future evolution given a certain control input. To characterize a stochastic process, we must specify the kind of dependence of $X(t)$ with respect to $X(\tau)$ for $\tau < t$. In the models we deal with, the state of the stochastic process has the same interpretation as in the case of deterministic dynamic systems, in the sense that only the present state can influence the future evolution; the memory of the past states is forgotten. This property does not hold in general, and we will see what conditions are required on the random variables involved.

The aim of stochastic models is to state properties of stochastic processes in terms of probability distributions; we may study the time evolution of the probability distribution of $X(t)$ given an initial condition or, alternatively, we may study the stochastic process in steady state, i.e., when $t \to \infty$. We will see that the second type of study is much more easily accomplished, and we will limit the treatment to long term probability distributions.

E.1.1 A refresher on Probability Theory and random variables

Consider a random experiment, whose **sample space** is denoted by S. Any subset of S is called an **event**. To each event E we associate a probability denoted by $P\{E\} \in [0, 1]$, in such a way that $P\{S\} = 1$ and $P\{\emptyset\} = 0$. The probability of the intersection of events E and F is denoted by $P\{E, F\}$; the conditional probability of E given F is denoted by $P\{E \mid F\}$ and can be computed as

$$P\{E \mid F\} = \frac{P\{E, F\}}{P\{F\}},$$

provided that $F \neq \emptyset$. Two events are independent if $P\{E, F\} = P\{E\}P\{F\}$; this implies that, in the case of independent events, $P\{E \mid F\} = P\{E\}$.

THEOREM E.1
Total Probability Theorem. *Given a collection of subsets $A_i \subset S$, $i = 1, \ldots, n$, such that*

$$A_i \bigcap A_j = \emptyset \qquad i, j = 1, \ldots, n;\ i \neq j,$$

and

$$\bigcup_{i=1}^{n} A_i = S,$$

for any event $B \subset S$ we have

$$P\{B\} = \sum_{i=1}^{n} P\{B \mid A_i\} P\{A_i\}.$$

If we associate a numeric value x to any event of a sample space, we obtain a **random variable** X. We speak of discrete or continuous random variables.

For a generic random variable X we define the **cumulative distribution function (CDF)** as

$$F_X(a) = P\{X \leq a\},$$

Motivation

for any real number a. In the case of discrete random variables we define a **probability mass function (PMF)** $p_X(x_i) = P\{X = x_i\}$. Clearly

$$\sum_{i=1}^{\infty} p_X(x_i) = 1.$$

The PMF can be related to the CDF

$$F_X(a) = \sum_{x_i \leq a} p_X(x_i).$$

In the case of a continuous random variable we define the **probability density function (PDF)** $f_X(x)$ such that

$$P\{X \in A\} = \int_A f_X(x)dx.$$

The PDF is related to the CDF

$$F_X(a) = \int_{-\infty}^{a} f_X(x)dx$$

$$f_X(a) = \frac{d}{da}F_X(a).$$

Given a random variable we define its **expected value** as

$$E[X] = \sum_i x_i p_X(x_i)$$

for the discrete case, and

$$E[X] = \int_{-\infty}^{\infty} x f_X(x)dx$$

for the continuous case.

We also define the **variance** of a random variable

$$\text{Var}[X] = E\left[(X - E[X])^2\right],$$

and the **standard deviation**

$$\alpha_X = \sqrt{\text{Var}[X]}.$$

Another useful quantity is the **coefficient of variation**

$$C_X = \frac{\alpha_X}{E[X]},$$

which gives an idea of the variability of the distribution.

Based on the total probability theorem, we may compute expectations and probabilities by conditioning. We have

$$E[X] = \begin{cases} \sum_y E[X \mid Y = y] P\{Y = y\} & \text{in the discrete case} \\ \int_{-\infty}^{\infty} E[X \mid Y = y] f_Y(y) dy & \text{in the continuous case,} \end{cases}$$

and, for any event E,

$$P\{E\} = \begin{cases} \sum_y P\{E \mid Y = y\} P\{Y = y\} & \text{in the discrete case} \\ \int_{-\infty}^{\infty} P\{E \mid Y = y\} f_Y(y) dy & \text{in the continuous case.} \end{cases}$$

Example E.3

Computing probabilities by conditioning is a most useful tool. Consider a set of independent and identically distributed continuous random variables $X_1, X_2, \ldots$ and an integer valued random variable N. Then we may define the random variable

$$Z = \sum_{i=1}^{N} X_i.$$

By conditioning on N, we can show the following properties:

$$E[Z] = E[N]E[X]$$
$$\text{Var}[Z] = E[N]\text{Var}[X] + (E[X])^2 \text{Var}[N].$$

See [162] for a detailed proof. ◻

E.1.2 Memoryless random variables

When coping with a deterministic dynamic system, we have introduced the concept of state as the information needed to predict future system behavior. The state represents the limited memory of the system; this concept of state is exploited by Dynamic Programming algorithms. Some stochastic systems enjoy a similar property. It is therefore useful to consider those probability distributions which are memoryless, both in the discrete and the continuous case.

Consider carrying out a set of independent and identical "trials", where the probability of success is p, and let X be the discrete random variable corresponding to the number of trials needed to obtain the first success; this type of random variable is called **geometric variable**, and its PFM is

$$P\{X = k\} = (1-p)^{k-1} p \quad k = 1, 2, 3, \ldots$$

Motivation

For a geometric variable X we have

$$E[X] = \frac{1}{p}, \qquad \text{Var}[X] = \frac{1-p}{p^2}.$$

We can also obtain the probability of having more than m trials before the success

$$P\{X > m\} = 1 - P\{X \leq m\} = 1 - \sum_{k=1}^{m} P\{X = k\} = (1-p)^m.$$

Now consider the following issue: if we know that m trials have already been carried out without success, does this information tell us something about the future trials? To ascertain whether this is the case or not, let us compute $P\{X = m+n \mid X > m\}$. We have

$$P\{X = m+n \mid X > m\} = \frac{P\{X = m+n, X > m\}}{P\{X > m\}}$$

$$= \frac{(1-p)^{m+n-1} p}{(1-p)^m} = (1-p)^{n-1} p$$

$$= P\{X = n\}.$$

Since $P\{X = m+n \mid X > m\} = P\{X = n\}$, we see that the variable is **memoryless**.

In the case of continuous random variables, the same role is played by the **exponential variable**, whose CDF is (for $x \geq 0$):

$$F_X(x) = 1 - e^{-\lambda x}.$$

The Probability Density Function (PDF) is

$$f_X(x) = \lambda e^{-\lambda x},$$

and we have

$$E[X] = \frac{1}{\lambda}, \qquad \text{Var}[X] = \frac{1}{\lambda^2}.$$

It is possible to show the following useful relations and properties:

1. $P\{X > x\} = e^{-\lambda x}$.
2. $P\{x \leq X \leq y\} = e^{-\lambda x} - e^{-\lambda y}$, where $0 \leq x \leq y$.
3. If X_1 and X_2 are independent exponential variables with parameters λ_1 and λ_2,

$$P\{X_1 < X_2\} = \frac{\lambda_1}{\lambda_1 + \lambda_2}.$$

4. If X_1 and X_2 are independent exponential variables with parameters λ_1 and λ_2, and we define $X = \min\{X_1, X_2\}$, then X is exponential with parameter $\lambda_1 + \lambda_2$.

To show that the exponential variable is memoryless, we proceed as in the geometric variable case:

$$P\{X > x+y \mid X > x\} = \frac{P\{X > x+y, X > x\}}{P\{X > x\}}$$
$$= \frac{P\{X > x+y\}}{P\{X > x\}} = \frac{e^{-\lambda(x+y)}}{e^{-\lambda x}} = e^{-\lambda x}$$
$$= P\{X > x\}.$$

E.2 Stochastic processes and Markov chains

DEFINITION E.4 A **stochastic process** *is a collection of random variables* $\{X(t) \mid t \in T\}$, *where* $X(t)$ *is a random variable for any* $t \in T$.

The set on which $X(t)$ takes its values is the **state space** of the stochastic process. There are different types of stochastic process depending on the state space, which may be discrete or continuous, and the parameter, which may be also discrete or continuous. We will be concerned with stochastic processes characterized by a countable state space; such processes are called **chains**. For instance the stochastic processes of Examples E.1 and E.2 are chains.

Stochastic processes are in general quite difficult to analyze, since they are characterized by the joint probability distribution of the random variables $X(t_i)$ for arbitrary collections of time instants $t_i, i = 1, \ldots, n$. However, an important class of stochastic processes, the **Markov processes**, is characterized by a limited memory, and their analysis is dramatically simplified. Before analyzing Markov stochastic processes, we recall the basic properties of an important type of stochastic process, the Poisson process.

E.2.1 The Poisson process

The Poisson process is a continuous-time stochastic process which plays a fundamental role in manufacturing systems: it is useful to model a memoryless stochastic process of job arrivals to the machines.

A stochastic process $N(t)$ is a **counting process** if $N(t)$ represents the number of events occurred up to time t. For a counting process we have:

1. $N(t) \geq 0$.
2. $N(t)$ takes integer values.
3. If $s < t$, then $N(s) < N(t)$.

4. If $s < t$, then $N(t) - N(s)$ is the number of events that occur during the interval (s,t).

A counting process has **independent increments** if the number of events occurring in disjoint time intervals are independent. A counting process has **stationary increments** if the number of events occurring in an interval depends only on its length, i.e., for an arbitrary s, the number of events during $(t_1 + s, t_2 + s)$ has the same distribution as the number of events during (t_1, t_2).

DEFINITION E.5 *A counting process $N(t)$ is called a **Poisson process** with rate λ if:*

1. $N(0) = 0;$
2. *the process has independent increments;*
3. *the number of events occurring during any interval of length t is a random variable with Poisson distribution, i.e.,*

$$P\{N(t+s) - N(t) = k\} = e^{-\lambda t}\frac{(\lambda t)^k}{k} \qquad k = 0, 1, 2, \ldots,$$

for arbitrary s, t.

A consequence of this definition is that a Poisson process has stationary increments and that $E[N(t)] = \lambda t$.

A Poisson process is typically used to model the number of job arrivals to a manufacturing system during a time interval $[0, t]$. It is interesting to consider the interarrival times of the Poisson process. For $k = 0$ and $t > 0$ we have

$$P\{X(t) = 0\} = e^{-\lambda t},$$

which can be interpreted as the probability of no arrival. If T is the time of the first arrival,

$$P\{T > t\} = e^{-\lambda t}.$$

It can be shown that the interarrival times between two generic arrivals of a Poisson process are independent and exponentially distributed with parameter λ.

The Poisson process enjoys the following interesting properties, illustrated in Figure E.3:

Superposition. Consider a set of part types arriving at a machining center. For each part type i we have a Poisson stream with rate λ_i, and the streams are joined at the machine. The arrival process at the machine is still Poisson, with a rate given by the sum of the rates λ_i.

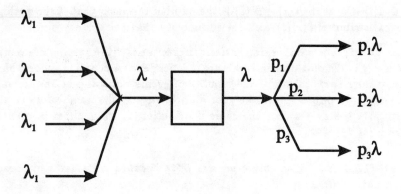

FIGURE E.3
Superposition and decomposition of Poisson streams

Decomposition. Consider the departure process from a machine, and assume it is a Poisson stream with rate λ. The parts are routed to other machines, and for each machine M_i there is a probability p_i. Then the arrival process at the machine i is Poisson with rate $p_i \lambda$.

E.2.2 Discrete-time Markov chains

We consider here a discrete-time process $X(t)$ for $t \in T = \{t_0, t_1, t_2, \ldots\}$; the state space S is countable, and we can set $S = \{0, 1, 2, \ldots\}$ without loss of generality. We use the standard notation $X_n = j$ to say that the state of the process at time t_n is j; X_0 is the initial state of the chain.

DEFINITION E.6 *A discrete-time chain is called a* **discrete-time Markov chain (DTMC)** *if*

$$P\{X_n = j \mid X_{n-1} = i, X_{n-2} = i_{n-2}, \ldots X_1 = i_1, X_0 = i_0\} = $$
$$P\{X_n = j \mid X_{n-1} = i\}$$

for all $i, j, i_0, i_1, \ldots, i_{n-2} \in S$.

We see that a discrete-time Markov chain has a strong similarity to a discrete-time dynamic system, in the sense that the future evolution depends only on the current state and not on the entire past history. The difference is that the evolution is not due to an applied control, but it is characterized by state transition probabilities. Furthermore, we cannot deal with the state trajectory $X(t)$, but with the evolution of the probability distributions $\pi_i(t)$, i.e., the probability that the system is in state i at time t.

Let $p_{ij}(m, m+n) = P\{X_{m+n} = j \mid X_m = i\}$ be the n-step transition probability from state i to state j. In general the transition probabilities are time-varying; if this is not the case, i.e., if

$$p_{ij}(m, m+n) = p_{ij}(n) \quad \forall i,j \in S;\ \forall m,n,$$

we speak of a **homogeneous** Markov chain. We will only consider homogeneous Markov chains; the reader wishing an example of a non-homogeneous Markov chain may think of the evolution of a Simulated Annealing algorithm.

The one-step transition probability, in the homogeneous case, is denoted by

$$p_{ij} = P\{X_n = j \mid X_{n-1} = i\}.$$

The one step transition probabilities are grouped in the **transition probability matrix**

$$\mathbf{P} = [p_{ij}] = \begin{bmatrix} p_{00} & p_{01} & p_{02} & \cdots \\ p_{10} & p_{11} & p_{12} & \cdots \\ p_{20} & p_{21} & p_{22} & \cdots \\ \vdots & \vdots & \vdots & \ddots \end{bmatrix}.$$

Note that $\mathbf{P}$ may be an infinite matrix. The matrix $\mathbf{P}$ is said to be a **stochastic matrix**. A stochastic matrix $\mathbf{P}$ is characterized by the following properties of its entries p_{ij}:

$$p_{ij} \in [0,1] \quad \forall i,j$$

$$\sum_j p_{ij} = 1 \quad \forall i.$$

A DTMC can be graphically represented as shown in Figure E.4.

We have introduced Markov chains as stochastic processes with a limited amount of memory. We can expect that this is reflected in the random variables pertaining to the process; indeed, it is easy to see that the mean **sojourn time** T_i in state i is a geometric random variable:

$$P\{T_i = n\} = P\{T_i > n-1\} - P\{T_i > n\} = p_{ii}^{n-1} - p_{ii}^n = p_{ii}^{n-1}(1 - p_{ii}).$$

This is a geometric variable with success probability $1 - p_{ii}$; the mean sojourn time in state i is therefore $1/(1 - p_{ii})$.

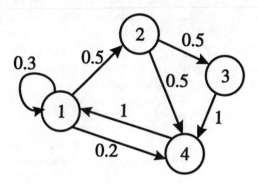

FIGURE E.4
A discrete-time Markov chain

E.2.3 Continuous-time Markov chains

DEFINITION E.7 *A continuous-time chain is called a* **continuous-time Markov chain (CTMC)** *if*

$$P\{X(t) = j \mid X(s) = i; X(u) = x(u), u \in [0, s)\} = P\{X(t) = j \mid X(s) = i\}$$

for $t \geq s$.

The standard notation for transition probabilities is

$$p_{ij}(s,t) = P\{X(t) = j \mid X(s) = i\}$$

We consider only homogeneous CTMC, for which

$$p_{ij}(s,t) = p_{ij}(t-s) = P\{X(u+t-s) = j \mid X(u) = i\} \qquad \forall u.$$

When considering a CTMC, an issue to consider is the time elapsing between state transitions. If the system has no memory apart from the current state, a natural consequence is that the sojourn time in a state must be an exponential random variable. To see this, consider

$$P\{T_i > s + x \mid T_i > s\} = h(x), \qquad s, x \geq 0.$$

This probability can be written as

$$h(x) = \frac{P\{T_i > s + x, T_i > s\}}{P\{T_i > s\}} = \frac{P\{T_i > s + x\}}{P\{T_i > s\}},$$

from which we get

$$P\{T_i > s + x\} = P\{T_i > s\}h(x). \tag{E.1}$$

If we set $s = 0$, noting that $P\{T_i > 0\} = 1$, we obtain

$$P\{T_i > x\} = h(x).$$

Therefore we may write (E.1) as

$$P\{T_i > s + x\} = P\{T_i > s\}P\{T_i > x\}.$$

This equation shows that the sojourn time is exponentially distributed.

E.3 Steady-state analysis of Markov chains

The evolution of state probabilities in a homogeneous DTMC or CTMC can be studied both in transient and steady state. Studying the steady state means analyzing the probability distribution

$$\pi_i = \begin{cases} \lim_{n \to \infty} \pi_i(n) & \text{in the DTMC case,} \\ \lim_{t \to \infty} \pi_i(t) & \text{in the CTMC case.} \end{cases}$$

In the following we will use the notation $\mathit{\Pi} = [\pi_0, \pi_1, \pi_2, \ldots]$ to denote the (row) vector of probabilities. Due to the lack of memory, we expect that, apart from pathological cases, a steady-state probability distribution exists and is independent of the initial state. The following treatment will be very limited and based on intuitive arguments; the reader is referred to the literature for a thorough analysis.

E.3.1 Steady-state analysis of DTMCs

We start from the equations expressing the multiple-step transition probabilities

$$p_{ij}(m + n) = P\{X_{m+n} = j \mid X_0 = i\},$$

with the aim of linking them to one-step transition probabilities.

If we consider every possible intermediate state X_m, and apply the total probability theorem, we may write

$$\begin{aligned} p_{ij}(m+n) &= \sum_{k \in S} P\{X_{m+n} = j, X_m = k \mid X_0 = i\} \\ &= \sum_{k \in S} P\{X_{m+n} = j \mid X_m = k, X_0 = i\}P\{X_m = k \mid X_0 = i\} \\ &= \sum_{k \in S} P\{X_{m+n} = j \mid X_m = k\}P\{X_m = k \mid X_0 = i\} \end{aligned}$$

$$= \sum_{k \in S} p_{ik}(m) p_{kj}(n). \qquad (E.2)$$

These are called the Chapman-Kolmogorov equations. Now we may easily link multiple-step transitions to one-step transitions. If we set $m = 1$ and $n = n - 1$ in (E.2) we get

$$p_{ij}(n) = \sum_{k \in S} p_{ik}(1) p_{kj}(n-1).$$

In matrix form we may write $\mathbf{P}(n) = \mathbf{P}\mathbf{P}(n-1)$, which, together with the initial condition $\mathbf{P}(0) = \mathbf{I}$, yields $\mathbf{P}(n) = \mathbf{P}^n$.

If the system starts in an initial state with a probability mass function $\Pi(0)$, we may ask what is the evolution of the probabilities $\pi_j(n)$. Invoking again the total probability theorem we see that

$$\pi_j(n) = P\{X_n = j\} = \sum_{i \in S} P\{X_n = j \mid X_0 = i\} P\{X_0 = i\} = \sum_{i \in S} \pi_i(0) p_{ij}(n),$$

or, in matrix form,

$$\Pi(n) = \Pi(0) \mathbf{P}^n.$$

To obtain the limiting probabilities π_j, we note that

$$\Pi(n) = \Pi(n-1) \mathbf{P};$$

in a steady-state condition, one would expect that we may find Π by solving

$$\Pi = \Pi \mathbf{P}, \qquad (E.3)$$

with the additional conditions

$$\sum_{j \in S} \pi_j = 1; \qquad \pi_j \in [0, 1].$$

Note that this implies that the memory of the initial state probability is lost.

Actually, finding the steady-state probability distribution is not trivial, since it is not guaranteed its existence and uniqueness. To insure that Π exists and is unique, we must rule out some "pathological" cases. One is shown in Figure E.5; the evolution of the chain is clearly periodic, and the state probability distribution does not settle to a steady-state value. Another pathological case is shown in Figure E.6: the memory of the initial state is not lost since we have non-communicating states. A DTMC where all states communicate is said to be **irreducible**. Another issue is the number of times a state will be visited: we say that a state i is **recurrent** if the expected number of times the process is in state i is infinite; furthermore, a state i is **positively recurrent** if, starting from i, the expected

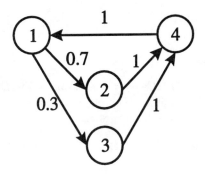

FIGURE E.5
A periodic DTMC

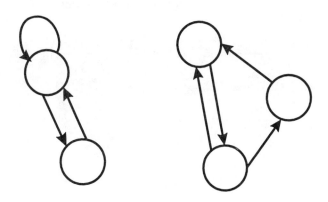

FIGURE E.6
A reducible DTMC

time to get back to i is finite. A DTMC is positive recurrent if all its states are positive recurrent; in a finite DTMC, all recurrent states are positive recurrent. An irreducible, aperiodic, positively recurrent DTMC is said to be **ergodic**. The existence and uniqueness of Π, independent of the initial state, is insured by the ergodicity of the chain.

E.3.2 Steady-state analysis of CTMCs

The steady-state analysis of CTMCs is trickier than in the DTMCs case, and it leads to differential Chapman-Kolmogorov equations. As in the discrete time case, we may use the total probability theorem to write

$$p_{ij}(s,t) = \sum_{k \in S} p_{ik}(s,u) p_{kj}(u,t) \qquad 0 \leq s \leq u \leq t,$$

or, in matrix form,

$$\mathbf{H}(s,t) = \mathbf{H}(s,u)\mathbf{H}(u,t).$$

We obtain differential equations by letting $u = t$ and $t = t + h$, where h is an infinitesimal time increment:

$$\mathbf{H}(s, t+h) = \mathbf{H}(s,t)\mathbf{H}(t, t+h).$$

Subtracting $\mathbf{H}(s,t)$ from both sides of this equation, dividing by h and letting $h \to 0$, we get

$$\lim_{h \to 0} \frac{\mathbf{H}(s,t+h) - \mathbf{H}(s,t)}{h} = \mathbf{H}(s,t) \left(\lim_{h \to 0} \frac{\mathbf{H}(t, t+h) - \mathbf{I}}{h} \right).$$

We obtain the forward Chapman-Kolmogorov equation

$$\frac{\partial \mathbf{H}(s,t)}{\partial t} = \mathbf{H}(s,t)\mathbf{Q}(t),$$

where

$$\mathbf{Q}(t) = \left(\lim_{h \to 0} \frac{\mathbf{H}(t, t+h) - \mathbf{I}}{h} \right)$$

is the **infinitesimal generator** of the CTMC. The initial condition is $\mathbf{H}(s,s) = \mathbf{I}$.

The evolution of the transition probabilities can be linked to the evolution of the state probabilities; using the total probability theorem,

$$p_j(t) = \sum_i P\{X(0) = i\} P\{X(t) = j \mid X(0) = i\}$$

$$= \sum_i p_i(0) p_{ij}(0, t).$$

By the same token, we obtain the backward Chapman-Kolmogorov equation

$$\frac{\partial \mathbf{H}(s,t)}{\partial t} = -\mathbf{Q}(s)\mathbf{H}(s,t).$$

It is important to interpret the elements of the infinitesimal generator $\mathbf{Q}$; from its definition, we have

$$1 - p_{ii}(t, t+h) = -hq_{ii}(t) + o(h),$$
$$p_{ij}(t, t+h) = hq_{ij}(t) + o(h).$$

Therefore $-q_{ii}(t)$ can be interpreted as the rate at which the system "departs" from state i, and $q_{ij}(t)$ is the rate of transitions from i to j. As a consequence

$$\sum_{j \in S} q_{ij}(t) = 0 \quad \forall i. \tag{E.4}$$

In the homogeneous case, the Chapman-Kolmogorov equations are simplified. In matrix form we have

$$\frac{d\mathbf{H}(t)}{dt} = \mathbf{H}(t)\mathbf{Q}, \qquad \mathbf{H}(0) = \mathbf{I}$$
$$\frac{d\mathbf{H}(t)}{dt} = \mathbf{Q}\mathbf{H}(t), \qquad \mathbf{H}(0) = \mathbf{I}.$$

In scalar form,

$$\frac{dp_{ij}(t)}{dt} = q_{jj}p_{ij}(t) + \sum_{k \neq j} q_{kj}p_{ik}(t)$$
$$\frac{dp_{ij}(t)}{dt} = q_{ii}p_{ij}(t) + \sum_{k \neq i} q_{ik}p_{kj}(t).$$

Note that the infinitesimal generator is in this case a constant matrix. In matrix form, the evolution of state probabilities is

$$\frac{d\mathbf{\Pi}(t)}{dt} = \mathbf{\Pi}(t)\mathbf{Q}; \tag{E.5}$$

or, in scalar form,

$$\frac{d\pi_j(t)}{dt} = q_{jj}\pi_j(t) + \sum_{k \neq j} q_{kj}\pi_k(t). \tag{E.6}$$

To obtain the steady-state probabilities, the same considerations hold as in the DTMC case. Under ergodicity conditions, we can find the steady-state probabilities $\mathbf{\Pi}$ by setting to zero the derivative in (E.5),

$$\mathbf{\Pi}\mathbf{Q} = 0$$

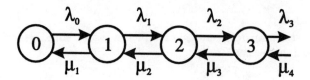

FIGURE E.7
A birth-death process

$$\sum_j \pi_j = 1$$

$$\pi_j \geq 0.$$

It is most useful to interpret (E.6) under stationary conditions:

$$q_{jj}\pi_j + \sum_{k \neq j} q_{kj}\pi_k = 0.$$

From (E.4) we have $q_{jj} = -\sum_{k \neq j} q_{jk}$, hence we may write

$$\pi_j \left(\sum_{k \neq j} q_{jk} \right) = \sum_{k \neq j} q_{kj}\pi_k. \qquad (E.7)$$

This equation can be interpreted as a *flow balance* equation: the probability flow into a state equals the probability flow out of that state.

E.3.3 An example: steady-state analysis of a birth-death process

A birth-death process is a simple CTMC with transitions rates:

$$q_{i,i+1} = \lambda_i \qquad i \geq 0,$$
$$q_{i,i-1} = \mu_i \qquad i \geq 1,$$
$$q_{ij} = 0 \qquad |i-j| > 1.$$

The graphical structure of the birth-death is shown in Figure E.7.

The system described in Example E.1 is a birth-death process, if the interarrival and service times are exponentially distributed with rates λ and μ.

To analyze the steady-state behavior of a birth-death process, we write the flow balance equations

$$\lambda_0 \pi_0 = \mu_1 \pi_1$$

$$(\lambda_k + \mu_k)\pi_k = \lambda_{k-1}\pi_{k-1} + \mu_{k+1}\pi_{k+1} \qquad k \geq 1.$$

The last equation can be slightly manipulated to obtain

$$\lambda_k \pi_k - \mu_{k+1}\pi_{k+1} = \lambda_{k-1}\pi_{k-1} - \mu_k \pi_k,$$

which, when recursively applied, results in

$$\lambda_k \pi_k - \mu_{k+1}\pi_{k+1} = \lambda_0 \pi_0 - \mu_1 \pi_1 = 0.$$

This yields

$$\pi_k = \frac{\lambda_{k-1}}{\mu_k}\pi_{k-1} = \frac{\lambda_0 \lambda_1 \cdots \lambda_{k-1}}{\mu_1 \mu_2 \cdots \mu_k}\pi_0.$$

To get the value of π_0 we can exploit the normalization condition

$$\sum_j \pi_j = 1,$$

which yields in this case

$$\pi_0 = \frac{1}{1 + \sum_{k \geq 1} \prod_{i=0}^{k-1} \frac{\lambda_j}{\mu_{j+1}}}.$$

For further reading

A suitable introduction to probability theory and stochastic processes can be found in [162]. Stochastic models at a more advanced level are treated in [185]. A thorough treatment of stochastic processes can be found in [111, 112].

Problems

PROBLEM 1 *Lot-sizing with make/buy decisions.*

We are given the demand forecasts for a set of items over a discrete-time horizon; for each item we have the typical data of a (big bucket) lot-sizing problem. It is possible to meet demand both by producing and purchasing items from an alternative source. Purchased items are characterized by a fixed cost and a linearly variable cost. Build a MILP model to minimize the overall costs.

PROBLEM 2 *Lot-sizing in a production-distribution environment.*

A factory produces a set of items which are sent to two decentralized stores. There is also a centralized store within the factory; the material flow is shown in Figure 1. The following assumptions are made:

- the inventory costs are the same for the three stores;
- we have to satisfy a forecasted demand for the two decentralized stores;
- for each time bucket we must plan the production in the factory and the transportation from the central store to the decentralized ones;
- the two transportation services are independent, and their capacity can be considered infinite;
- the transportation cost has a linearly variable component depending on the weight of the transported items; there is a fixed cost depending on whether or not the service is activated for a certain time bucket (note that the fixed cost is paid *once* for *all* the items transported during a period).

Build a MILP model to minimize the overall costs.

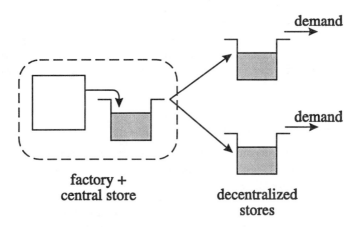

FIGURE 1
Lot-sizing in a production-distribution environment.

PROBLEM 3 *Lot-sizing and scheduling in a press shop.*

A press shop consists of M nonidentical machines, on which plastic components are produced. The machines can be considered as parallel machines (one machine visit is required). For each item, we know on which machines it can be produced, and the quantity produced per hour. The shop operates on three eight-hour daily shifts (0-8; 8-16; 16-24). Starting a new batch implies a sequence-independent setup cost and a setup time ranging from five to seven hours, depending on the item to be produced and the difficulty of mounting its die.
There are different constraints on the setup operations:

- The setup requires a skilled worker. There are two of them, and they are available only during the 8-16 shift.
- Each die is available in a single copy.
- Certain dies are shared by a family of items.
- Before being mounted, a die requires some maintenance, which is carried out within a maintenance department. This is also necessary to change production from a certain product to another one requiring the same die. Furthermore, it is necessary to send a die to the maintenance department at least 24 hours before its use.

Build a MILP model to meet the daily demand, minimizing inventory and setup costs.

PROBLEM 4 *Single-item lot-sizing with quantity discounts.*

Consider a purchasing problem characterized by a weekly demand to be met at minimum cost. We have an inventory cost and a fixed cost associated with each purchase order. Adapt the Wagner-Within procedure to the case of a piecewise-linear purchase cost, accounting for discounts.

PROBLEM 5 *Lot-sizing with perishable items.*

Consider a food production system consisting of a single line. We want to adopt a DLSP modeling framework, but we have to cope with some complications.

- Apart from inventory and setup costs (assumed sequence-independent), we must consider shutdown costs; we pay a certain cost when turning off the line.
- We must plan one maintenance period during the scheduling horizon; the maintenance period takes one time bucket, and it must be carried out not earlier than time bucket t_1 and not later than time bucket t_2.
- The items are perishable and cannot be stored for more than $\hat{t}$ time buckets.

Build a MILP model for the problem; then modify the model to cope with a maintenance period taking two *consecutive* time buckets.

PROBLEM 6 *Lot-sizing in an assembly plant.*

We have a plant for the assemlby of different end items. End items are characterized by a flat bill of materials (one level of components); let α_{ij} be the quantity of component j needed to assemble one unit of end item i. The raw components are purchased from two vendors; then they are processed and assembled. There is an inventory for raw components and an inventory for end items; there is *no* inventory for processed components. Given the weekly demand for end items, we must plan both the production and the transportation of raw components. The objective is to minimize inventory, setup and transportation costs.

There are different (finite-capacity) workcenters for processing components and one workcenter for assembly. The assembly workcenter is very flexible, and there is no setup issue with it. Components are partitioned into families; there are *family* setup costs and times for components processing.

There are two *independent* transportation services from the two vendors. Transportation is planned on a weekly basis; the components transported during a week are available for processing at the beginning of the next week. Transportation is capacity-constrained; assume that capacity is measured in terms of the *volume* of transported components. The transportation cost is a piecewise constant function of the capacity required, i.e.,

$$C = \begin{cases} 0 & \text{if } V = 0 \\ C_1 & \text{if } 0 < V \leq V_1, \\ C_2 & \text{if } V_1 < V \leq V_2, \\ C_3 & \text{if } V_2 < V \leq V_3, \end{cases}$$

where C is the cost and V is the volume.

Build a MILP model to solve a problem. Consider how the model should be changed to take an inventory of processed components into account.

PROBLEM 7 *Load balancing with tooling constraints.*

A set of jobs must be processed on a set of identical parallel machines. Each job must undergo a sequence of operations, which are consecutively carried out during a single machine visit. Each operation requires a tool, tools are stored on each machine in a limited capacity tool magazine, and for each tool type there is a limited amount of tool copies.

- Build a MILP model for the assignment of parts and tools to machines in order to minimize the maximum load.
- Modify the model in order to minimize the maximum load unbalance (i.e., the maximum load minus the minimum load).
- Define a neighborhood structure in order to apply a tabu search procedure for this problem.
- How can the model be modified in order to take tool wear into account?

Hint: merge the models illustrated in Section 4.6.5 for $R//C_{\max}$ and FMS part type selection.

PROBLEM 8 *Strenghtening MILP models.*

Consider the following knapsack-like constraint on binary variables:

$$15x_1 + 10x_2 + 8x_3 + 7x_4 + 5x_5 \leq 24.$$

Given this constraint, we may add the following constraints (among others) to strengthen the formulation:

$$x_1 + x_2 + x_3 + x_4 \leq 3$$
$$x_1 + x_2 + x_3 \leq 2$$
$$x_1 + x_2 \leq 1.$$

Which one is the best constraint to add, if we want to solve the problem by an LP-based Branch and Bound procedure?

PROBLEM 9 *Part type selection with family setup times.*

In Example 4.3 we have considered a single machine part type selection problem with family setup times. Adapt the model to the case of parallel identical machines.

PROBLEM 10 *Disjunctive graphs and the Johnson algorithm.*

Consider an orientation of the disjunctive graph for a $F2//C_{\max}$ problem, and use it to justify the Johnson algorithm.

PROBLEM 11 *Job shop scheduling with external setups.*

Consider the following variant of the $J//C_{\max}$ problem. There are M machines. $M - 1$ have negligible setup times; one is characterized by considerable, though sequence-independent, setup times. As a consequence, this machines is the bottleneck. To ease the problem, a possible strategy is to adopt an *external* setup: the setup is done on a part, while the machine is working on another part. In other words, setup and processing are separated, and only processing requires the machine. Assume that there is always someone able to carry out the setup. Build a disjunctive graph representation to solve the problem by local search or by the shifting bottleneck procedure.

PROBLEM 12 *Job shop scheduling with sequence-dependent setup times.*

Consider a $J/r_i, s_{ij}/L_{\max}$ problem, i.e., a job shop scheduling problem with non-zero release times, sequence-dependent setup times and a max lateness objective.
- Devise a possible priority rule for solving the problem.
- Consider adopting a tabu search approach; discuss a possible neighborhood structure.

- In Section 4.8.7 we have considered the shifting bottleneck procedure for a $J//C_{\max}$ problem. Try to adapt the method to our problem, under the assumption that only one machine is characterized by very large and sequence-dependent setup times.

PROBLEM 13 *Tabu search for $R//\sum_i w_i C_i$* [15].

Consider the problem of minimizing the total weighted completion time on a set of unrelated parallel machines. Devise a neighborhood structure for the problem.
Hint: given an assignment of parts to machines, is coping with sequencing issues difficult?

PROBLEM 14 *Single-machine scheduling with precedence constraints* [191].

Consider a $1/prec/\sum_i C_i$ problem; we have five jobs, whose processing times are 3, 10, 1, 3, 2, respectively. There are two precedence constraints: job 2 must be scheduled before job 3, and job 4 before job 5 (note: before does *not* mean *immediately* before).

- Build a MILP model for the problem.

- Adopt a Lagrangian relaxation strategy and dualize the complicating constraints.

- Consider a vector of Lagrangian multipliers $\hat{\mu}$ with all the components set to 0.5. For this $\hat{\mu}$ compute the value of the dual function and find a subgradient. Is the dual function differentiable in $\hat{\mu}$?

Hint: see Section 4.7.4.

PROBLEM 15 *The $M/M/1$ queue with machine breakdowns.*

Consider an $M/M/1$ queue with an infinite capacity buffer. Let λ and μ be the arrival rates (the usual exponential distributions are assumed). When processing a part, the machine may break down; the time between machine breakdowns is exponentially distributed with rate f. The time to repair the machine is also exponentially distributed, with rate r. After repairing the machine, the part is checked; with a probability α, the part is scrapped.
Draw the Markov Chain diagram under the following alternative policies concerning a part continuing its processing after a machine breakdown:

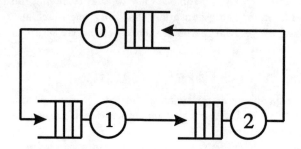

FIGURE 2
A cyclic queueing network.

- Preemptive resume policy: the processing is resumed from the point at which it was interrupted.

- Preemptive repeat policy: the processing is restarted from scratch.

- Separate setup and processing: there is an exponentially distributed setup time with rate μ_s and an exponentially distributed processing time with rate μ_p. After repair, the setup need not be repeated, but the processing is restarted from scratch.

PROBLEM 16 *The convolution algorithm for closed queueing networks.*

The convolution algorithm requires solving the traffic equations, to obtain the relative utilizations ρ_i for each station i. For a closed network, there are infinite solutions of the traffic equations: actually, they are of the form $\alpha \rho_i$, where α is any strictly positive real number. Show that the choice of α is irrelevant for the performance measures we compute with the convolution algorithm.

PROBLEM 17 *Throughput in a balanced cyclic queueing network.*

Consider a closed queueing network such as the one depicted in Figure 2; such an arrangement is called a *cyclic* network. The network consists of M machines and it is balanced, in the sense that the machines are characterized by the same exponential service time distribution with rate μ. Let N be number of available fixtures. Show that the system throughput is

$$\Theta(N) = \frac{\mu N}{N + M - 1}.$$

Hint: use Mean Value Analysis.

Bibliography

[1] J. Adams, E. Balas, D. Zawack. 1988. The Shifting Bottleneck Procedure for Job Shop Scheduling. *Management Science* **34**, 391-401.

[2] H.H. Adelsberger, J.J. Kanet. 1989. The Leitstand - A New Tool in Computer-Aided Manufacturing Scheduling. In: K.E. Stecke, R. Suri (eds.). *Proceedings of the Third ORSA/TIMS Conf. on FMS*. Elsevier, Amsterdam. 231-236.

[3] P. Afentakis, B. Gavish, U.S. Karmarkar. 1984. Computationally Efficient Optimal Solutions to the Lot-Sizing Problem in Multistage Assembly Systems. *Management Science* **30**, 222-239.

[4] R.K. Ahuja, T.L. Magnanti, J.B. Orlin. 1993. *Network Flows: Theory, Algorithms, and Applications*. Prentice Hall, Englewood Cliffs, NJ.

[5] R. Akella, Y. Choong, S.B. Gershwin. 1984. Performance of a Hierarchical Production Scheduling Policy. *IEEE Trans. on Components, Hybrids and Manufacturing Technology* **7**, 225-240.

[6] R.G. Askin, C.R. Standridge. 1993. *Modeling and Analysis of Manufacturing Systems*. Wiley, New York.

[7] L.H. Avonts, L.N. Van Wassenhove. 1988. The Part Mix and Routing Mix Problem in FMS: a Coupling Between an LP Model and a Closed Queueing Network. *International Journal of Production Research* **26**, 1891-1902.

[8] F. Baccelli, G. Cohen, G.J. Olsder, J.-P. Quadrat. 1992. *Synchronization and Linearity: an Algebra for Discrete Event Systems*. Wiley, Chichester.

[9] H.C. Bahl. 1983. Column Generation Based Heuristic Algorithm for Multi-Item Scheduling. *IIE Transactions* **15**, 136-141.

[10] H.C. Bahl, L.P. Ritzman, J.N.D. Gupta. 1987. Determining Lot Sizes and Resource Requirements: a Review. *Operations Research* **35**, 329-345.

[11] K.R. Baker. 1974. *Introduction to Sequencing and Scheduling*. Wiley, Chichester.

[12] K.R. Baker. 1975. A Comparative Study of Flow Shop Algorithms. *Operations Research* **23**, 62-73.

[13] E. Balas. 1969. Machine Sequencing via Disjunctive Graphs: an Implicit Enumeration Algorithm. *Operations Research* **17**, 941-957.

[14] Barnes, J.W., and M. Laguna. 1993. A Tabu Search Experience in Production Scheduling. *Annals of Operations Research* **41**, 141-156.

[15] Barnes, J.W., and M. Laguna. 1993. Solving the Multiple-Machine Weighted Flow Time Problem Using Tabu Search. *IIE Transactions* **25**, 121-128.

[16] C. Barnhart. 1993. Dual-Ascent Methods for Large-Scale Multicommodity Flow Problems. *Naval Research Logistics* **40**, 305-324.

[17] S. Bashyam, M.C. Fu. 1994. Application of Perturbation Analysis to a Class of Periodic Review (s, S) Inventory Systems. *Naval Research Logistics* **41**, 47-80.

[18] E.M.L. Beale, J.A. Tomlin. 1970. Special Facilities in a General Mathematical Programming Systems for Non-Convex Problems Using Ordered Sets of Variables. In: J. Lawrence (ed.). *Proceedings of the 5th International Conf. on Operational Research*. Tavistock Publications, London.

[19] B.J. Berkley. 1992. A Review of the Kanban Production Control Research Literature. *Production and Operations Management* **1**, 393-411.

[20] J.W.M. Bertrand, J.C. Wortmann, J. Wijngaard. 1990. *Production Control: a Structural and Design Oriented Approach*. Elsevier, New York.

[21] D.P. Bertsekas. 1987. *Dynamic Programming: Deterministic and Stochastic Models*. Prentice-Hall, Englewood Cliffs, NJ.

[22] D.P. Bertsekas. 1992. *Linear Network Optimization: Algorithms and Codes*. MIT Press, Cambridge, MA.

[23] P.J. Billington, J.O. McClain, L.J. Thomas. 1983. Mathematical Programming Approaches to Capacity-Constrained MRP Systems: Review, Formulation and Problem Reduction. *Management Science* **29**, 1126-1141.

[24] G.R. Bitran, S. Dasu. 1992. A Review of Open Queueing Network Models of Manufacturing Systems. *Queueing Systems* **12**, 95-134.

[25] G.R. Bitran, D. Tirupati. 1993. Hierarchical Production Planning. In: S.C. Graves, A.H.G. Rinnooy Kan, P.H. Zipkin (eds.). *Logistics of Production and Inventory*. North Holland, Amsterdam. 523-568.

[26] J. Blażewicz, W. Cellary, R. Slowiński, J. Weglarz. 1986. *Scheduling under Resource Constraints - Deterministic Models*. Annals of Operations Research **7**. Baltzer, Amsterdam.

[27] J. Błażewicz, K. Ecker, G. Schmidt, J. Weglarz. 1993. *Scheduling in Computer and Manufacturing Systems*. Springer-Verlag, Berlin.

[28] J. Błażewicz, M. Dror, J. Weglarz. 1991. Mathematical Programming Formulations for Machine Scheduling: a Survey. *European Journal of Operational Research* **51**, 283-300.

[29] Bowman, V.J. 1975. On the Relationship of the Tchebycheff Norm and the Efficient Frontier of Multiple-Criteria Problems. In: H. Thiriez, and S. Zionts (eds.). Multiple Criteria Decision Making. *Lecture Notes in Economics and Mathematical Systems* **130**, 76-86.

[30] P. Brandimarte. 1992. Neighbourhood Search-Based Optimization Algorithms for Production Scheduling: a Survey. *Computer-Integrated Manufacturing Systems* **5**, 167-176.

[31] P. Brandimarte. 1993. Routing and Scheduling in a Flexible Job Shop by Tabu Search. *Annals of Operations Research* **41**, 157-183.

[32] P. Brandimarte, M. Calderini. 1995. A Hierarchical Bicriterion Approach to Integrated Process Plan Selection and Job Shop Scheduling. *International Journal of Production Research* **33**, 161-181.

[33] P. Brandimarte, W. Ukovich, A. Villa. 1995. Continuous Flow Models for Batch Manufacturing: a Basis for a Hierarchical Approach. To appear in *International Journal of Production Research*.

[34] P. Brandimarte, A. Villa (eds.). 1995. *Optimization Models and Concepts in Production Management*. Gordon and Breach, Reading, UK.

[35] J.A. Buzacott, J.G. Shanthikumar. 1992. Design of Manufacturing Systems Using Queueing Models. *Queueing Systems* **12**, 135-214.

[36] J.A. Buzacott, J.G. Shanthikumar. 1993. *Stochastic Models of Manufacturing Systems*. Prentice-Hall, Englewood Cliffs, NJ.

[37] J.P. Buzen. 1973. Computational Algorithms for Closed Queueing Networks with Exponential Servers. *Communications of the ACM* **16**, 527-531.

[38] G. Buxey. 1990. The Myth of Aggregate Planning. *Production Planning and Control* **1**, 222-234.

[39] G.M. Campbell, V.A. Mabert. 1991. Cyclical Schedules for Capacitated Lot-Sizing with Dynamic Demands. *Management Science* **37**, 409-427.

[40] M. Caramanis, G. Liberopoulos. 1992. Perturbation Analysis for the Design of Flexible Manufacturing Systems Flow Controllers. *Operations Research* **40**, 1107-1126.

[41] J. Carlier. 1982. The One-Machine Sequencing Problem. *European Journal of Operational Research* **11**, 42-27.

[42] J. Carlier, E. Pinson. 1989. An Algorithm for Solving the Job Shop

Problem. *Management Science* **35**, 164-176.

[43] A. Carrie. 1988. *Simulation of Manufacturing Systems*. Wiley, Chichester.

[44] D.G. Cattrysse, J. Maes, L.N. Van Wassenhove. 1990. Set Partitioning and Column Generation Heuristics for Capacitated Dynamic Lotsizing. *European Journal of Operational Research* **46**, 38-47.

[45] D.G. Cattrysse, M. Salomon, R. Kuik, L.N. Van Wassenhove. 1993. A Dual Ascent and Column Generation Heuristic for the Discrete Lotsizing and Scheduling Problem with Setup Times. *Management Science* **39**, 477-486.

[46] D.G. Cattrysse, M. Salomon, L.N. Van Wassenhove. 1994. A Set Partitioning Heuristic for the Generalized Assignment Problem. *European Journal of Operational Research* **72**, 167-174.

[47] D.G. Cattrysse, L.N. Van Wassenhove. 1992. A Survey of Algorithms for the Generalized Assignment Problem. *European Journal of Operational Research* **60**, 260-272.

[48] Y.-L. Chang, H. Matsuo, R.S. Sullivan. 1989. A Bottleneck-based Beam Search for Job Scheduling in a Flexible Manufacturing System. *International Journal of Production Research* **27**, 1949-1961.

[49] O. Charalambous, K.S. Hindi. 1991. A Review of Artificial Intelligence-Based Job Shop Scheduling Systems. *Information and Decision Technologies* **17**, 189-202.

[50] W.-H. Chen, J.-M. Thizy. 1990. Analysis of Relaxations for the Multi-Item Capacitated Lot-Sizing Problem. *Annals of Operations Research* **26**, 29-72.

[51] H.C. Co, J.S. Biermann, S.K. Chen. 1990. A Methodical Approach to the Flexible Manufacturing System Batching, Loading and Tool Configuration Problems. *International Journal of Production Research* **28**, 2171-2186.

[52] E.G. Coffman, M.R. Garey, D.S. Johnson. 1978. An Application of Bin Packing to Multiprocessor Scheduling. *SIAM Journal of Computing* **7**, 1-17.

[53] D. Connors, G. Feigin, D. Yao. 1992. Scheduling Semiconductor Lines Using a Fluid Network Model. *Proc. 3rd IEEE International Conf. on Computer Integrated Manufacturing*, Troy NY, 174-183.

[54] H.E. Crowder, E.L. Johnson, M.W. Padberg. 1983. Solving Large-Scale Zero-One Linear Programming Problems. *Operations Research* **31**, 803-834.

[55] R.L. Daniels, R.J. Chambers. 1990. Multiobjective Flow Shop Scheduling. *Naval Research Logistics* **37**, 981-995.

[56] R. De Matta, M. Guignard. 1994. Studying the Effects of Production Loss due to Setup in Dynamic Production Scheduling. *European Journal of Operational Research* **72**, 62-73.

[57] M. Dell'Amico, M. Trubian. 1993. Applying Tabu-Search to the Job Shop Scheduling Problem. *Annals of Operations Research* **41**, 231-252.

[58] E.V. Denardo. 1982. *Dynamic Programming: Models and Applications.* Prentice-Hall, Englewood Cliffs, NJ.

[59] M. Di Mascolo. Y. Frein, Y. Dallery, R. David. 1991. A Unified Modeling of Kanban Systems Using Petri Nets. *International Journal of Flexible Manufacturing Systems* **3**, 275-307.

[60] M. Diaby, H.C. Bahl, M.H. Karwan, S. Zionts. 1992. A Lagrangean Relaxation Approach for very Large-Scale Capacitated Lot-Sizing. *Management Science* **38**, 1329-1340.

[61] F. DiCesare, G. Harhalakis, J.M. Proth, M. Silva, F.B. Vernadat. *Practice of Petri Nets in Manufacturing.* Chapman & Hall, London.

[62] G. Dobson, U.S. Karmarkar. 1989. Simultaneous Resource Scheduling. *Operations Research* **37**, 592-600.

[63] G. Dobson, U.S. Karmarkar. 1995. Large Scale Job SCheduling by Lagrangian Decomposition. In: P. Brandimarte, A. Villa (eds.). 1995. *Optimization Models and Concepts in Production Management*, 45-69. Gordon and Breach, Reading, UK.

[64] R.A. Dudek, S.S. Panwalkar, M.L. Smith. 1992. The Lessons of Flowshop Scheduling Research. *Operations Research* **40**, 7-13.

[65] J. Duggan, J. Browne. 1991. Production Activity Control: a Practical Approach to Scheduling. *International Journal of Flexible Manufacturing Systems* **4**, 79-103.

[66] R.W. Eglese. 1990. Simulated Annealing: a Tool for Operational Research. *European Journal of Operational Research* **46**, 271-281.

[67] E.A. Elsayed, T.O. Boucher. 1994. *Analysis and Control of Production Systems (2nd ed.).* Prentice-Hall, Englewood Cliffs, NJ.

[68] G.D. Eppen, R. Kipp Martin. 1987. Solving Multi-Item Capacitated Lot-Sizing Problems Using Variable Redefinition. *Operations Research* **35**, 832-848.

[69] L.F. Escudero. 1987. A Mathematical Formulation of a Hierarchical Approach for Production Planning in FMS. In: A. Kusiak (ed.). *Modern Production Management Systems.* North Holland, Amsterdam. 231-245.

[70] L.F. Escudero, A. Sciomachen. 1987. The Job Sequencing Ordering Problem on a Card Assembly Line. In: T.A. Ciriani, R.C. Leachman (eds.). *Optimization in Industry.* Wiley, Chichester. 251-262.

[71] J.R. Evans, E. Minieka. 1992. *Optimization Algorithms for Networks and Graphs.* Dekker.

[72] A.V. Fiacco, G.P. McCormick. 1968. *Nonlinear Programming: Sequential Unconstrained Minimization Techniques.* Wiley, Chichester.

[73] M.L. Fisher. 1981. The Lagrangian Relaxation Method for Solving Integer Programming Problems. *Management Science* **27**, 1-18.

[74] M.L. Fisher, B.J. Lageweg, J.K. Lenstra, A.H.G. Rinnooy Kan. 1983. Surrogate duality relaxation for job shop scheduling. *Discrete Applied Mathematics* **5**, 65-75.

[75] M.L. Fisher, R. Jaikumar, L.N. Van Wassenhove. 1986. A Multiplier Adjustment Method for the Generalized Assignment Problem. *Management Science* **32**, 1095-1103.

[76] B. Fleischmann. 1990. The Discrete Lot-Sizing and Scheduling Problem. *European Journal of Operational Research* **44**, 337-348.

[77] B. Fleischmann. 1994. The Discrete Lot-Sizing and Scheduling Problem with Sequence-Dependent Setup Costs. *European Journal of Operational Research* **75**, 395-404.

[78] R. Fletcher. 1987. *Practical Methods of Optimization.* Wiley, Chichester.

[79] S. French. 1982. *Sequencing and Scheduling: an Introduction to the Mathematics of the Job Shop.* Ellis Horwood, Chichester, UK.

[80] T.D. Fry, J.F. Cox, J.H. Blackstone. 1992. An Analysis and Discussion of the Optimized Production Technology Software and its Use. *Production and Operations Management* **1**, 229-242.

[81] M.C. Fu. 1994. Optimization by Simulation: a Review. *Annals of Operations Research* **53**, 199-247.

[82] M.R. Garey, D.S. Johnson. 1979. *Computers and Intractability: a Guide to the Theory of NP-completeness.* W.H. Freeman and Company, San Francisco.

[83] S.B. Gershwin. 1994. *Manufacturing Systems Engineering.* Prentice Hall, Englewood Cliffs, NJ.

[84] C.A. Glass, J.N.D. Gupta, C.N. Potts. 1994. Lot Streaming in Three-Stage Production Processes. *European Journal of Operational Research* **75**, 378-394.

[85] C.R. Glassey, F. Markgraf, H. Fromm. 1993. Real Time Scheduling of Batch Operations. In: T.A. Ciriani, R.C. Leachman (eds.). *Optimization in Industry.* Wiley, Chichester. 113-137.

[86] A.M. Geoffrion. 1970. Elements of Large-Scale Mathematical Programming. *Management Science* **16**, 653-675.

[87] A.M. Geoffrion. 1974. Lagrangian Relaxation for Integer Program-

ming. *Mathematical Programming Study* **2**, 82-114.

[88] P. Glasserman. 1991. *Gradient Estimation via Perturbation Analysis.* Kluwer, Boston.

[89] P. Glasserman. 1994. Perturbation Analysis of Production Networks. In: D.D. Yao (ed.). *Stochastic Modeling and Analysis of Manufacturing Systems*, 233-280. Springer-Verlag, Berlin.

[90] F. Glover. 1989. Tabu Search Part I. *ORSA Journal on Computing* **1**, 190-206.

[91] F. Glover, M. Laguna, E. Taillard, D. de Werra (eds.). 1993. *Tabu Search.* Baltzer, Amsterdam.

[92] F. Glover, E. Taillard, D. de Werra. 1993. A User's Guide to Tabu Search. *Annals of Operations Research* **41**, 1-28.

[93] D.E. Goldberg. 1989. *Genetic Algorithms in Search, Optimization and Machine Learning.* Addison-Wesley, Reading, MA.

[94] R.L. Graham. 1966. Bounds for Certain Multiprocessing Anomalies. *Bell System Technical Journal* **45**, 1563-1581.

[95] S.C. Graves, A.H.G. Rinnooy Kan, P.H. Zipkin (eds.). 1993. *Logistics of Production and Inventory.* North Holland, Amsterdam.

[96] H.J. Greenberg, F.H. Murphy. 1992. A Comparison of Mathematical Programming Modeling Systems. *Annals of Operations Research* **38**, 177-238.

[97] M. Guignard, and M.B. Rosenwein. 1989. An Application-Oriented Guide for designing Lagrangean Dual Ascent Algorithms. *European Journal of Operational Research* **43**, 197-205.

[98] H.O. Günther. 1986. The Design of an Hierarchical Model for Production Planning and Scheduling. In: S. Axsäter, C. Schneeweiss, E. Silver (eds.). Multi-Stage Production Planning and Inventory Control. *Lecture Notes in Economics and Mathematical Systems* **266**, 227-260. Springer-Verlag, Berlin.

[99] M.-H. Han, Y.K. Na, G.L. Hogg. 1989. Real-Time Tool Control and Job Dispatching in Flexible Manufacturing Systems. *International Journal of Production Research* **27**, 1257-1267.

[100] P. Hansen, B. Jaumard, V. Mathon. 1993. Constrained Nonlinear 0-1 programming. *ORSA Journal on Computing* **5**, 97-119.

[101] A.C. Hax, D. Candea. 1984. *Production and Inventory Managament.* Prentice-Hall, Englewood Cliffs, NJ.

[102] M. Held, R.M. Karp. 1962. A Dynamic Programming Approach to Sequencing Problems. *Journal of SIAM* **10**, 196-210.

[103] J.-B. Hiriart-Urruty, C. Lemaréchal. 1993. *Convex Analysis and Minimization Algorithms I.* Springer-Verlag, Berlin.

[104] Y.-C. Ho, X.-R. Cao. 1991. *Perturbation Analysis of Discrete Event Dynamic Systems.* Kluwer, Boston.

[105] J.H. Holland. 1975. *Adaptation in Natural and Artificial Systems.* University of Michigan Press, Ann Arbor.

[106] Horst, Tuy. 1993. *Global Optimization: Deterministic Approaches.* Springer-Verlag, Berlin.

[107] Y. Ibaraki. 1987. Enumerative Approaches to Combinatorial Optimization - Part I. *Annals of Operations Research* **10**. Baltzer, Amsterdam.

[108] E. Ignall, L. Schrage. 1965. Application of the Branch and Bound Technique to Some Flow Shop Scheduling Problems. *Operations Research* **13**, 400-412.

[109] E.L. Johnson, M.M. Kostreva, U.H. Suhl. 1985. Solving 0-1 Integer Programming Problems Arising from Large Scale Planning Models. *Operations Research* **33**, 803-819.

[110] K. Kant. 1992. *Introduction to Computer System Performance Evaluation.* McGraw-Hill, New York.

[111] S. Karlin, H.M. Taylor. 1975. *A First Course in Stochastic Processes.* Academic Press, Boston, MA.

[112] S. Karlin, H.M. Taylor. 1981. *A Second Course in Stochastic Processes.* Academic Press, Boston, MA.

[113] U.S. Karmarkar. 1981. Equalization of the Runout Times. *Operations Research* **29**, 757-762.

[114] Y.-D. Kim. 1990. A Comparison of Dispatching Rules for Job Shops with Multiple Identical Jobs and Alternative Routeings. *International Journal of Production Research* **28**, 953-962.

[115] Y.-D. Kim, C.A. Yano. 1993. Heuristic Approaches for Loading Problems in Flexible Manufacturing Systems. *IIE Transactions* **25**, 26-39.

[116] J.G. Kimemia, S.B. Gershwin. 1983. An Algorithm for the Computer Control of Production in Flexible Manufacturing Systems. *IIE Transactions* **15**, 353-362.

[117] R. Kipp Martin. 1987. Generating Alternative Mixed-Integer Programming Models Using Variable Redefinition. *Operations Research* **35**, 820-831.

[118] S. Kirkpatrick, C.D. Gelatt, M.P. Vecchi, 1983. Optimization by Simulated Annealing. *Science* **220**, 621-680.

[119] J.P.C. Kleijnen. 1992. Simulation and Optimization in Production Planning: a Case Study. *Decision Support Systems* **8**.

[120] P.J. Kuehn. 1979. Approximate Analysis of General Queuing Networks by Decomposition. *IEEE Trans. on Communications* **27**, 113-

126.

[121] A. Kusiak, S.S. Heragu. 1989. Expert Systems and Optimization. *IEEE Trans. Software Engineering* **15**, 1017-1020.

[122] P. L'Ecuyer, N. Giroux, P.W. Glynn. 1994. Stochastic Optimization by Simulation: Numerical Experiments with the $M/M/1$ Queue in Steady State. *Management Science* **40**, 1245-1261.

[123] B.J. Lageweg, J.K. Lenstra, A.H.G. Rinnooy Kan. 1978. A General Bounding Scheme for the Permutation Flow Shop Problem. *Operations Research* **26**, 53-67.

[124] M. Laguna, J.L.G. Velarde. 1991. A Search Heuristic for Just-in-Time Scheduling in Parallel Machines. *Journal of Intelligent Manufacturing* **2**, 253-260.

[125] L.S. Lasdon. 1970. *Optimization theory for large systems*. MacMillan, London, UK.

[126] L.S. Lasdon, R.C. Terjung. 1971. An Efficient Algorithm for Multi-Item Scheduling. *Operations Research* **19**, 946-969.

[127] A.M. Law, W.D. Kelton. 1991. *Simulation Modeling and Analysis (2nd ed.)*. McGraw-Hill, New York.

[128] E.L. Lawler, J.K. Lenstra, A.H.G. Rinnooy Kan, D.B. Shmoys. 1993. Sequencing and Scheduling: Algorithms and Complexity. In: S.C. Graves, A.H.G. Rinnooy Kan, P.H. Zipkin (eds.). *Logistics of Production and Inventory*. North Holland, Amsterdam. 445-522.

[129] E.L. Lawler, J.K. Lenstra, A.H.G. Rinnooy Kan, D.B. Shmoys (eds.). 1985. *The Traveling Salesman Problem*. Wiley, Chichester.

[130] C.-Y. Lee, J.D. Massey. 1988. Multiprocessor Scheduling: Combining LPT and Multifit. *Discrete Applied Mathematics* **20**, 233-242.

[131] C.-Y. Lee, R. Uzsoy, L.A. Martin-Vega. 1992. Efficient Algorithms for Scheduling Semiconductor Burn-In Operations. *Operations Research* **40**, 764-775.

[132] S. Lozano, J. Larraneta, L. Onieva. 1991. Primal-Dual Approach to the Single Level Capacitated Lot-Sizing Problem. *European Journal of Operational Research* **51**, 354-366.

[133] A.D. Luber. 1991. *Solving Business Problems with MRP II*. Digital Press, Bedford, MA.

[134] P.B. Luh, D.J. Hoitomt. 1993. Scheduling of Manufacturing Systems using the Lagrangian Relaxation Technique. *IEEE Trans. on Automatic Control* **38**, 1066-1079.

[135] M. Lundy, A. Mees. 1986. Convergence of an Annealing Algorithm. *Mathematical Programming* **34**, 111-124.

[136] J. Maes, J.O. McClain, L.N. Van Wassenhove. 1991. Multilevel Ca-

pacitated Lotsizing Complexity and LP-based Heuristics. *European Journal of Operational Research* **53**, 131-148.

[137] A.S. Manne. 1958. Programming of Economic Lot Sizes. *Management Science* **4**, 115-135.

[138] S. Martello, P. Toth. 1990. *Knapsack Problems: Algorithms and Computer Implementations.* Wiley, Chichester.

[139] S.V. Maturana. 1994. Issues in the Design of Modeling Languages for Mathematical Programming. *European Journal of Operational Research* **72**, 243-261.

[140] J.O. McClain, L.J. Thomas, J.B. Mazzola. 1992. *Operations Management.* Prentice Hall, Englewood Cliffs, NJ.

[141] K.N. McKay, J.A. Buzacott, F.R. Safayeni. 1989. The Scheduler's Knowledge of Uncertainty: the Missing Link. In: J. Browne (ed.). *Knowledge Based Production Management Systems.* Elsevier, NY. 171-189.

[142] G. McMahon, M. Florian. 1975. On Scheduling with Ready Times and Due Dates to Minimize Maximum Lateness. *Operations Research* **23**, 475-482.

[143] J. Miltenburg. 1989. Level Schedules for Mixed-Model Assembly Lines in Just-in-Time Production Systems. *Management Science* **35**, 192-207.

[144] M. Minoux. 1986. *Mathematical Programming: Theory and Algorithms.* Wiley, Chichester.

[145] Y. Mondem. 1983. *Toyota Production System.* Institute of Industrial Engineers Press, Norcross, GA.

[146] M. Montazeri, L.N. Van Wassenhove. 1990. Analysis of Scheduling Rules for an FMS. *International Journal of Production Research* **28**, 785-802.

[147] T.E. Morton, D.W. Pentico. 1993. *Heuristic Scheduling Systems.* Wiley, Chichester.

[148] K.G. Murty. 1983. *Linear Programming.* Wiley, Chichester.

[149] A. Nagar, J. Haddock, S. Heragu. 1995. Multiple and Bicriteria Scheduling: a Literature Survey. *European Journal of Operational Research* **81**, 88-104.

[150] G.L. Nemhauser, L.A. Wolsey. 1988. *Integer Programming and Combinatorial Optimization.* Wiley, Chichester.

[151] G.L. Nemhauser, A.H.G. Rinnooy Kan, M.J. Todd (eds.). 1989. *Optimization.* North Holland, Amsterdam.

[152] P.S. Ow, T.E. Morton. 1988. Filtered Beam Search in Scheduling. *International Journal of Production Research* **26**, 35-62.

[153] I.H. Osman, C.N. Potts. 1989. Simulated Annealing for Permutation Flow Shop Scheduling. *OMEGA International Journal of Management Science* **17**, 551-557.

[154] C.H. Papadimitriou, K. Steigliz. 1982. *Combinatorial Optimization: Algorithms and Complexity.* Prentice-Hall, Englewood Cliffs, NJ.

[155] R.G. Parker, R.L. Rardin. 1988. *Discrete Optimization.* Academic Press, Boston, MA.

[156] J.M. Proth, H.P. Hillion. 1990. *Mathematical Tools in Production Management.* Plenum Press, NY.

[157] M. Reiser, S.S. Lavenberg. 1980. Mean Value Analysis of Closed Multichain Queueing Networks. *Journal of the ACM* **27**, 312-322.

[158] A.H.G. Rinnooy Kan. 1976. *Machine Scheduling Problems.* Martinus Nijhoff, The Hague.

[159] T.G. Robertazzi. 1990. *Computer Networks and Systems: Queueing Theory and Performance Evaluation.* Springer-Verlag, Berlin.

[160] T.R. Rohleder, G. Scudder. 1993. A Comparison of Order-Release and Dispatch Rules for the Dynamic Weighted Early/Tardy Problem. *Production and Operations Management* **2**, 221-238.

[161] Y. Roll, R. Karni. 1991. Multi-Item, Multi-Level Lot SIzing with an Aggregate Capacity Constraint. *European Journal of Operational Research* **51**, 73-87.

[162] S.M. Ross. 1993. *An Introduction to Probability Models (5th ed.).* Academic Press, Boston, MA.

[163] M.H. Safizadeh, R. Signorile. 1994. Optimization of Simulation via Quasi-Newton Methods. *ORSA Journal on Computing* **6**, 398-408.

[164] M. Salomon. 1991. *Deterministic Lot-Sizing Models for Production Planning.* Lecture Notes in Economics and Mathematical Systems, Vol. 355. Springer-Verlag, Berlin.

[165] M. Salomon, L.G. Kroon, R. Kuik, L.N. Van Wassenhove. 1991. Some Extensions of the Discrete Lot-Sizing and Scheduling Problem. *Management Science* **37**, 801-812.

[166] M.W.P. Savelsbergh. 1994. Preprocessing and Probing Techniques for Mixed-Integer Programming Problems. *ORSA Journal on Computing* **6**, 445-454.

[167] A.-W. Scheer, A. Hars. 1991. The Leitstand - A New Tool for Decentral Production Control. Competitive Manufacturing. In: G. Fandel, G. Zäpfel (eds.). *Modern Production Concepts*, 371-385. Springer-Verlag, Berlin.

[168] P. Serafini, W. Ukovich. 1989. A Mathematical Model for Periodic Scheduling Problems. *SIAM Journal on Algebraic and Discrete Meth-*

ods **2**, 550-581.

[169] J.G. Shanthikumar, J.A. Buzacott. 1980. On the Approximations to the Single Server Queue. *International Journal of Production Research* **18**, 761-773.

[170] J.G. Shanthikumar, J.A. Buzacott. 1981. Open Queueing Network Models of Dynamic Job Shops. *International Journal of Production Research* **19**, 255-266.

[171] A. Sharifnia. 1995. Optimal Production Control Based on Continuous Flow Models. In: P. Brandimarte, A. Villa (eds.). 1995. *Optimization Models and Concepts in Production Management*, 153-185. Gordon and Breach, Reading, UK.

[172] B. Shetty, H.K. Bhargava, R. Krishnan (eds.). 1992. Model Management in Operations Research. *Annals of Operations Research* **38**. Baltzer, Amsterdam.

[173] J.P. Sousa, L.A. Wolsey. 1992. A Time Indexed Formulation of Non-Preemptive Single Machine Scheduling Problems. *Mathematical Programming* **54**, 353-367.

[174] M.G. Speranza, W. Ukovich. 1992. Models for Periodic Production. *Proc. 3rd IEEE International Conf. on Computer Integrated Manufacturing*, Troy, NY, 348-354.

[175] M.G. Speranza, C. Vercellis. 1993. Hierarchical Models for Multi-Project Planning and Scheduling. *European Journal of Operational Research* **64**, 312-325.

[176] R.E. Steuer. 1985. *Multiple Criteria Optimization: Theory, Computation, and Application*. Wiley, Chichester.

[177] U.H. Suhl, R. Szymanski. 1994. Supernode Processing of Mixed-Integer Models. *Computational Optimization and Applications* **3**, 317-331.

[178] R. Suri, R.R. Hildebrandt. 1985. Modeling Flexible Manufacturing Systems Using Mean-Value Analysis. *Journal of Manufacturing Systems* **3**, 27-38.

[179] R. Suri, S. de Treville. 1993 Rapid Modeling: the Use of Queueing Models to Support Time-Based Competitive Manufacturing. In: G. Fandel, T. Gulledge, A. Jones (eds.). *Operations Research in Production Planning and Control*, 21-30. Springer-Verlag, Berlin.

[180] R. Suri, J.L. Sanders, M. Kamath. 1993. Performance Evaluation of Production Networks. In: S.C. Graves, A.H.G. Rinnooy Kan, P.H. Zipkin (eds.). *Logistics of Production and Inventory*. North Holland, Amsterdam. 199-286.

[181] M.T. Tabucanon, S. Mukyangkoon. 1985. Multi-Objective Micro-

computer-Based Interactive Production Planning. *International Journal of Production Research* **23**, 1001-1023.

[182] U. Tetzlaff. 1990. *Optimal Design of Flexible Manufacturing Systems.* Springer-Verlag, Berlin.

[183] J.M. Thizy, L.N. Van Wassenhove. 1985. Lagrangean Relaxation for the Multi-Item Capacitated Lot-Sizing Problem: A Heuristic Implementation. *IIE Transactions* **17**, 308-313.

[184] L.J. Thomas, J.O. McClain. 1993. An Overview of Production Planning. In: S.C. Graves, A.H.G. Rinnooy Kan, P.H. Zipkin (eds.). *Logistics of Production and Inventory.* North Holland, Amsterdam. 333-370.

[185] H.C. Tijms. 1986. *Stochastic Modeling and Analysis: a Computational Approach.* Wiley, Chichester.

[186] E. Toczylowski et al. 1989. Multi-Level Production Scheduling for a Class of Flexible Machining and Assembly Systems. *Annals of Operations Research* **17**, 163-180.

[187] P. Toth. 1980. Dynamic Programming Algorithms for the Zero-One Knapsack Problem. *Computing* **25**, 29-45.

[188] D. Trietsch, K.R. Baker. 1993. Basic Techniques for Lot Streaming. *Operations Research* **41**, 1065-1076.

[189] W.W. Trigeiro, L.J. Thomas, J.O. McClain. 1989. Capacitated Lot Sizing with Setup Times. *Management Science* **35**, 353-366.

[190] R. Uzsoy, C.-Y. Lee, L.A. Martin-Vega. 1992. Scheduling Semiconductor Test Operations: Minimizing Maximum Lateness and Number of Tardy Jobs on a Single Machine. *Naval Research Logistics* **39**, 369-388.

[191] S. Van de Velde. 1990. Dual Decomposition of Single Machine Scheduling Problems. *Proc. 1st Conference on Integer Programming and Combinatorial Optimization*, University of Waterloo, Canada, May 1990, 495-507.

[192] S. Van de Velde. 1990. Minimizing the Sum of the Job Completion Times in the Two-Machine Flow Shop by Lagrangian Relaxation. *Annals of Operations Research* **26**, 257-268.

[193] S. Van de Velde. 1991. *Machine Scheduling and Lagrangian Relaxation.* Ph.D. Thesis. CWI, Amsterdam.

[194] S. Van de Velde. 1993. Duality-based Algorithms for Scheduling Unrelated Parallel Machines. *ORSA Journal on Computing* **5**, 192-205.

[195] S. Van de Velde. 1995. Deterministic Machine Scheduling. In: P. Brandimarte, A. Villa (eds.). 1995. *Optimization Models and Concepts in Production Management*, 1-43. Gordon and Breach, Reading, UK.

[196] N.M. van Dijk. 1993. *Queueing Networks and Product Forms. A Sys-*

tems Approach. Wiley, Chichester.

[197] P.J.M. Van Laarhoven, E.H.L. Aarts. 1987. *Simulated Annealing: Theory and Applications.* Reidel Publishing Company, Dordrecht, The Netherlands.

[198] P.J.M. Van Laarhoven, E.H.L. Aarts, J.K Lenstra. 1992. Job Shop Scheduling by Simulated Annealing. *Operations Research* **40**, 113-125.

[199] T.J. Van Roy, L.A. Wolsey. 1987. Solving Mixed Integer Programming Problems Using Automatic Reformulation. *Operations Research* **35**, pp. 45-57.

[200] C.J. Van Wyk. 1990. *Data Structures and C Programs.* Addison-Wesley, Reading, MA.

[201] A.P.J. Vepsalainen, T.E. Morton. 1987. Priority Rules for Job Shops with Weighted Tardiness Costs. *Management Science* **33**, 1035-1047.

[202] N. Viswanadham, Y. Narahari. 1992. *Performance Modeling of Automated Manufacturing Systems.* Prentice-Hall, Englewood Cliffs, NJ.

[203] T.E. Vollmann, W.L. Berry, D.C. Whybark. 1992. *Manufacturing Planning and Control Systems.* Irwin, Homewood, IL.

[204] H.M. Wagner. 1975. *Principles of Operations Research.* Prentice-Hall, Englewood Cliffs, NJ.

[205] J. Walrand. 1988. *An Introduction to Queueing Networks.* Prentice-Hall, Englewood Cliffs, NJ.

[206] C.D.J. Waters. 1992. *Inventory Control and Management.* Wiley, Chichester.

[207] S. Webster, F.R. Jacobs. 1993. Scheduling a Flexible Machining System with Dynamic Tool Movement. *Production and Operations Management* **2**, 38-54.

[208] W. Whitt. 1983. The Queueing Network Analyzer. *The Bell System Technical Journal* **62**, 2779-2815.

[209] M. Widmer, A. Hertz. 1989. A New Heuristic Method for the Flow Shop Sequencing Problem. *European Journal of Operational Research* **41**, 186-193.

[210] H.P. Williams. 1993. *Model Building in Mathematical Programming.* Wiley, Chichester.

[211] H.P. Williams. 1993. *Model Solving in Mathematical Programming.* Wiley, Chichester.

[212] J.M. Wilson. 1989. Alternative Formulations of a Flow Shop Scheduling Problem. *Journal of the Operational Research Society* **40**, 395-399.

[213] R.J. Wittrock. 1985. Scheduling Algorithms for Flexible Flow Lines. *IBM Journal of Research and Development* **29**, 401-412.

[214] R.J. Wittrock. 1990. Scheduling Parallel Machines with Major and Minor Setup Times. *International Journal of Flexible Manufacturing Systems* **2**, 329-341.

[215] D.L. Woodruff, M.L. Spearman. 1992. Sequencing and Batching for Two Classes of Jobs with Deadlines and Setup Times. *Production and Operations Management* **1**, 87-102.

[216] M. Zhou, F. DiCesare. 1993. *Petri Net Synthesis for Discrete Event Control of Manufacturing Systems*. Kluwer, Boston.

Index

A
algorithm
    enumerative 45
    exponential 38
    heuristic 156
    polynomial 38
    pseudopolynomial 36, 49, 89
APP (Aggregate Production Planning) 61, 70
arc 281
    conjunctive 132
    directed 281
    disjunctive 132
    fixed-charge 301
    undirected 281
arrival theorem 214
Artificial Intelligence 179
aspiration criteria 343
assemble-to-order 12
assembly 60, 70, 72, 80, 88
assignment
    Generalized Assignment Problem 187, 313, 318, 327
    problem 304, 314
    Quadratic Assignment Problem 319
    variables 142
ATSP *see* Traveling Salesman Problem

B
backlog 16, 55
cost 237
balance equation 200, 374
    global 226
    local 226
batch
    processor 105, 115, 180, 311
    size, sizing 84
    transfer 84, 104, 236
beam search 176, 336
best-first 333
best-improving 343
bin packing 173, 311
binary search 89, 173, 181
birth-death process 196, 374
block 146
blocking 9, 105, 161, 200
bottleneck 132, 149, 176, 177
bound
    lower 40, 64, 266, 271, 321, 323
    upper 31, 64
boundary condition 24
Branch and Bound 32, 37, 38, 45, 52, 61, 61, 73, 77, 124, 144, 164, 319, 321, 328
    truncated 337
branching 147, 148, 320, 328

C
capacity constraint 3, 7, 57, 63, 70, 73, 87, 184
buffer 84, 88

399

tool magazine 38
chain (*see* Markov chain)
Chapman-Kolmogorov equations
    370, 372
circuit 183, 283
class
    $\mathcal{NP}$ 37
    $\mathcal{P}$ 33, 45
clique 133, 183, 282
CLSP (Capacitated Lot-Sizing
    Problem) 60, 70, 74, 95
coefficient of variation 199, 361
column generation 69, 318
combination
    convex 49, 248, 275, 279,
        316, 317
    linear 29
complementary slackness 260,
    262, 267
completion time 16, 34
    maximum (*see* makespan)
    total 86, 110
    total weighted completion
        time 110, 139, 153
computational complexity 21,
    37, 45, 65, 211
    average-case 33
    worst-case 33
concave
    function 248, 269, 287, 316,
        325
    problem 250
constraint
    active 260
    dualized 59
    elimination 148, 323
    equality 22, 254-256, 267
    inequality 22, 254-255
    integrality 22
    linear 283
    logical 314
    method 49, 183, 186
    qualification 256-258, 260

control
    feedback 79, 356
    policy 350
    variable 55, 84, 349
convex
    combination 49, 248, 275,
        279, 316, 317
    function 248-250, 252, 273,
        315
    hull 40, 97, 248, 274, 281,
        325, 327
    problem 250, 257, 268, 287
    region 63
    set 49, 247, 250, 279
convexity 31, 247, 266-267, 273,
    287, 296
convolution algorithm 209, 236
cooling schedule 164, 341
cost
    backlog 237
    fixed 2, 61
    inventory 15, 54, 60, 61, 94,
        237
    production 58, 90, 94
    reduced 290, 298
    resource 14, 59
    setup 2, 15, 52, 60, 61, 94
cost-to-go 65, 238, 349-350, 354
    (*see also* value
    function)
counting process 193, 206, 364
criterion vector 29
CTMC *see* Markov chain,
    continuous-time
cumulative
    demand 99
    production 84
cumulative distribution function
    234, 360
customer order 12
customer service 112
cycle
    directed 283
    Hamiltonian 303, 311

INDEX

D

deadline 109
decision problem 45, 335
decision process
    sequential 349
decomposition 44, 347, 350
    dual (Lagrangian,
        price-directed) 52, 53,
        62, 64, 63, 74, 77, 270
    hierarchical 335
    one-way 176
    two-way 176
    primal (resource-directed) 271
DEDS (Discrete-Event Dynamic System) 8
demand
    dependent 3, 72
    independent 3, 72
    uncertainty 70
depth-first 332
difference
    central 231, 254
    finite 231, 253
differentiability 247
direction
    ascent 275
    descent 253
    search 253, 268
directional derivative 272, 274
disaggregation 63, 71
discrete-time
    model 7, 140
    dynamic system 55
dispatching rule *see* priority rule
disturbance 79
diversification 344
DLSP (Discrete Lot-Sizing and Scheduling Problem) 52, 74, 91
    single-item 99
dominance condition 321, 333
dominant schedule 68

DTMC *see* Markov chain, discrete-time
dual
    feasible, feasibility 299
    function 43, 57, 58, 63, 265, 268, 270, 274, 298, 325
    heuristic 64, 77, 161, 335
    problem *see* problem
    variable *see* variable
duality 246, 264
    in Linear Programming 278, 296
    Lagrangian 31, 246
    strong 266, 268, 270, 296
    weak 63, 266, 267, 271, 323, 324
dualization 56, 265 (*see also* constraints, dualized)
due date 15, 16, 34, 48, 84, 159
Dynamic Programming 28, 32, 37, 48, 52, 65, 75, 77, 99, 125, 131, 180, 238, 334, 337, 346, 362
    backward 349
    forward 126, 353
dynamic system 347
    discrete-event *see* DEDS
    discrete-time 349

E

earliness 16, 88, 110, 118
EDD rule *see* priority rule
efficient
    set 30
    solution 30
engineer-to-order 12
EOQ (Economic Order Quantity) 2, 8, 11, 60
epigraph 248-249
error
    modeling 79, 90
    tracking 79, 82
event 360
    discrete *see* DEDS

independent 360
expected value 28, 361

F
family
   of products, items 70
   setup 70
fathoming 321
feasibility problem 89, 181
feasible
   set 22, 323
   solution 22, 332
first-improving 338
fixed-charge 62, 67
   arc *see* arc
   problem 301, 312
flexible
   flow line (FFL) 83
   manufacturing system (FMS) 38, 143, 235
flow
   continuous 44
   minimum cost 302, 312
   time 16, 110, 183
form
   canonical 285
   standard 285
formulation
   modal 44, 68, 313, 317
   strengthening 146
   strong 63, 95
frequency-based experimentation 233
function
   additive 16, 270
   affine 25, 269, 283
   barrier 256
   concave 248, 269, 287, 316, 325
   convex 248-250, 252, 273, 315
   dual 43, 57, 58, 63, 265, 268, 270, 274, 298, 325
   differentiable 249
   Lagrangian 257, 261, 264
   linear 283
   minmax 16
   minsum 16
   nondifferentiable 31, 246, 272, 325
   penalty 55, 238, 254, 255, 344
      exterior 255
      interior 255
   piecewise linear 315, 325, 330
functional equation 65, 100, 182, 347, 351
   backward 349
   forward 349

G
Gantt chart 9, 103
gradient method 31, 231, 253, 273, 337
graph 278, 281
   acyclic 283
   bipartite 219, 283, 301
   complete 282
   directed 281
   disjunctive 132, 137, 171, 182
   optimization 278
   undirected 281
greedy (rule or algorithm) 82, 89, 335
Group Technology 180 (*see also* layout)

H
half-space 279
hedging point 240
heuristic algorithm, strategy 43, 53, 156, 334
   general purpose 334
   iterative 334
   one-shot 334, 335
   special purpose 334

INDEX

heuristic
    dual 64, 77, 161, 335
    LP-based 73, 335
HPP (Hierarchical Production Planning) 70, 72
hyperplane 249, 279
    support 249, 273

I

independent increments 206, 365
infinitesimal generator (of a CTMC) 338, 372
integrality property 64, 327
interchange argument 120
inventory
    control 1
        fixed period 3
        fixed quantity 2
    cost 15, 54, 60, 61, 94, 237
    level 94

J

Johnson algorithm 122
just in time 55, 81

K

knapsack 35, 45, 48, 117, 308, 311, 328, 334, 353
Kuhn-Tucker conditions 261

L

Lagrangian
    decomposition *see* decomposition
    multiplier 57, 59, 63, 77, 151, 246, 254, 257, 268, 274, 324
    problem *see* problem
    relaxation *see* relaxation
lateness 16, 34, 110
    maximum 16, 87, 89, 110, 178
layout
    group technology 14
    process-oriented 14, 108
    product-oriented 12, 108
lead time 2, 12, 15, 112, 115, 180, 194, 205
Least Squares 232
leitstand 180
likelihood ratio method 233
Linear Programmming *see* LP
list scheduling 113, 157
Little's law 195, 197, 205, 212
load
    balancing 91, 142, 171
    profile 70
local improvement 338
local search 163, 337
lot size 3
lot-sizing 4, 52
    in assembly systems 70
    capacitated lot-sizing *see* CLSP
    discrete lot-sizing and scheduling *see* DLSP
    multi-level 60, 72
    single-item 77
    single-item, uncapacitated 64, 68
lot streaming 104
LP (Linear Programming) 31, 49, 52, 235, 238, 278
    model 58, 131
    problem 25, 54, 283

M

machine
    down state 359
    failure 79, 237
    loading 185
    up state 359
    utilization 112, 194, 236
make-to-order 11, 15, 16, 53
make-to-stock 11, 15, 53
makespan 9, 16, 88, 110, 112, 174

Markov chain
    continuous-time (CTMC) 194, 223, 238, 368
    discrete-time (DTMC) 366
    ergodic 372
    homogeneous 367
    irreducible 370
    periodic 370
Markov process 238, 364
material handling 74
Material Requirements Planning (MRP) 4
Mathematical Programming 24
matrix
    incidence 59, 305, 312
    routing 199, 236
    stochastic 367
    transition probability 367
    unimodular 40, 278, 304, 327
Mean Value Analysis (MVA) 210, 236
    approximate 216
memory, long-term 344
meta-heuristics 338
MILP (Mixed-Integer Linear Programming) 32, 52, 60
    model 43, 60, 135, 149, 182, 241, 328
    problem 323
minimizer 250
model
    analytical (evaluative) 191
    continuous-flow 237
    continuous-time 7, 237
    discrete-time 7, 140
    evaluative 5
    experimental (evaluative) 191
    generative 5
    (mixed-integer) linear programming 43, 60, 135, 149, 182, 241, 328
    nominal 79
    queueing 5
    simulation see simulation
    stochastic 28
MPS file 240
Multifit algorithm 173, 174, 176
multi-stage system 72
multiplier see Lagrangian multiplier
MVA see Mean Value Analysis

N

neighborhood structure 164, 338
network 281, 299
    acyclic 346
    directed 130
    flow 58, 67
    layered 351
    optimization 31, 59, 278
    project 130, 140
Newton method 253
node 281
    decomposition 216
    demand 300
    down-child 329
    potential 132, 182
    source 300
    transit 300, 301
    up-child 329
nondeterministic computer 45
nondifferentiability point 273, 275
norm
    Euclidean 80
    Tchebycheff 50, 188
NP–complete (problem) 47, 61
NP–completeness 37, 45
NP–hard (problem) 21, 47, 61, 78, 89, 144
number of late jobs 86
numerical method 246, 253

O

objective function 23

INDEX
405

additive (minsum) 86, 89, 350
minmax 87
nonregular 88, 118
regular 110, 113
OPT (Optimized Production Technology) 84, 177
optimal control 24, 79
   continuous-time 354
   discrete-time 350
optimality
   global 252, 260, 267
   local 31, 251, 252, 257, 260, 261
   principle of (Bellman) 351
optimization method
   exact 20, 31
   heuristic 20, 31, 37
optimization (method, problem)
   combinatorial 26, 307
   constrained 254
   discrete 26, 49, 247, 268, 307, 346
   dynamic 24
   finite-dimensional 23
   global 322
   infinite-dimensional 24, 32, 238, 346, 354
   multiobjective 29, 49, 185
   nonsmooth 272
   pseudoboolean 27
   stochastic 28, 32, 237, 346
   unconstrained 246, 251, 253
   vector 29, 49, 185
optimum
   global 250
   local 250, 253
orientation 133

P

parallel machines
   identical 83, 108
   uniform 108
   unrelated 108

part type selection 38, 105, 117, 143
path
   critical (see path, longest)
   directed 283
   longest 131, 171, 303, 349
   shortest 64, 67, 96, 303, 335, 346, 351
penalty function see function
performance measure 14
permutation 34, 135, 148, 164, 330
   variable 135
perturbation 263, 298
perturbation analysis 233
   infinitesimal 234
Petri net 192, 219
planning
   Aggregate Production Planning see APP
   Hierarchical Production Planning see HPP
point, extreme 279, 287, 304
Poisson
   arrival 194
   process 206, 364
polyhedron 279, 286
   bounded (see polytope)
polynomial
   algorithm 63, 89, 118, 181
   problem 21
   reduction 46
polytope 279, 287
precedence constraint 103, 130, 151
preemption 104
primal
   decomposition 271
   problem 57, 63, 265, 324
   variable 257
priority rule 78
   ATC (Apparent Tardiness Cost) 159
   EDD (Earliest Due Date)

34, 36, 121, 146, 159, 181
dynamic 158
LPT (Longest Processing Time) 161, 172, 176
LWKR (Least Work Remaining) 159
MWKR (Most Work Remaining) 159
SLACK 159
SPT (Shortest Processing Time) 119, 159, 176
S/RMOP (Slack per Remaining Operation) 159
S/RMWK (Slack per Remaining Work) 159
SRPT (Shortest Remaining Processing Time) 124
static 158
WSPT (Weighted Shortest Processing Time) 120, 151, 159
probability 360
   conditional 360
   density function 361
   mass function 361
problem
   dual 57, 77, 266, 298, 325
   instance 33
   Lagrangian 274
   primal 57, 63, 265, 324
   relaxed 265, 267-270, 323
   unconstrained 272
process plan 11, 103
processing time 34, 48
product form 200, 209, 226
production
   batch 12
   control 11
   cumulative 84
   job shop 12
   mass 12
   mix 12, 81, 116, 183, 235, 237
   one-of-a-kind 12
   order 2, 12
   planning 10, 52, 53 (*see also* APP, HPP)
   repetitive 12, 80, 116, 183, 237
   scheduling 11, 78
programming
   binary (0/1) 320
   integer 26, 145, 307
   linear 267
   mixed-integer 27
   nonlinear 25, 246
   nonlinear 0/1 319
   pure integer 27
   quadratic 26, 257
projection 254, 272

Q

queueing
   model 5
   network 194
      closed 199, 207, 227, 235
      multiclass 194, 216
      open 199
      single-class 194
      tandem 200
   system
      $GI/G/1$ 198
      $M/M/1$ 196
      $M/M/1/N$ 197
      $M/M/m$ 197
   theory 191

R

random variable 360
   exponential 363, 365, 368
   geometric 362, 367
   memoryless 362
ray 281
   extreme 281

real time
  control (*see also* production control)
  rescheduling 78
recursion 210, 351
recursive equation 211, 347
  backward 65
reentrant flow 88, 105, 180
regression 232
relaxation
  continuous 38, 40, 43, 53, 61, 73, 97, 145, 241, 323, 326, 332
    Lagrangian 43, 60, 97, 149, 324, 326, 332
  surrogate 162, 326
release time 16, 36, 48, 84, 88, 138, 147
reorder level (ROL) 2
resource allocation 353
Response Surface Methodology 231
rolling (time) horizon 28, 72, 112
routing 11, 44, 175, 235
  Markovian 199
  matrix 199, 236
RSM *see* Response Surface Methodology
runout time 72

S

safety stock 54
satisfiability problem 48
scalarization 29, 49, 186
schedule
  active 113, 157
  nondelay 115, 157
  semiactive 113
scheduling 52, 323
  classification of problems 106
  dynamic 106, 112
  FJS (Flexible Job Shop) 175
  flow shop 108, 148
  permutation 108, 135
  FMS (Flexible Manufacturing System) 175
  job shop 103, 108, 171
  open shop 108
  parallel machine 106, 155, 171, 174
    identical 108
    uniform 108
    unrelated 108
  periodic 116, 182
  preemptive 123
  single-machine 106
  static 106, 112
  with simultaneous use of resources 153
scheduling problem
  $1//L_{\max}$ 34, 36, 121
  $1/r_i/L_{\max}$ 36, 48, 121, 146, 177
  $1//T_{\max}$ 121
  $1/prec/\max_i \gamma_i(C_i)$ 121
  $1//\sum_i C_i$ 119, 145
  $1//\sum_i w_i C_i$ 120
  $1/r_i/\sum_i C_i$ 124
  $1/pmptn, r_i/\sum_i C_i$ 124
  $1/prec/\sum_i w_i C_i$ 137, 151
  $1//\sum_i T_i$ 125
  $1/r_i/\sum_i T_i$ 138
  $1/s_{ij}/C_{\max}$ 129
  $1/s_{ij}/\sum_i w_i C_i$ 139
  $F2//C_{\max}$ 122
  $F//C_{\max}$ 135, 148, 163, 167
  $F//\sum_i C_i$ 135
  $F//\sum_i w_i C_i$ 152
  $J2//C_{\max}$ 123
  $J//C_{\max}$ 135, 139, 148, 171, 177
  $J//\sum_i w_i T_i$ 159
  $P//C_{\max}$ 171
  $R//C_{\max}$ 161
search
  strategy 332

tree 320, 332
sensitivity 263
  in LP 298
sequencing 34
service level 16
Set
  Covering 312, 318
  Packing 313
  Partitioning 313, 318
setup 8, 13
  cost 2, 15, 52, 60, 61, 94
  family 117
  major-minor 70, 117, 174
  sequence-dependent 15, 105, 116, 127, 139, 160, 179
  sequence-independent 15
  time 14, 15, 52, 61, 64, 76, 309
shadow price 264
shifting bottleneck 177
Simplex 54, 58
  algorithm 38
  dual 298
  method 32, 43, 236, 278, 324, 346
simulated annealing 163, 338, 367
simulation 5, 191, 231
sojourn time 367
solution
  basic 288
  basic feasible 289
  incumbent 322, 332
special ordered set (SOS) 241, 317, 329
  type 1 (SOS1) 329
  type 2 (SOS2) 329
  type 3 (SOS3) 330
SPT rule see priority rule
standard deviation 361
state 359, 362
  equation 6, 24, 55
  positively recurrent 370
  recurrent 370

space 207, 364
trajectory 350
variable 55, 78, 81, 84, 100, 349
stationarity 251, 253, 257, 258, 262-264
stationary increments 365
steady state 194, 360, 369, 372
steepest descent 253
step length 59, 253, 268, 275
stochastic
  modeling 357
  process 358, 364
strategic oscillation 344
subdifferential 273, 274, 325
subgradient 58, 59, 64, 77, 161, 273, 325
subgraph 281
subtour 309
supply 300

T
tabu
  attribute 342
  list 343
  search 167, 176, 342
tardiness 16, 110
  maximum 180
  quadratic 155
  total 16, 86, 110
  total weighted 110
temperature 339
throughput 15, 183, 194, 211
time
  bucket 52, 54
    large 60
    small 74
time-sharing 84
tool 84
  magazine 38, 143
total probability theorem 360, 369, 372
Toyota Goal Chasing method 82
tracking 79

traffic
  equation 204
  intensity 196
trajectory 78, 82
  reference (or target) 79
  state 350
transportation problem 64, 301
Traveling Repairman Problem 139
Traveling Salesman Problem (TSP) 116, 139, 309
  Asymmetric (ATSP) 303
  Symmetric 303
truncated exponential scheme 336

U
uncertainty 78
unimodularity 304
utilization 15 (*see also* machine utilization)

V
valid inequality 42
value function 100, 182, 349 (*see also* cost-to-go)
variable
  artificial 294
  assignment 142
  basic 69, 289
  binary 27
  control 55, 84, 349
  decision 22
  dual 59, 257, 261
  (general) integer 27
  logical 27
  nonbasic 289
  permutation 135
  primal 257
  semicontinuous 329
  slack 286
  state *see* state
  time-indexed 140, 155
variance 361
  visit ratio 206, 208

W
Wagner-Within property 65
waiting time 159
WIP (Work-in-Process) 6, 15, 84, 91, 105, 112, 119, 183, 194, 205, 234, 357
workforce 56
worst-case bound 172
WSPT rule *see* priority rule